PARALLEL COMPUTATIONAL
FLUID DYNAMICS '92

PARALLEL COMPUTATIONAL FLUID DYNAMICS '92

Proceedings of the Conference on
Parallel CFD '92
Implementations and Results Using Parallel Computers
New Brunswick, NJ, USA, 18–20 May, 1992

Edited by

R.B. PELZ
*Department of Mechanical and
Aerospace Engineering
Rutgers University
Piscataway, NJ, USA*

A. ECER
*Department of Mechanical Engineering
Indiana University Purdue
University Indianapolis
Indianapolis, IN, USA*

J. HÄUSER
*Institute of Turbomachinery
Department of Mechanical Engineering
Technical College of Braunschweig-Wolfenbüttel
Wolfenbüttel, Germany*

1993

NORTH-HOLLAND
AMSTERDAM • LONDON • NEW YORK • TOKYO

ELSEVIER SCIENCE PUBLISHERS B.V.
Sara Burgerhartstraat 25
P.O. Box 211, 1000 AE Amsterdam, The Netherlands

ISBN: 0 444 89986 3

© 1993 ELSEVIER SCIENCE PUBLISHERS B.V. All rights reserved.

No part of this publication may be reproduced, stored in a retrieval system or transmitted, in any form or by any means, electronic, mechanical, photocopying, recording or otherwise, without the prior written permission of the publisher, Elsevier Science Publishers B.V., Copyright & Permissions Department, P.O. Box 521, 1000 AM Amsterdam, The Netherlands.

Special regulations for readers in the U.S.A. - This publication has been registered with the Copyright Clearance Center Inc. (CCC), Salem, Massachusetts. Information can be obtained from the CCC about conditions under which photocopies of parts of this publication may be made in the U.S.A. All other copyright questions, including photocopying outside of the U.S.A., should be referred to the copyright owner, Elsevier Science Publishers B.V., unless otherwise specified.

No responsibility is assumed by the Publisher for any injury and/or damage to persons or property as a matter of products liability, negligence or otherwise, or from any use or operation of any methods, products, instructions or ideas contained in the material herein.

This book is printed on acid-free paper.

Printed in The Netherlands

PREFACE

The Parallel Computational Fluid Dynamics conference held in New Brunswick in June 1992 has shown a healthy trend continuing. Massively parallel systems are gaining wider acceptance in science and technology. Major computer vendors like Convex, Cray, DEC, and IBM have announced their entry in the parallel market. This is a clear sign that parallel computing is becoming a mature field. Exciting parallel systems from Intel, Maspar, Ncube, Parsytec, Thinking Machines and other manufacturers are already available and more vendors will be on the market in 93.

However, this rapid development in parallel hardware has to be complemented by the respective parallel software. Major efforts are needed to develop parallel application software to use the powerful platforms available. In particular, application software for industrial computations is needed. Comparing the procedings of Parallel CFD 91 with the present one, it is obvious that progress has been made, but envisaging the next stage in the sofware development process a closer cooperation between research institutions, universities and industry as well as manufacturers of parallel machines is needed to solve large scale, applied problems of complex geometry on massively parallel systems on a routinely basis. Special emphasis has to be given to the questions of problem scaling and load balancing of these applications.

In general, the engineer is not interested in the parallel architecture and will use any configuration that will provide the proper speedup. Codes therefore have to be portable and must not be dependent upon a certain architecture.

Also, since new parallel systems are entering the market, the question of portability of parallel codes is of even higher importance. Some kind of standard in processor communication will be needed. Parallel software engineering along with new software design concepts like object-oriented programming have to be considered too. Also many users are still working in Fortran, it seems to be the right time that the latest hardware is complemented by the new software design techniques that were developed during the last decade. Languages like C++ seem to have a high potential in numerical applications, which is worthwhile to investigate.

In addition, teaching and training programs for engineers and scientists in parallel programming as well as modern software design techniques are of great importance.

To summarize the present situation, it is save to say that parallel computing offers an enormous potential for CFD applications, but massive support from industry and government is required to make parallel systems a valuable production tool and to provide the resources for education and training of a new class of scientists and engineers.

Jochem Häuser

on hehalf of the organizing committee

ACKNOWLEDGEMENT

Parallel CFD 92 was sponsored by:

Computer Aids for Industrial Productivity (CAIP) of Rutgers University
Convex Computer Corporation
Cray Research Corporation
Intel Corporation, Supercomputers Division
International Business Machines
National Aeronautics and Space Administration, Ames Research Center
Thinking Machines Corporation

TABLE OF CONTENTS

Finite Element Simulation of Memory Fluids on Message Passing Parallel Computers
R. Aggarwal and R. Keunings — 1

A Database Management System for Parallel Processing of CFD Algorithms
H.U. Akay, R. Blech, A. Ecer, D. Ercoskun, B. Kemle, A. Quealy and A. Williams — 9

A Multiblock Hypersonic Flow Solver for Massively Parallel Computer
S. Borelli, A. Matrone and P. Schiano — 25

Applications of a Parallel Pressure-Correction Algorithm to 3-D Turbomachinery Flows
M.E. Braaten — 39

The Calculation of 3-D Compressible Flow through a Rectangular Nozzle Using a Data Parallel Finite Element Model
F.P. Brueckner D.W. Pepper T.H. Sobota and R.H. Chu — 51

A Comparison of Lattice Gas Automata Implementations on the MasPar MP-1
J. Butterworth and J.F. Prins — 63

Solving CFD Methods Involving Global Communication on Distributed Memory MIMD
A. Chalmers and S.P. Fiddes — 75

A Fast Vortex Method for the Simulation of 3-D Flows on Parallel Computers
K. Chua and T.R. Quackenbush — 87

CFD on the 1-K Node nCUBE/2
D.D. Cline — 99

Multidomain Computations of Compressible Flows in a Parallel Scheduling Environment
F. Dellagiacoma, S. Paoletti, F. Poggi and M. Vitaletti — 111

CFD Experiences on a Range of Novel Architecture Systems
D.R. Emerson, R.J. Blake and R.J. Allan — 123

Programming Paradigms for Spectral Element Methods in DM-MIMD Architectures
N. Floros, J. Reeve and O. Tutty   133

Application of a Parallel CFD Code to Large-Scale Practical Systems
E.R. Galea, A. Chan, M. Cross, N. Hoffmann, C. Ierotheou, S. Johnson and K. Pericleous   147

Sparse Grid Multilevel Methods, their Parallelization and their Application to CFD
M. Griebel   161

Unsteady Fluid Flow Calculations Using a Machine Independent Parallel Programming Environment
A. Gursoy, L.V. Kale and S.P. Vanka   175

Algorithm Modifications for Parallel Operation of a Multigrid Navier-Stokes Solver
C.S. Gwilliam and J.S. Rollett   187

Features of Architecture Independent Parallel CFD Software
J. Häuser, H.D. Simon and H.-G. Paap   199

Data Parallel Finite Element Techniques for CFD on the Connection Machine Systems
Z. Johan, T.J.R. Hughes, K.K. Mathur and S.L. Johnsson   215

Mapping Unstructured Mesh CFD Codes onto Local Memory Parallel Architectures
B.W. Jones, M.F. Everett and M. Cross   231

Parallel Adaptive Navier-Stokes Algorithm on the Cray Y-MP-8
Y. Kallinderis and A. Vidwans   241

Challenges Posed for Parallel Processing on iPSC-860 by DNS Schemes for Supersonic Flows
F. Ladeinde   253

Block-Structured Multigrid for the Navier-Stokes Equations: Experiences and Scalability Questions
J. Linden G. Lonsdale, H. Ritzdorf and A. Schüller   267

Parallel Algorithms for Gas Dynamics
L.N. Long 279

Evaluation of Different Approaches to Parallel Processing for CFD
C. de Nicola G. Pietro and L. Paparone 291

Compressible Vortex Reconnection on the Connection Machine
R.B. Pelz, T. Scheidegger, N.J. Zabusky and O.N. Boratav 299

Numerical Simulation of Complex Fluid Flows on MIMD Computers
M. Perić, M. Schäfer and E. Schreck 311

Direct Numerical Simulation of Turbulence on the Connection Machine
J.B. Perot 325

Parallel CFD in a Networked Environment
G. Prisco, D. Ferrer-Pellicer, J.P. Huot, R. Molina and M. Roest 337

Large Eddy Simulations of Turbulence on a Massively Parallel Computer
J. Robichaux, S.P. Vanka and D.K. Tafti 349

Parallel CFD on Unstructured Meshes Using Algebraic Multigrid
G. Robinson 359

Order Finite Difference Scheme
C. Shu and B.E. Richards 371

Efficient Parallelisation of Implicit and Explicit Solvers on a MIMD Computer
D.M. Smith and S.P. Fiddes 383

The Uniform Boundary Algorithm for Supersonic and Hypersonic Flows
Y.S. Weber J.W. Weber, J.D. Anderson, E.S. Oran and C. Li 395

Parallelising Explicit and Fully Implicit Navier-Stokes Solutions for
Compressible Flows
X. Xu, N. Qin and B.E. Richards 407

Parallelization of KIVA-II on the iPSC-860 Supercomputer
O. Yasar and C.J. Rutland 419

On the Implementation Issues of Domain Decomposition Algorithms
for Parallel Computers
J. Zhu 427

Finite Element Simulation of Memory Fluids on Message-Passing Parallel Computers

R. Aggarwal and R. Keunings

Division of Applied Mechanics, Université Catholique de Louvain,
B-1348 Louvain-la-Neuve, Belgium

Abstract

We describe a parallel algorithm for the numerical simulation of memory fluids on message-passing, distributed memory computers. The algorithm has been implemented within the general-purpose finite element package POLYFLOW. The resulting code has been ported on a 128 processor Intel iPSC/860 hypercube. We report high levels of parallel efficiency in the simulation of extrusion flow. Further work is needed towards automatic load balancing.

1. INTRODUCTION

It has been known for many years that the Navier-Stokes equations are unable to describe the elastic effects associated with the flow of memory fluids such as polymer solutions or melts. Available macroscopic constitutive equations for viscoelastic fluids take the form of either partial differential equations for the polymer contribution to the stress, or time-integrals of a memory-weighted deformation kernel along the particle paths. Modeling, mathematical and numerical issues associated with the simulation of memory fluids are discussed in [1-2].

One of the difficulties associated with the simulation of memory fluids is that of computing resources. Indeed, realistic time-dependent simulations of three-dimensional viscoelastic flows cannot be performed today within a reasonable timeframe, even with the help of vector supercomputers. The objective of our work is to explore the potential of massively-parallel computers in the simulation of non-Newtonian flows. The present paper focuses on the class of viscoelastic fluids described by integral constitutive equations. Our related work on differential models is reported in [3-5].

The outline of the paper is as follows. We first review in Section 2 the basic equations governing the flow of integral memory fluids. In Section 3, we briefly describe the numerical discretization technique as well as the iterative scheme used in sequential computations. The corresponding parallel algorithm is presented in Section 4. Results obtained on a 128 processor Intel iPSC/860 are reported and discussed in Section 5.

2. GOVERNING EQUATIONS

The macroscopic mathematical description of viscoelastic flow consists of conservation laws, constitutive equations, and suitable boundary and initial conditions. In the present paper, we consider steady-state, isothermal, incompressible flows, for which the conservation laws read

$$\nabla \cdot (-p\mathbf{I} + \mathbf{T}) + \rho \mathbf{f} = \rho \ \mathbf{v} \cdot \nabla \mathbf{v} \ ,$$

$$\nabla \cdot v = 0. \tag{1}$$

Here, p is the pressure, I is the unit tensor, T is the extra-stress tensor, v is the velocity vector, f is the body force per unit mass of fluid, and ρ is the fluid density.

In macroscopic simulations, the set of conservation equations (1) is closed with a constitutive model that relates the extra-stress T to the deformation experienced by the fluid. In this work, we shall consider single-integral constitutive models. First, we decompose the extra-stress into two components, i.e.

$$T = T_N + T_V, \qquad T_N = 2\mu_N D, \tag{2}$$

where T_N and T_V are respectively the Newtonian and viscoelastic contributions to the extra-stress. The symbol μ_N is a viscosity coefficient, while D is the rate of strain tensor.

In order to describe integral constitutive equations for the viscoelastic contribution T_V, let us consider a fluid particle whose position at present time t is given by $x(t)$. The fluid motion is assumed to be described by the vector relation

$$x(t') = \chi(x(t), t, t'), \tag{3}$$

which gives the particle position $x(t')$ at historical time t' ranging between $-\infty$ and t. We then define the relative deformation gradient F_t and the right Cauchy-Green strain tensor C_t by

$$F_t(t') = \frac{\partial \chi}{\partial x}, \qquad C_t(t') = F_t^T(t') F_t(t'). \tag{4}$$

Single-integral constitutive equations give the viscoelastic extra-stress T_V at a fluid particle through a time integral of the deformation history. Their generic form is given by

$$T_V(x(t)) = \int_{-\infty}^{t} M(t-t') S_t(t') \, dt', \tag{5}$$

where the operator $\int \cdot dt'$ is a time integral taken along the particle path parameterized by the historical time t'. The kernel S_t is a deformation-dependent tensor of the generic form

$$S_t(t') = \phi_1(I_1, I_2) [C_t^{-1}(t') - I] + \phi_2(I_1, I_2) [C_t(t') - I], \tag{6}$$

where C_t^{-1}, the inverse of C_t, is known as the Finger strain tensor. The scalars ϕ_1 and ϕ_2 are given functions of the invariants I_1 and I_2, defined respectively as the trace of the Finger and Cauchy-Green strain tensors. Finally, the factor $M(t-t')$ appearing in (5) is a memory function.

In the present paper, we use a particular case of the popular K.B.K.Z. model, for which we have $\phi_1 = 1$, $\phi_2 = 0$, and

$$M(t-t') = \sum_{i=1}^{n} \frac{\mu_i}{\lambda_i^2} \exp[-(t-t')/\lambda_i] \, H(t-t') \;,\quad H(t-t') = \frac{\alpha}{\alpha-3+I_1\beta+I_2(1-\beta)} \;. \qquad (7)$$

The above model includes two material constants α, β, and a spectrum of n relaxation times λ_i and viscosity coefficients η_i. Its use in numerical simulations of actual polymer melt flows is described in [6].

3. NUMERICAL TECHNIQUE

Available techniques for solving integral viscoelastic flows are based on a decoupled approach, whereby the computation of the viscoelastic extra-stress is performed separately from that of the flow kinematics. From known kinematics, one first calculates the viscoelastic extra-stress by means of the constitutive equation (5); the kinematics are then updated by solving the conservation equations (2) using the viscoelastic stresses as a known pseudo body force term. The procedure is then iterated upon. The update scheme is usually akin to Picard's iterative algorithm. We only indicate here the major steps of the above procedure. For details, see [6] and the references therein.

One iteration of the numerical procedure involves the following steps :

Step 1 : Integrate the constitutive equation (5) to compute the viscoelastic extra-stress, using the kinematics calculated at the previous iteration.

Step 2 : Update the kinematics by solving the conservation laws (2); the viscoelastic stress computed in Step 1 is treated as a pseudo-body force.

Step 2 amounts to a classical Navier-Stokes problem, which is solved in [6] by means of a standard Galerkin, finite element velocity-pressure formulation. This so-called u-v-p problem leads to an algebraic set of equations for the nodal unknowns of velocity and pressure; it is tackled by means of a direct frontal solver. In order to compute the generalized load vector of the u-v-p problem, we need the values of the viscoelastic extra-stress at *all* Gauss integration points of the finite element mesh. Step 1 thus consists of three sub-tasks, to be performed for each Gauss point :

Tracking : On the basis of the velocity field computed at the previous iteration, determine the upstream trajectory and the travel time of the integration point.

Strain : At a finite number of past times, compute the deformation gradient, and from it the integrand of (5).

Stress : Compute the memory integral (5) numerically.

The main numerical difficulties associated with the use of integral models lie in the tracking of past particle positions, the computation of the strain history, and the overall iterative scheme. Details on those issues can be found in [6].

The above sequential numerical algorithm has been implemented by Goublomme [6] in the commercial package POLYFLOW developed in Louvain-la-Neuve for the simulation of polymer processing flows [7].

4. PARALLEL IMPLEMENTATION

Finite element computations generically involve two potentially expensive phases, namely (i) the computation of the element contributions to the stiffness matrix and load vector, and (ii) the solution of the finite element equation set. In many applications, including viscoelastic flow described by differential models, the solution phase consumes a major fraction of the *CPU* time. With integral viscoelastic fluids, however, the *CPU* time needed for the solution phase (i.e. Step 2) is only a very small fraction of that for the computation of the load vector (i.e. Step 1). In typical two-dimensional steady-state simulations of integral fluids, we have observed that Step 1 takes more than 95% of the the complete sequential *CPU* timing. Figures as high as 99.6% have even been obtained in some flow problems, as we shall see below. In view of these observations, we shall concentrate here on the parallel treatment of Step 1. Our work on the parallelization of the direct solver phase is reported elsewhere [3-5].

The programming model that we have selected is that of a *Multiple Instruction Multiple Data* computer, endowed with P processors capable of simultaneous and independent processing on their own data. We currently use in our experiments the 16 processor Intel iPSC/2, the 128 processor Intel iPSC/860, as well as a heterogeneous network of engineering workstations. Programming is done in FORTRAN 77 with calls to a suitable communication library.

It is conceptually simple to devise a parallel algorithm for the computation of the stiffness matrix and load vector in integral viscoelastic calculations (Step1). Indeed, computation of the viscoelastic stress can be done for each Gauss point independently of the others. One can thus distribute the elements to the available processors, and have them work in parallel at the evaluation of the contribution of "their" elements to the finite element equation sets. Thus, at each iteration, the P available processors compute in parallel the viscoelastic stresses at their allocated Gauss points. One of those processors is also in charge of the frontal assembly and elimination of the pseudo-Navier-Stokes problem; this small task is currently performed *sequentially*. The P processors compute in parallel the resulting local finite element matrices which are communicated to the "frontal solver" processor. In view of the non-local character of tracking and memory integral operations, all processors must have at their disposal the whole velocity field computed at the previous iteration. The velocity field is broadcast by the "frontal solver" processor to the other processors at the end of each non-linear iteration.

With this approach, the same copy of the code is loaded on all processors, together with the input data on finite element mesh, material properties and boundary conditions. We have implemented the above parallel algorithm in an experimental version of the POLYFLOW package.

Speedup is defined as the ratio between the total elapsed time for the sequential algorithm and that obtained with the parallel algorithm using P processors. *Efficiency* is then defined as the speedup divided by the number of processors. In the present context, possible sources of loss of efficiency are (i) the load imbalance between processors, (ii) the inter-processor communication overheads, and (iii) the occurrence of sequential bottlenecks. The issue of load balancing is a rather peculiar one in the present application. At first sight, it would seem that equipartition of the finite element mesh into P sub-domains would guaranty perfect load balance. Actually, the compute load involved in the computation of the viscoelastic stress at a given Gauss point depends on the material parameters (e.g. spectrum of relaxation times), on the current kinematics, and on the location of the Gauss point within the overall flow domain. It is indeed impossible to predict *a priori* the compute load required by the stress calculation, and this load can vary significantly from one Gauss point to the other. (A similar issue arises in non-linear structural mechanics involving elasto-plastic materials.) These points will be illustrated in Section 5. The results described in the present paper are based upon a *static* allocation of workload among the available processors. Dynamic allocation schemes are currently under development in our group.

Inter-processor communication overheads and sequential bottlenecks also lead to losses in efficiency. In particular, Amdhal's law states that an upper-bound for parallel speedup is the inverse of the time fraction taken by the non-parallelized part of a sequential algorithm. In the present case, the u-v-p direct solver phase is performed sequentially by a particular processor. Although the corresponding Amdhal fraction is found to be less than 5% in problems of sufficiently large sizes, it is clear that this sequential bottleneck will be of increasing importance as the number of processors increases. Again, we address this issue further in the next section.

5. RESULTS AND DISCUSSION

We have evaluated the above parallel algorithm in the simulation of *stick-slip* flow of a K.B.K.Z. fluid. This two-dimensional, steady-state flow problem is an idealization of extrusion from a planar die [1]. The computational domain is rectangular, and it is discretized by means of a structured finite element mesh. We have used a sequence of meshes with increasing refinement. In this section, we review the results obtained on the 128 processor Intel iPSC/860. The focus is put on the issue of parallel efficiency. Details on the physics of the stick-slip flow can be found in [1] and the references therein.

Observed values of the parallel speedup are listed in Table 1 as a function of mesh refinement (i.e. number of finite elements) and of the number of available processors. We refer to those values as *global* speedup results in the sense that the *CPU* time spent in the sequential solver phase is taken into account. Note that the speedup results for the 256 and 512 element meshes correspond to one non-linear iteration, while those for the coarser meshes have been obtained for a complete simulation.

P =	2	4	8	16	32	64	128
16 elts	1.9	3.5	6.3	11.6	-	-	-
80 elts	1.9	3.5	6.8	13.5	19.4	25.9	-
180 elts	2.0	3.9	7.3	13.1	24.0	41.1	48.4
256 elts	2.0	3.7	6.6	13.0	24.7	46.5	78.7
512 elts	1.9	3.6	6.4	12.6	24.1	43.4	70.0

Table 1. Stick-slip flow of a K.B.K.Z. integral fluid. Observed global speedup as a function of mesh refinement and number of processors (sequential solver *CPU* time included).

These speedup results are rather satisfactory in view of the fact that the algorithm contains a purely sequential part (i.e. the frontal solver for the pseudo Navier-Stokes problem), and that static allocation of elements to processors has been used. Let us consider these two issues separately.

Even though the sequential solver phase is very short, it does have a major impact on global speedup when the number of processors becomes large. In order to illustrate this point, let us focus on the 256 and 512 element meshes.

For the 256 element mesh, the solver phase takes 0.4% of the total sequential run on one processor. With 128 processors, the measured global speedup is 78.7, and the efficiency is 62%. Let us now assume that the sequential solver phase takes 0% of the total sequential run, instead of the measured 0.4%. It is then easy to compute the speedup of the parallel computation of viscoelastic stress alone. We find a value of 117.6, namely an efficiency of 91%. This simple calculation shows that only 9% efficiency have actually been lost in load imbalance and inter-processor communications, with the 256 element mesh and 128 processors. This is a very good result, considering the complexity of the application, and the fact that distribution of workload to the processors has been performed statically.

We list in Table 2 the computed speedup values for the viscoelastic stress integration, for the 256 and 512 element meshes. With the latter, the sequential solver phase is 0.5% of the total sequential run on one processor. Comparison of Tables 1 and 2 shows that parallelization of the solver phase, even with a small number of processors, will boost the global speedup to very high levels.

P =	2	4	8	16	32	64	128
256 elts	2.0	3.7	6.8	13.7	27.5	57.7	117.6
512 elts	2.0	3.7	6.6	13.3	27.2	54.6	104.9

Table 2. Stick-slip flow of a K.B.K.Z. integral fluid. Speedup of the viscoelastic stress computation as a function of mesh refinement and number of processors (sequential solver *CPU* time neglected).

Finally, let us focus on the issue of load balancing. We have already pointed out that the compute load within each element cannot be predicted *a priori* with integral viscoelastic models. A static allocation scheme of elements to processors is thus likely to induce load imbalance. We show in Fig.1 the *CPU* time spent for the computation of viscoelastic stress in each element of the 256 element mesh, as a function of the element index.

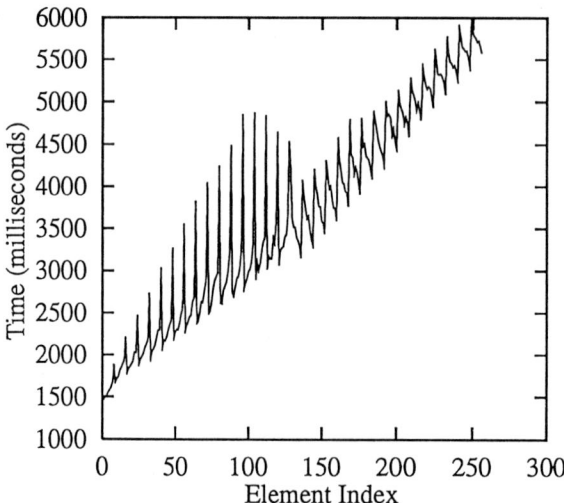

Fig.1 Stick-slip flow of a K.B.K.Z. integral fluid. *CPU* time for computation of viscoelastic stress as a function of element index (256 element mesh; first non-linear iteration).

The element numbering starts at the upstream, inlet section. This explains why the compute load within an element increases with the element index. The oscillatory character of the curve is due to the fact that computation of viscoelastic stress takes longer in the elements located near the die wall than in the elements close to the plane of symmetry. Note that the curve shown in Fig.1 is for the first non-linear iteration; it changes from one iteration to the next.

The speedup results presented in Tables 1-2 have been obtained with a static allocation scheme. The extent of load imbalance induced by this scheme can be estimated by inspection of Table 3. Here, we show the measured maximum relative difference in compute load within each processor, relative to the average processor load. These figures are given for the 256 and 512 element meshes, for the first non-linear iteration, as a function of the number of processors.

P =	2	4	8	16	32	64	128
256 elts	1.3	4.1	10.6	12.9	17.5	24.6	38.6
512 elts	0.2	2.5	8.1	9.6	12.6	18.1	30.5

Table 3. Stick-slip flow of a K.B.K.Z. integral fluid. Maximum load imbalance (in percent) as a function of mesh refinement and number of processors (figures based on viscoelastic stress computation only).

For example, load imbalance for the 256 element mesh and 128 processors is 38.6%. We observe that, for a given mesh, load imbalance induced by the static allocation scheme increases with the number of processors. There is clearly a need for a dynamic allocation procedure which would preserve load balance during the non-linear iteration procedure. This is the subject of our current work.

ACKNOWLEDGMENTS
The post-doctoral research of R. Aggarwal is supported financially by the BRITE/EURAM European Programme. We wish to thank A. Goublomme for his help. We also wish to thank M. Cosnard (ENSL, France), P. Leca (ONERA, France), and D. Roose (KUL, Belgium) for providing us with ample computer time on the iPSC/860 hypercubes located at their institutions. The results presented in this paper have been obtained within the framework of Interuniversity Attraction Poles initiated by the Belgian State, Prime Minister's Office, Science Policy Programming. The scientific responsibility rests with its authors.

REFERENCES
1. R. Keunings, in: C.L Tucker III (Ed.), Fundamentals of Computer Modeling for Polymer Processing, Carl Hanser Verlag, 1989, 402-470
2. M.J. Crochet, Amer. Chem. Soc., 62, 1989, 426-455
3. O. Zone, R. Keunings, D. Roose, in : A. Bode (Ed.), Distributed Memory Computing, Lecture Notes in Computer Science, Vol. 487, Springer Verlag, 1991, 294-303
4. O. Zone and R. Keunings, in: M. Durand and F. El Dabaghi (Eds.), High Performance Computing II, North-Holland, 1991, 333-344
5. R. Aggarwal, P. Henriksen, R. Keunings, D. Vanderstraeten, O. Zone, in: C. Hirsch (Ed.), Proc. First Eur. Comp. Fluid Dyn. Conf., Elsevier, 1992, in press
6. A. Goublomme, Ph.D. Thesis, Université Catholique de Louvain, Belgium, 1992
7. M.J. Crochet, B. Debbaut, R. Keunings, J.M. Marchal, in: K.T. O'Brien (Ed.), Computer Modeling for Extrusion and Other Continuous Polymer Processes, Carl Hanser Verlag, 1992, 25-50

A DATABASE MANAGEMENT SYSTEM FOR PARALLEL PROCESSING OF CFD ALGORITHMS

H.U. Akay[*], R. Blech[†], A. Ecer[*], D. Ercoskun[*], B. Kemle[*], A. Quealy[§] and A. Williams[†]

[*] Department of Mechanical Engineering, Indiana University Purdue University Indianapolis, Indianapolis, Indiana 46202, USA

[†] NASA Lewis Research Center, Computational Technologies Branch, Cleveland, Ohio 44135, USA

[§] Sverdrup Technology, Inc., NASA Lewis Research Center Group, Brookpark, Ohio 44142, USA

Abstract

A database management system for parallel processing of CFD algorithms was developed. For CFD problems with computational grids, a data base was defined in terms of blocks of grid points and their interfaces. A portable parallel library was employed to provide portability and efficiency. The data was managed for parallel algorithms which were defined in terms of block and interface solvers. An application of the capability was demonstrated on Intel iPSC/860 and IBM RS6000/320 computers using an explicit time-integration scheme.

1. INTRODUCTION

Parallel computation of flow problems has rapidly become popular during the last five years. The ultimate objective of these efforts is to solve large problems with complex geometries and with large number of grid points on many processors. The term "massively parallel", which is mentioned quite often, implies that a large set of data has to be managed on a large number of processors.

Activities in parallel CFD, leading to the present effort, have evolved through the following steps:

- A block-structured grid generation scheme was developed [1], where the computational grid was generated as an assembly of blocks. The connectivity of these blocks was recorded and managed during the grid generation process. Such block structured grid generation schemes are commonly used for solving complex CFD problems [2,3].

- A block structured solution scheme [4,5] was developed for the solution of potential and Euler equations, for solving large problems with complex geometries. This implicit solution scheme involved the solution of a set of equations in each block with explicit updating of the boundary conditions between the blocks. Such domain decomposition schemes have become popular over the last decade [6-8].

- Following these developments, it was natural to implement these schemes on parallel computers, since the problem was already defined and solved as a series of smaller problems [9-12]. Three-dimensional flow problems were tested on different types of computers, which involved distributed and shared memory implementations of the same algorithm.

Through these applications the following key conclusions were drawn:

- Most of the effort in these applications involved the grid generation. Thus, activities in parallel computing and grid generation have to be tied very closely for employing parallel computing efficiently.

- The parallel computing hardware and related software tools are developing and changing very rapidly. We expect the parallel computing environment to be dynamic during the coming years, where hardware configurations will be changing continuously. The portability issues in such an environment are critical.

- At this time, developing parallel codes requires familiarity with grid generation, the computational algorithm and the operation of the computer. The algorithm developers can afford to develop expertise on a computer only up to a certain level and therefore are not very enthusiastic about parallel computers.

The present activity was originated to treat the above issues.

Every CFD code reads a computational grid and performs computations on a set of data as described by this grid. Most of the activities related to parallel computing involve management of such "grid-oriented" data on an array of processors. For simplifying such operations, we have developed a "grid-oriented data base (GPAR)" for parallel processing of CFD algorithms. A three-dimensional computational grid is divided into a series of "blocks (a cluster of grid points)" connected to each other with "interfaces (another cluster of grid points)". GPAR creates and manages the data defined in terms of such blocks and interfaces. It employs "a machine portable parallel library (APPL)" [13]. This library was developed at NASA Lewis Research Center for the purpose of providing portability at high communication speeds on different machines. Since APPL calls are used inside GPAR, the parallel CFD user only needs to learn GPAR calls. In this fashion, issues related to parallel computers are confined to the development of GPAR and APPL and are not directly addressed by the CFD user. The parallel CFD user has to deal with grid generation, algorithm development and the use of GPAR. As it will be discussed later, the term "block" can loosely be used to define any cluster of grid points. The terminology defined in GPAR is general enough to include different types of algorithms involving structured and unstructured grids, explicit and implicit schemes, etc. The relationship between three levels of algorithms envisioned for parallel computing in CFD are shown in Figure 1.

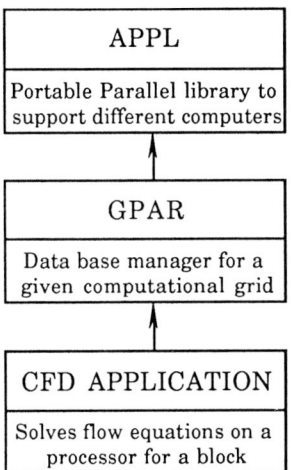

Figure 1. Three levels of algorithms for parallel computing in CFD.

In the present development, the parallelization of an algorithm is defined as implementing the same solution algorithm on different sets of data rather than different portions of an algorithm on the same set of data. Since the main objective is to solve very large problems, the key issues involve addressing large sets of data and the proximity of such sets on parallel computers. By using the developed parallel computing environment, one can generate portable CFD codes and test them on different hardware configurations.

2. AN APPLICATION PORTABLE PARALLEL LIBRARY (APPL)

APPL is a subroutine-based communication library [13], that is callable from applications written in FORTRAN or C. It provides a consistent programmer interface to a variety of distributed and shared-memory multiprocessor MIMD machines. A programming approach based on APPL is intended to minimize the learning curve and programming effort required to move application codes to different and/or networked machines.

Using the APPL approach, an application program can be developed by the user thinking in terms of communication processes rather than a particular hardware architecture. The processes execute in parallel on separate processors and exchange information as needed. Basic message passing primitives, comprising APPL, are used to communicate between processes. The primitives are the same regardless of the machine, thus making the application program portable. These primitives include:

- PBEGIN: Initialize the environment
- PEND: Clean up the environment
- MYID: Identify the calling process
- NPROCS: Number of user processes
- SSEND: Synchronous send of a message
- SRECV: Synchronous receive
- ASEND: Asynchronous send of a message
- ARECV: Asynchronous receive
- PROBE: Determine if a message exists to receive
- WHAT_TIME: Current value of system clock
- INFO_PID, INFO_LEN, INFO_TYPE: Display information from last receive or probe operation
- BCAST: Broadcast
- Global Arithmetic Operations: Global sum, product, max. and min.

During the present study, the portability of this library was tested on Intel iPSC/860 parallel processors, a network of IBM RS/6000 workstations and Alliant FX/80. Once the CFD code is written by using GPAR calls, it becomes portable on all machines which are supported by the APPL library. In the case of a network of RS6000 workstations, one is able to specify the number of available workstations and the number of blocks to be solved on each workstation. In this case, several blocks can be assigned to a single workstation. This provides more flexibility in terms of load balancing.

3. A GRID ORIENTED DATABASE FOR PARALLEL COMPUTING (GPAR)

The GPAR database is simple and contains only two entities: A **block** and an **interface**. A block consists of an assembly of grid points. It can be derived from a structured or unstructured grid. For structured grids, only the coordinates of the grid points are in this data set. In the case of unstructured grids, a connectivity list for the grid points is also maintained. An interface is defined for a pair of neighboring blocks between which boundary conditions are exchanged. It includes two sets of grid points. It includes a set of grid points from the two neighboring blocks which define the interface boundary. The grid points as defined from both sets can be matching or unmatching. They can be overlapping or non-overlapping. Simple, two-dimensional examples of these combinations are shown in Figure 2. for structured grids. Unstructured grids are also supported in GPAR. Each interface is assigned to a block. It also contains a pointer indicating the neighboring interface on the adjacent block. The number of interfaces for each block is equal to the number of neighboring blocks connected to the particular block. Depending on the algorithm, usually, the number of grid points forming an interface is smaller than the number of grid points in a single block.

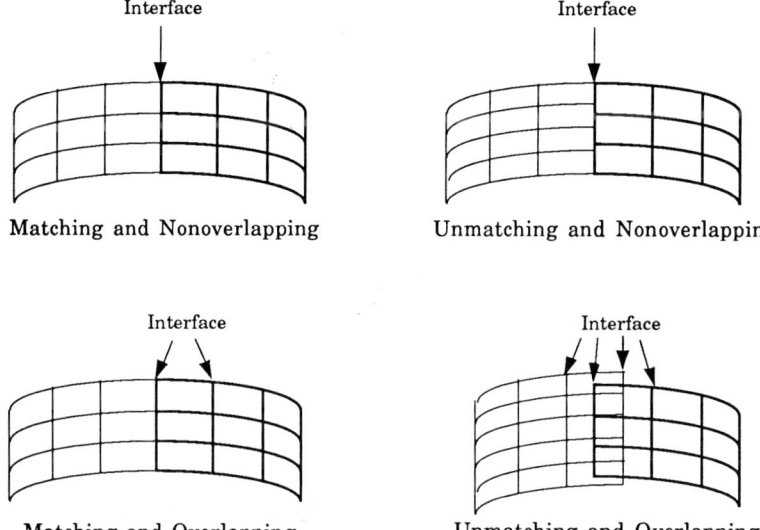

Figure 2. Different types of interfaces for structured grids.

The main features of GPAR can be summarized as follows:
- It acts as a block and interface manager during the parallel computations,
- It utilizes a machine portable library (APPL),
- It consists of library functions for:
  - Reading and generating block data,
  - Reading and generation interface data,
  - Exchanging block and interface information during the computations,
  - Updating block and interface data during the computations.

GPAR supports the following list of primitives for manipulating the data base:
- *dbSetup*: initialize the data base,
- *dbCheck*: check the consistency of the data base,
- *ReadandGenerateBk*: read the contents of a block from a file and generate corresponding interface entry in the database
- *ReadandGenerateIf*: read the contents of an interface from a file and generate corresponding interface entry in the database
- *GenerateBk*: generate block entry in the database
- *GenerateIf*: generate interface entry in the database
- *SendBk*: send a block to a specific processor
- *SendIf*: send an interface to a specific processor
- *RecvBk*: receive a block
- *RecvIf*: receive an interface
- *SendNeighIf:* send an interface to the processor of the neighboring interface

- *RecvNeighIf*: receive the incoming neighboring interface
- *UpdateIffBk*: update the values at the interface with the values at the block
- *UpdateBkfIf*: update the values at the block with the values at the interface

As can be seen from the above list, GPAR provides two types of conveniences to the CFD algorithm developer:

- Send and receive tasks necessary for parallel computing have to be planned only for two data sets: blocks and interfaces.
- The origin and destination of data are described in terms of grid blocks and their interfaces rather then a collection of processors and communication channels.

4. PARALLELIZING A CFD CODE BY USING GPAR

The main objective of the present activity was to enable a CFD code developer to write a parallel code without being penalized for working with a parallel computer. When using GPAR one has to follow the steps described below:

1. Create **block** and **interface** files: As mentioned above, many block-structured grid generation schemes are being used by CFD researchers. The creation of such files from these schemes is usually straight forward. In the case of a single-block grid, one has to divide the grid into a series of blocks to generate such data.

2. Provide a **setup** procedure to read and generate blocks and interfaces by using the GPAR primitives:

   ReadandGenerateBlocks (Blockfilename)
   ReadandGenerateInterfaces (Interfacefilename)

3. Provide a **block solver**: This is the CFD algorithm which operates on a set of given grid points. It includes no read or write statements. At each time step, the block solver sets up the equations for a given block of grid points, uses the boundary conditions established by the interface solver and solves the equations for a block. This block solver is duplicated on each processor. Of course, one can have more than one block solver, e.g., a potential, Euler or Navier-Stokes, in different blocks. The process involves parallelization of the data set and the algorithm at a block level. We assume that one can, for example, "vectorize" a block solver at a much lower level which will not be addressed here.

4. Provide an **interface solver**: Again the user has to provide an algorithm to interchange boundary conditions between the neighboring blocks. At each time step, the interface solver sends computed boundary conditions to neighboring blocks, receives boundary conditions from adjacent blocks, updates block boundary conditions. In the case of an explicit scheme with matching grids, the process is simple. For example, for a two-dimensional

grid, the interface involves three sets of grid points, one set is on the boundary which is common to both blocks while the other two sets are defined by each block. At the end of a block solution, the values of an interface which are common to that particular block are automatically updated. The user has to call only the GPAR routine, *SendInterface (Interface number)*, to send the values of the interface to its neighbor. Also, by calling, *ReceiveInterface (InterfaceNumber)*, it receives and updates the interface values as calculated by its neighbor. In the case of explicit schemes, the grid points which lie on the boundary are calculated twice, once in each block. If one uses many small blocks, there is an overhead due to this duplication [12].

The above steps have to be repeated for each processor. A host program is optional if one reads the block and interface data directly into the processors without going through a host. Further details of the implementation are described below for a specific application.

5. PARALLEL IMPLEMENTATION OF A CFD CODE USING GPAR

A three-dimensional, unsteady, compressible potential solver was parallelized using GPAR. The algorithm is based on the finite element method and involves a two-step, explicit solver [12]. The governing equations for density and velocity potential are written as follows:

$$\frac{\partial \rho}{\partial t} + \frac{\partial}{\partial x_i}(\rho u_i) = 0 \qquad (1)$$

$$\frac{\partial \phi}{\partial t} = H_o - \frac{1}{2} u_i u_i - \frac{\gamma}{\gamma+1} \frac{p}{\rho} \qquad (2)$$

$$u_i = \frac{\partial \phi}{\partial x_i} \qquad (3)$$

$$p = \kappa \rho^\gamma \qquad (4)$$

where, u_i is the velocity component, ρ is the mass density, ϕ is the velocity potential function, p is the static pressure, H_O is the total enthalpy, γ is the ratio of specific heats and κ is a constant. The solution scheme starts, as a first step, with the forward time computation of density with upwinding. As a second step, the equation for the velocity potential is calculated explicitly and with forward time-differencing. However in this case, the recently calculated values of density for the forward time step is utilized. The flow chart for the application program is shown in Figure 3. As can be seen from this chart, each time step requires a pair of block and interface solvers. This explicit CFD algorithm was employed to demonstrate the applicability of GPAR as will be discussed.

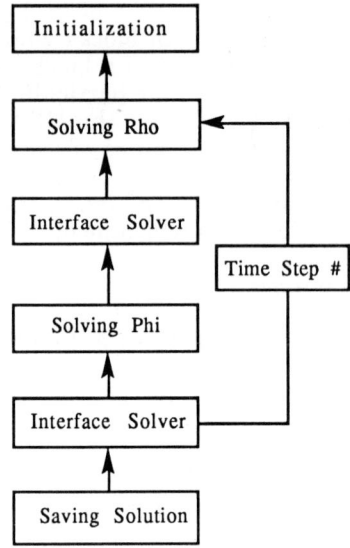

Figure 3. The flow-chart of flow analysis program.

6. PERFORMANCE STUDIES

6.1 Introduction

The selected numerical example is the computation of transonic potential flow around a circular-arc airfoil with M =0.675 at the inlet[12]. The computational grid employed in the analysis is an H-type grid as shown in Figure 4. The employed finite element algorithm assumes that the grid is unstructured even though the particular grid used is structured. Figure 5 shows a typical six block configuration of this grid for which the blocks are not exactly equal in size. Although it is desirable to have blocks of equal size, it is not always practical. For this particular grid, the same problem was solved by dividing the same grid into different numbers of blocks. Since the scheme is explicit the numerical results are identical, however, there is more repetition of computations at the interfaces for larger numbers of blocks.

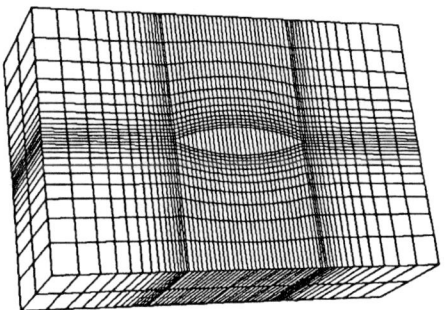
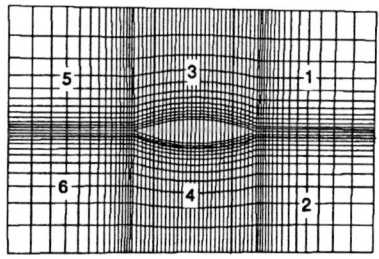

Figure 4. Computational grid (3x25x61 grid points).

Figure 5. A six-block model of the computational grid.

6.2 Timing Studies on INTEL iPSC/860

The real time required for each iteration step on an Intel iPSC/860 machine is shown in Figure 6. The number of blocks in each case is equal to the number of processors utilized. As can be seen from this figure, the efficiency decreases with decreasing size of blocks. The interface solver, which includes the communication between the processors, requires a negligible amount of time. Up to twelve blocks, with 500 grid points per block, the communication cost is small in comparison with the computational effort required to solve the equations in each block, even for this explicit scheme. The overall speedup of the parallelization is shown in Figure 7, where the relationship is almost linear yet less than the ideal speedup.

The details of the observed efficiency is further studied. The efficiency is defined as,

$$E = \text{speedup/number of processors}$$

A source of loss in efficiency is due to the overhead in computing the grid points located at the block interfaces. The efficiency due to such an overhead is defined as,

$$E_0 = 1 - O$$

where O is the ratio of the number of elements located at the interfaces of a block to the number of elements inside a block. The comparison of actual computed efficiency, E, and the efficiency due to the above overhead is shown in Figure 8. The difference between these two curves is due to the differences between the sizes of blocks assigned to each processor. Thus, even for the chosen simple example, where the size of the blocks are not drastically different, the load balancing becomes important.

The same grid was further refined to obtain the solution of a larger problem. The grid with 3x25x61 grid points and fifteen blocks was refined to obtain a 22x25x61 grid, while keeping the number of blocks the same. The ratios of the sizes of different blocks were also kept constant. The obtained results are summarized in Table 1. One can observe that while 420 grid points on a processor required less than one second per time step, 3000 grid points required 10 seconds. Such computational speeds are acceptable, considering that 3000 grid points for unstructured grids may fully exploit the memory of a single processor. The interface solvers take appreciable time only due to the wait between the blocks with uneven sizes. It was observed that by using GPAR and APPL, communication costs were kept at a negligible level with minimum effort in programming.

In Figure 9, the cost of solving a block with a given number of grid points is shown. The computation time again varies almost linearly up to 3500 grid points. This figure shows the breakdown of the computation time for two block and two interface solvers and their sum. One can utilize such a curve for balancing the loads between the different processors.

Table 1. CPU time requirements in seconds for solution of each block of the fifteen-block case.

	Solve Rho	Boundary Solver 1	Solve Phi	Boundary Solver 2	Total Time	Grid Nodes IxJxK	Number of Nodes
Block 1	2.455	0.468	7.431	0.055	10.409	22x10x14	3,080
Block 2	2.456	0.457	7.430	0.063	10.406	22x10x14	3,080
Block 3	2.183	0.749	6.605	0.870	10.407	22x9x14	2,772
Block 4	2.455	0.419	7.422	0.104	10.400	22x10x14	3,080
Block 5	2.460	0.019	7.430	0.110	10.019	22x10x14	3,080
Block 6	2.182	0.701	6.597	0.928	10.408	22x9x14	2,772
Block 7	2.456	0.414	7.446	0.092	10.408	22x10x14	3,080
Block 8	2.462	0.401	7.480	0.064	10.407	22x10x14	3,388
Block 9	2.183	0.696	6.614	0.914	10.407	22x9x14	2,772
Block 10	2.458	0.037	7.463	0.450	10.408	22x10x14	3,080
Block 11	2.462	0.029	7.454	0.463	10.408	22x10x14	3,388
Block 12	2.185	0.320	6.631	1.272	10.408	22x9x14	2,772
Block 13	2.225	0.832	6.867	0.647	10.571	22x10x13	2,860
Block 14	2.253	0.821	6.849	0.671	10.594	22x10x13	3,146
Block 15	1.999	1.092	6.105	1.399	10.595	22x9x13	2,574
Average	2.325	0.497	7.055	0.540	10.417		

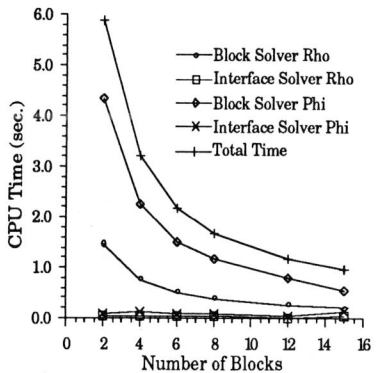

Figure 6. Computation times (seconds per step) for the test problem with different number of blocks.

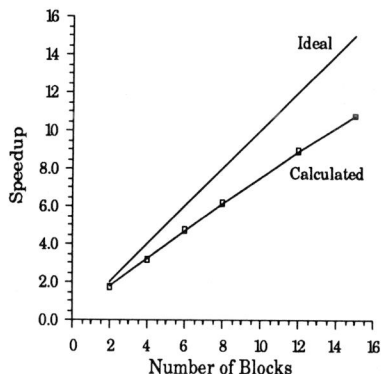

Figure 7. Speedup obtained for the test problem by increasing the number of processors.

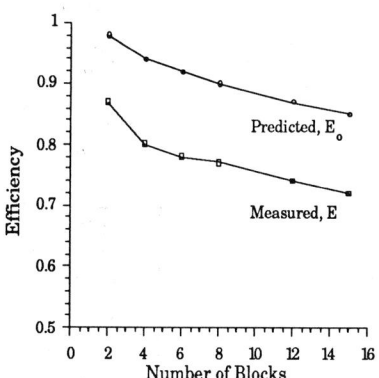

Figure 8. Efficiency of the computations for the test case (3x25x61 grid points).

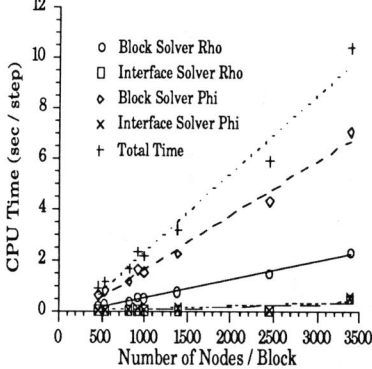

Figure 9. Computation times for a block with different number of grid points.

6.3 Timing Studies on a Network of RS6000 Workstations

For the cases tested on Intel iPSC/860 machine, it was observed that the communication costs were small and the computation time was acceptable. However, unbalanced loading contributed to the wait time and decreased the efficiency with increasing number of blocks. The main restriction, in terms of load balancing, at this time, was due to the fact that there was always a single block on a single processor. The same basic problem of 3x25x61 grid points was studied on a network of IBM RS6000/320 workstations. In this case, there were three workstations with peak single-precision performance ratings of 11.7, 9.2 and 11.7 MFLOPS, respectively. For this application, the load balancing was dependent on the machine speeds, as well as the load distribution. The three workstations were connected with Ethernet which was considerably slower than the Intel Interprocessor Communication Network. With APPL, GPAR can distribute any given number of blocks on a given number of workstations. The block solver, which remained the same on both Intel and IBM machines, and the associated interface solvers were assigned as separate processes on a UNIX workstation. In the case of a workstation with several blocks, these processes were executed simultaneously and competed with each other.

The original grid of 3x25x61 grid points was analyzed. This grid was studied on one, two and three machines with different numbers of blocks on each machine. The obtained results are summarized in Table 2. The computation time for the entire grid as a single block on a single machine is 6.1 sec/time step. This compares to 4.6 sec/time step on an iPSC/860.

Table 2. CPU requirements in seconds for solutions with a cluster of IBM RS6000/320 computers and different combination of blocks (3x25x61 grid points).

Machine 1	Machine 2	Machine 3	Elapsed Time (sec/time step)	Speedup S_n	Efficiency S_n/n
1 Block			6.1	1.00	1.00
8 Blocks			6.4	0.95	0.95
4 Blocks	4 Blocks		4.0	1.53	0.77
6 Blocks	6 Blocks		4.5	1.36	0.68
1 Block	1 Block	1 Block	3.0	2.03	0.67
4 Blocks	4 Blocks	4 Blocks	3.3	1.85	0.62
3 Blocks	3 Blocks	2 Blocks	3.4	1.79	0.60

Eight blocks on a single machine produced an efficiency of 95%. This value, in fact, is higher than the predicted efficiency of $E_0 = .90$ in Figure 11. It suggests that eight processes sharing a single processor may be an efficient combination. When two machines, with four blocks on each, were employed, the efficiency went down to 77%. This is mainly due to the communication costs, waiting and the differences between the speeds of the machines. Two machines, with six blocks each, demonstrated a similar behavior. The use of three machines further deteriorated the efficiency. In these cases, the need for better communication and load balancing became apparent. However, these are the particular cases one would like to optimize load balancing for improving the efficiency. When the number of blocks are larger than the number of processors such an optimization becomes practical. As can be seen from the above examples, without such an optimization, increasing the number of blocks deteriorates the efficiency. The case of massively parallel computers requires considerable care in terms of load balancing for complex problems. The tools such as GPAR and APPL may be useful in such an environment.

6.4 MFLOP Performance

In order to measure the MFLOP performance of each processor in double-precision, we have performed simple benchmark cases involving additions, subtractions, multiplications, divisions and exponentiations. These cases did not require the use of any arrays and IF statements. The operations were performed within a FORTRAN DO loop, and no I/O was required. Shown in Table 3 is a summary of the measured performance for each operation. Also, the slowness of division and exponentiation operations on iPSC is rather excessive. We have further compared our measured ideal MFLOP performances with the performance of our parallel flow code. As may be observed from Table 4, we reach to about 45% of the ideal performance on iPSC/860 and 74% on RS6000. The slowness of division and exponentiation operations accounts for the lower performance reached on iPSC/860. The performance calculation of the parallel code was based on a count of all arithmetic operations converted to an equivalent number of double-precision multiplications.

Table 3. Measured MFLOP performance of each machine in double-precision arithmetic operations (under ideal conditions).

Machine	Addition	Subtraction	Multiplication	Division	Exponentiation
iPSC/860	6.633	6.638	6.633	0.765	0.326
IBM RS6000/ 320H	8.333	8.333	8.333	1.163	1.250

Table 4. Comparison of measured MFLOP performance in double-precision multiplications.

Machine	Ideal	Parallel Flow Code
iPSC/860	6.63	2.99
IBM RS6000/320H	8.33	6.13

7. CONCLUSIONS

A database management system was developed specifically for the parallel processing of CFD algorithms which operate on a computational grid. The objectives achieved through this development are: portability, efficiency and ease in programming in a parallel environment. The important features of the development were the use of an efficient and portable parallel library and a data base defined for computational grids. The portability and efficiency issues were tested on an iPSC/860 and a network of RS6000/320 computers. For an explicit scheme, the communication cost between the nodes on the iPSC/860 was reduced to a negligible level. The same algorithm was readily portable to the distributed RS6000 environment. By using the developed database management system, it is possible to test several load balancing issues in a parallel environment without any extra effort.

8. ACKNOWLEDGMENTS

This research was supported by NASA Lewis Research Center under grant No. NAG3-1246.

9. REFERENCES

1. Ecer, A., Spyropoulos, J.T. and Maul, J., "A Block-Structured Finite Element Grid Generation Scheme for the Analysis of Three-Dimensional Transonic Flows," *AIAA Journal,* Vol. 23, No. 10, October 1985.

2. Thompson, J.F., Warsi, Z.U.A. and Mastin, C.W., *Numerical Grid Generation: Foundations and Applications*, North-Holland, The Netherlands, 1985.

3. The GRIDGEN 3D Multiple Block Grid Generation System, Volume II: User's Manual, *WRDC-TR-90-3022,* Flight Dynamics Laboratory, Wright Research and Development Center, Wright-Patterson, Ohio, 1990.

4. Ecer A. and Spyropoulos, J.T., "Block-Structured Solution Scheme for Analyzing Three-Dimensional Transonic Potential Flows," *AIAA Journal*, Vol. 25, No. 10, October 1987.

5. Ecer, A., Spyropulos, J.T. and Rubek V., "Block-Structured Solution of Euler Equations for Transonic Flows," *AIAA-86-1080, 4th Fluid Mechanics, Plasma Dynamics and Lasers Conference,* Atlanta, Georgia, May 1986.

6. Glowinski, R., Dihn, O.V. and Periaux, J., "Domain Decomposition Methods for Nonlinear Problems," *Computer Methods in Applied Mechanics and Engineering*, Vol. 40, 1983.

7. Benek, J.A., Dougherty, F.C. and Buning, P.G., "Chimera: A Grid Embedding Technique," *AEDC-TR-85-64*, December 1985.

8. Gropp, W.D. and Keyes, D.E., "Domain Decomposition Methods in Computational Fluid Dynamics," *International Journal for Numerical Methods in Fluids*, Vol. 14, 1992.

9. Ecer, A., Akay, H.U. and Erwin, S., "A Parallel Algorithm for the Solutions of 3D Euler Equatiions in Turbomachinery," *Proceedings of the Fourth Conference on Hypercube Concurrent Computers and Applications,* Monterey, California, March 6-8, 1989.

10. Ecer, A., Chang, S.M. and Spyropoulos, J.T., "Parallel Computation of Three-Dimensional Transonic Flows," *Proceedings of the AIAA Aerospace Sciences Meeting*, Nevada, January 1990.

11. Akay, H.U. and Beskok, A., "A Parallel Algorithm for Compressible Flows Through Rotor-Stator Combinations," *Parallel CFD '90, Research Directions in Parallel CFD,* edited by H.D. Simon, MIT Press, Cambridge, Mass., 1991.

12. Akay, H.U., Ecer, A. and Kemle, W.B., "A Parallel Explicit Solver for Unsteady Compressible Flows, Parallel Computational Fluid Dynamics '91, edited by K.G. Reinsch et al., Elsevier Science Publishers, The Netherlands, 1992.

13. Quealy, A., "Portable Programming on Parallel Networked Computers Using Aplication Portable Parallel Library (APPL), *NASA Technical Reports*, 1992 (in preparation).

A MULTIBLOCK HYPERSONIC FLOW SOLVER FOR MASSIVELY PARALLEL COMPUTER

Salvatore Borrelli, Alfonso Matrone & Pasquale Schiano

C.I.R.A., *Centro Italiano Ricerche Aerospaziali*, 81043 Capua, Italy

Abstract

A parallel implementation on a MIMD computer of a non equilibrium hypersonic flow solver, based on a Domain Decomposition Technique (D.D.T.), is showed.

The 2D physical domain is subdivided in "slices" according to one or both of the axes of the domain reference frame. The different types of domain decomposition are compared in accordance with the arising of load balancing problems that occur for the evaluation of production terms. The capability to handle, in efficient way, the domain shows a solution of this problem. Such a solution is investigated to get a dinamically assigned computational load for each processor. Finally the handling of multiblock grids is performed. An algorithm, called "Masked Multiblock", independent of the physical subdivision in blocks of the grid, is developed.

1 Introduction

Parallel computing is a way to reach better results in the aerodynamic field. The growth of the H/W and S/W of the computer systems and, consequently, the improvement in terms of computational performances has permitted to develop more accurate analysis of flows around complex aerodynamics configurations. In any case, the computational power of state-of-art computers is still inadequate to solve with sufficient accuracy a large part of aerodynamics flow fields around an aircraft. Especially hypersonic flows that involve chemical reactions are quite heavy from the computational point of view.

To reply at these needs of the aerospace community, and in general the whole scientific world, there has been a proliferation of new computer architectures that are different from classical serial computers: the massively parallel computers. These machines seem to be the most promising to achieve orders-of-magnitude increases in computational power at reasonable costs, even if, at this time, these computers are quite hard to use by an end-user.

One of the most effective techniques to use efficiently this kind of machines consists of a decomposition of the physical domain of the problem, where the calculations and the data are spread among different processors. The paper is concerned with a parallel implementation of a 2D non equilibrium reacting flow solver using an upwind methodology based on the FDS formulation. Different types of domain decomposition, meeting the problem of the dynamic load balancing that occurs with the evaluation of the production terms, are implemented. Moreover an approach for the multiblock solver is showed.

The target architecture is a MEIKO COMPUTING Surface composed by INMOS T800 processors.

2 The domain decomposition technique

The idea to reduce the solution of differential problems defined on a domain Ω to the concurrent solution of problems of same type corresponding to subdomains Ω_i was early introduced by Schwartz [1]. He used a Domain Decomposition Technique (D.D.T.) to demonstrate the existence of harmonic functions on regions with irregular boundaries. The D.D.T. can be regarded as a "divide and conquer" algorithm, and it is now largely diffused in parallel C.F.D.. It allows to use efficiently parallel computers, keeping the consolidate numerical schemes used in C.F.D.[2]. The basic idea of the D.D.T. is showed below:
let consider a general conservation law in integral form:

$$\frac{\partial}{\partial t}\int_D \vec{W} d\tau + \int_{\delta D} \vec{F}(\vec{W}) ds = \int_D \vec{P}(\vec{W}) d\tau \qquad (1)$$

where D is the global domain which the problem is defined on.
If $D = \bigcup_{h=1}^{k} D_h$ the equation (1) can be replaced with a set of k independent equations

$$\frac{\partial}{\partial t}\int_{D_h} \vec{W}_h d\tau + \int_{\delta D_h} \vec{F}(\vec{W}_h) ds = \int_{D_h} \vec{P}(\vec{W}_h) d\tau \qquad (2)$$

where $\vec{W}_h = \vec{W}/D_h$. The problems (2) can be solved concurrently: the only difficulty comes from the numerical evaluation of flux terms at the interfaces of subdomains D_h.

3 The fluid dynamic model

The physics of hypersonic flows is rather complicated. The Euler equations well describe the basic fluid-dynamics. Thermodynamical and chemical models (we adopt the model given in [3]) close the system of the governing equations. The matching and the interaction of the fluid-dynamics with the relaxations characterize the hypersonic flow. The complete system of equations in integral form is:

$$\frac{\partial}{\partial t}\int_{Vol} \rho_i\, dVol + \int_{\partial Vol} \rho_i \mathbf{V}\cdot d\mathbf{S} = \int_{Vol} \Omega_i\, dVol \qquad i=1,3$$

$$\frac{\partial}{\partial t}\int_{Vol} \rho\, dVol + \int_{\partial Vol} \rho\, \mathbf{V}\cdot d\mathbf{S} = 0 \qquad (3)$$

$$\frac{\partial}{\partial t}\int_{Vol} \rho \mathbf{V}\, dVol + \int_{\partial Vol} \rho \mathbf{V}(\mathbf{V}\cdot d\mathbf{S}) + \int_{\partial Vol} p\, d\mathbf{S} = 0$$

$$\frac{\partial}{\partial t}\int_{Vol} e\, dVol + \int_{\partial Vol} (p+e)\, \mathbf{V}\cdot d\mathbf{S} = 0$$

where the concentrations Y of the molecular oxygen O_2 and of the molecular nitrogen N_2 follow from the conservation of the atomic species:

$$Y_{O_2} = .233 - \frac{\mu_{O_2}}{2}\left(\frac{Y_O}{\mu_O} + \frac{Y_{NO}}{\mu_{NO}}\right) \; ; \; Y_{N_2} = .767 - \frac{\mu_{N_2}}{2}\left(\frac{Y_N}{\mu_N} + \frac{Y_{NO}}{\mu_{NO}}\right)$$

The equation of state of the gas mixture and the enthalpy are given by:

$$\frac{p}{\rho} = \sum_{i=1}^{5} R_i Y_i T \; ; \; h = \sum_{i=1}^{5} c_{pi} Y_i T + \sum_{i=1}^{3} Y_i h_i^o + \sum_{i=3}^{5} Y_i e_i^{vib}$$

being h_i^o the heat of formation of the i species, e_i^{vib} the vibrational energy and $c_{pi} = \frac{5}{2}R_i$ (for O and N) and $\frac{7}{2}R_i$ (for NO, O_2 and N_2). In the present work we use the Lighthill approximation $e_i^{vib} = .5 c_{pi} T$.

The discretization occurs according to a standard *Finite Volume* technique applied straightforward to the integral form of the governing equations. The computational domain is divided in quadrilater cells (volumes) and the net flux is evaluated according with "Flux Difference Splitting Formulation (FDS)" as it will shown afterwards. Increasing the Damköholer number and approaching the equilibrium conditions, the source terms in the chemical equations induce severe numerical instabilities. The remedy is well known and founded by carrying out an implicit evaluation of the source terms.

As mentioned above the interface flux is evaluated according with FDS formulation whose basic steps are the definition and the solution of appropriate Riemann problems. The solution of the Riemann problem, that is the collapse of the initial discontinuity, is described on the basis of the quasi linear form of Euler equations. Indicating with a,b,c,d the four regions delimited from the three waves of a Riemann problem, through the wave *I* the following relationship are verified:

$$\begin{aligned} Y_{ic} &= Y_{ia} & (i=1,..,3) \\ p_c + (\rho_a\, a_{fa})\, u_c &= p_a + (\rho_a\, a_{fa})\, u_a \\ v_c &= v_a \\ h_c - p_c/\rho_a &= h_a - p_a/\rho_a \end{aligned}$$

On the contact surface (wave *II*), the usual continuity of the pressure and velocity are imposed:

$$p_c = p_d \quad ; \quad u_c = u_d$$

finally, through the wave *III*:

$$\begin{aligned} Y_{id} &= Y_{ib} & (i=1,..,3)\\ p_d - (\rho_b\, a_{fb})\, u_d &= p_b - (\rho_b\, a_{fb})\, u_b\\ v_d &= v_b\\ h_d - p_d/\rho_b &= h_b - p_b/\rho_b \end{aligned}$$

The frozen speed of sound, that appears in the continuity equation, is defined by:

$$a_f^2 = \frac{h_\rho}{1/\rho - h_p} \tag{4}$$

On the basis of this equations, it is possible to evaluate all the flow properties. To evaluate the boundary condition, since there are no volumes behind the wall, a Riemann problem in the same way as in the interior volumes cannot be defined. So, an *half* Riemann problem with the flow properties evaluated at inner point is defined and a replacement of those unavailable values with the boundary conditions is performed. Since the flow in this point does not respect, in general, those boundary conditions, a wave flowing backward from the boundary in the cell is expected. In the new region generated behind this wave, both the signals flowing on characteristics crossing the wave, and the boundary conditions are imposed. We refer the reader to [4-6] for the details on this matter.

4 The parallel architecture

The distribuited memory MIMD machines are the most diffused and the most promising massively parallel computers. In this classification falls the transputer network that we have used for the parallel implementation. The machine is a MEIKO Computing Surface composed by up to 256 Inmos T800 - 25 Mhz processors hosted on a Sun Sparc 4/330, installed at CNR/IRSIP and accessed by a link at 64Kb/sec. These types of systems permit to investigate the MIMD parallelism at a moderate cost and with a scalable number of processors.

The code is entirely written in ANSI FORTRAN with calls to the MEIKO communication library, named CS-TOOLS [7], for the message passing between the processors.

5 The parallel implementation

A single subdomain and the strips of the cells belonging to the contiguous subdomains (overlapping areas) are assigned to each processor [8-9]. The overlapping areas are need to compute the fluxes on the interfaces between the blocks. The values of the solution in the overlapping areas are updated at each time step.

A Master-Slave programming model is adopted. The master process is concerned with the data inizialization, the global time step calculation and the evaluation of maximum of the residual. The slave processes, instead, execute the main part of the work operating on their computational blocks.

The network interconnection topologies adopted are a double ring in the 1D decomposition cases and a torus in the 2D decomposition case. The total communication work is divided in global communications (master to slaves and viceversa) and local communications (between neighbour slaves).

At the first, a comparative study about the partitioning ways of the computational domain was made. Two 1D decompositions (along I axis and along J axis) and a 2D decomposition were implemented (fig. 1-3).

The best performances were obtained in the 1D partitioning along I axis because it permits the best load balancing. It depends on the following considerations:

- in the points where the production terms are evaluated ($T > 2000$) more computational work is needed;

- the reactive region is concentrated around the body surface (fig. 4).

consequently in the 1D decomposition along J axis and in the 2D decomposition cases some processors work always on whole subdomains where $T > 2000K$ and others on whole subdomains where $T < 2000K$. In the figure (5) a comparison, in terms of speed-up, between 1D along I decomposition and 2D decomposition after 5000 iterations is showed.

6 The load balancing problem

Generally, when a D.D.T. is used, the partitioning of the computational domain is performed in a uniform way (some number of grid points per each subdomain). For reactive fluxes this kind of partitioning results inadequate. As a matter of fact it is not possible to predict when and where the numerical evaluation of the production terms is performed; such evaluation produces an additional computational work. In our case the ratio between the computational work to evaluate the solution in a point where is necessary to calculate the production terms and in a point where it is not necessary is almost 10. In conclusion, a fixed domain decomposition produces a bad load balancing.

A procedure to balance the load in a dynamic way (same number of floating point operations in each subdomain per each time iteration step) was developed. A brief description of it is showed in the following.

Let D be the whole computational domain and $\{D_1^{(i)}, D_2^{(i)}, ...D_k^{(i)}\}$ a partition of it at the time step i. A discrete function

$$f : D \to \Re$$

measuring the computational work in each point of D is defined. If the f_h $h = 1,..k$ are the restrictions of f on $D_h^{(i)}$, the total computational work in each $D_h^{(i)}$ is

$$l_h^{(i)} = \sum_{x \in D_h} f_h(x)$$

moreover, the global computational work and the average load are, respectively:

$$L^{(i)} = \sum_{h=1}^{k} l_h$$

$$\langle L^{(i)} \rangle = \frac{L^{(i)}}{k}$$

By comparison between the $l_h^{(i)}$ and the average load, a new partition

$$\{D_1^{(i+1)}, D_2^{(i+1)}, ... D_k^{(i+1)}\}$$

so that $l_h^{(i+1)} \sim \langle L^{(i)} \rangle$ is determined.
In our case the function f is:

$$f : x \in D \longrightarrow \begin{cases} 1 & \text{if } T(x,y,t) > 2000K \\ 10 & \text{if } T(x,y,t) \leq 2000K \end{cases}$$

where T is the temperature.

The figure (6) shows a comparison between the times per cycle of the balancing and the non balancing versions a 192 X 48 grid on 4 processors, in picture (7) the improvements, in terms of speedup, are showed.

7 The masked multiblock

In the last years, the increasing interest in the simulation of flows over complex configurations, such as a complete aircraft, leaded to new computational techniques; one of these are the multiblock methods.

Some flow regions are too complex to be mapped into a single block grid, with reasonable skewness and grid smoothness restrictions; so the flow region is splitted into a set of blocks where each block is composed by a several number of grid points. The advantage is not only that the grid smoothness properties are improved, but also that it is possible to concentrate more grid points in critical regions and to allocate a less number of them in others flow regions.

This decomposition implies that instead of one structured grid, a number of complete subgrids are applied. The difficulties arise from the coupling of the solutions on the interfaces of contiguous blocks where a proper handling of the conditions is needed.

The parallel implementation of our hypersonic multiblock flow solver is complicated from the load balancing problems coming from the different computational work assigned at each cell for each time iteration step. The Masked Multiblock Algorithm (M.M.A.) is independent of the physical subdivision of the grid in blocks and assigns, according to

the dynamic load balancing algorithm, at each processor one, more or a part of subgrid. The basic idea of the M.M.A. is showed in the following:
let G be the grid composed by p blocks

$$\{G_1, G_2, \ldots, G_p\}$$

and $C^{(i)}$ the set of the computational subdomains at the time step i

$$\{C_1, C_2, \ldots, C_k\}$$

coming from the load balancing procedure. An operator, called Masking Operator,

$$\Gamma : C^i \longrightarrow G$$

so that

$$\bigcup_{j=1}^{k} \Gamma(C_j^{(i)}) = G$$

and

$$\Gamma(C_j^{(i)}) \subseteq \bigcup_{h \in H} G_h$$

with

$$H \subseteq \{1, 2 \ldots, k\}$$

is defined. In particular, if $G = C^{(i)}$ and $p = k$, Γ maps $C_j^{(i)}$ into G_j (this is the case when each block is assigned at one processor). The sets $C^{(i)}$ are choosen so that

$$P(C_j^{(i)}) \approx P(G)/k$$

where P is the function measuring the computational work and k is the number of processors (or subdomains). The Masking Operator enables for each processor, working on one of the C_j, to know what are the blocks or the subset of blocks assigned to it. In this way the multiblock structure of the solver is preserved.

The figg. (9-11) show how the masked multiblock approach works. Finally fig. (8) represents some results coming from the masked multiblock approach.

8 Conclusions

A parallel implementation of the multiblock hypersonic flow solver was presented. Using a well known slicing partitioning, different strategies of cutting were tested. Two algorithms were developed: the first, called LB, was introduced to contrast the decay of parallel performances induced by chemistry calculation: the second, called MMA, was developed to take into accont the two different stategies to decompose the domain:

1. geometrical multiblock partitioning

2. parallel slicing

Our conclusions are that parallel computer are able to well answer to the fluidynamicists requests of powerful calculation, even if in hypersonic a new class of problems arises.

References

[1] H.A.Schwarz, *Uber einige Abbildungsaufgaben*, Ges. Math. Abh. Vol.11, 1869.

[2] Chan, Glowinski, Periaux, Widlund, *Domain Decomposition Methods for Partial Differential Equation*, SIAM, 1990.

[3] C.Park, On Convergence of Computation of Chemically Reacting Flows, AIAA Paper-85-0247, Jan. 1985.

[4] M.Pandolfi, *On the Flux Difference Splitting Formulation*, Notes on Numerical Fluid Mechanics, Vol.24, Vieweg, 1989.

[5] M.Pandolfi and S.Borrelli *An upwind Formulation for Hypersonic Non-equilibrium Flows*, Modern Research Topics In Aerospace Propulsion, Springer & Verlag, 1991.

[6] S.Borrelli and M.Pandolfi *A Contibution to the Prediction of Hypersonic Non-equilibrium Flows*, Workshop on Hypersonic Flows for Reentry Problems, INRIA, Antibes, Jan. 1990.

[7] Meiko Computer, *CS-TOOLS Manual*, 1990.

[8] C.De Nicola, G.De Pietro and P.Schiano, *CFD Equations Solutions on Massively Parallel Computer*, Proc. of V SIAM Conf. on Parallel Computing, Prentice-Hall, 1992.

[9] S.Borrelli and P.Schiano, *A Non Equilibrium Hypersonic Flow Calculation on Massively Parallel Computer*, Proc. of Parallel CFD '91 Conf., Stuttgart, June 1991, North-Holland, 1992.

[10] M.Willebeek-LeMair, A.P.Reeves, *A Distribuited Dynamic Load Balancing Strategy for Higly Parallel Multicomputer Systems*, proc. of IV conf. on Parallel Processing for Scientific Computing, SIAM, 1989.

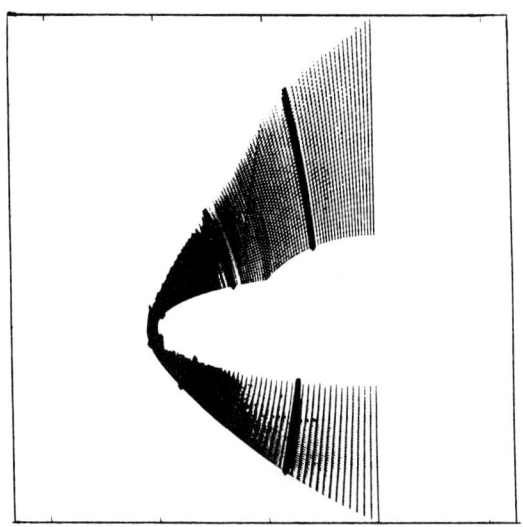

figure 1. 1D decomposition along I axis

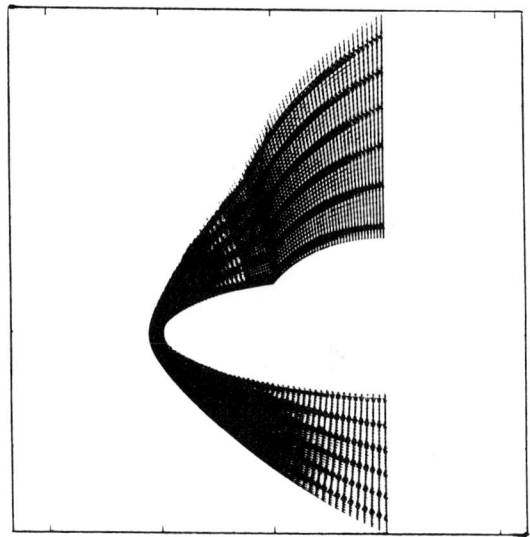

figure 2. 1D decomposition along J axis

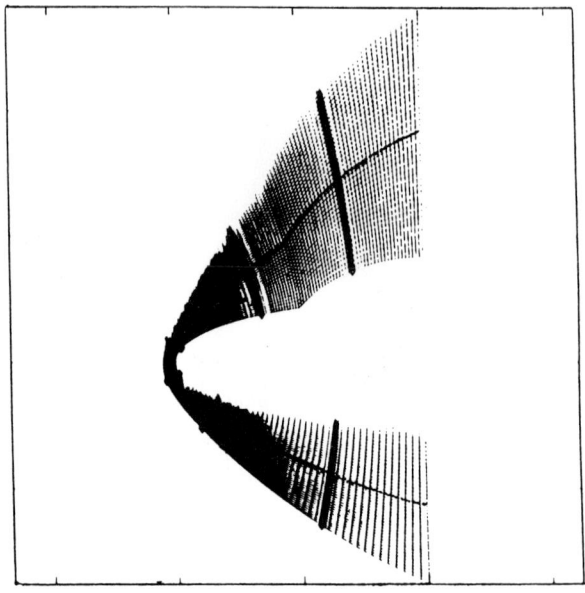

figure 3. 2D decomposition

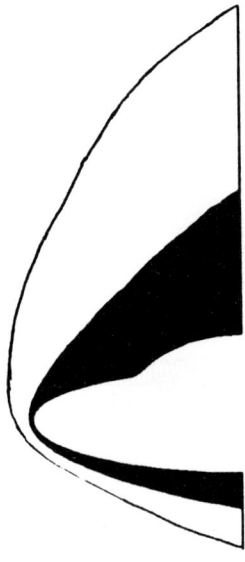

figure 4. Reactive region (black colour)

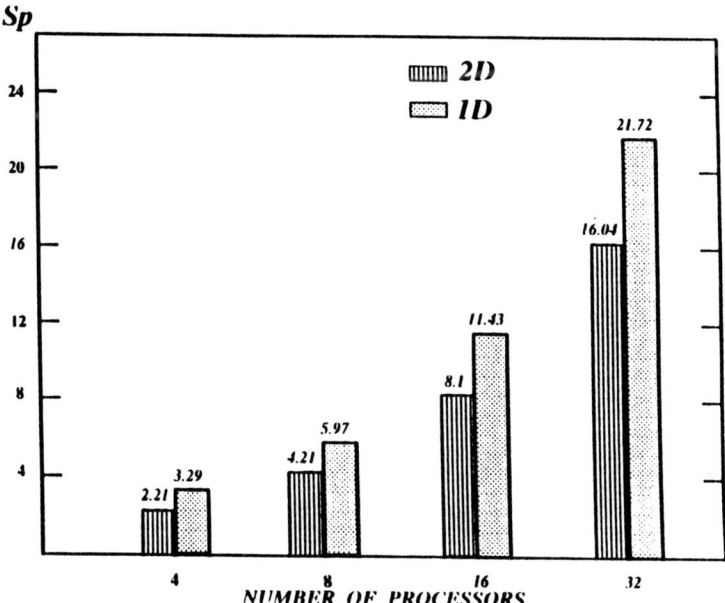

figure 5. Comparison between 1D dec. along I and 2D dec.

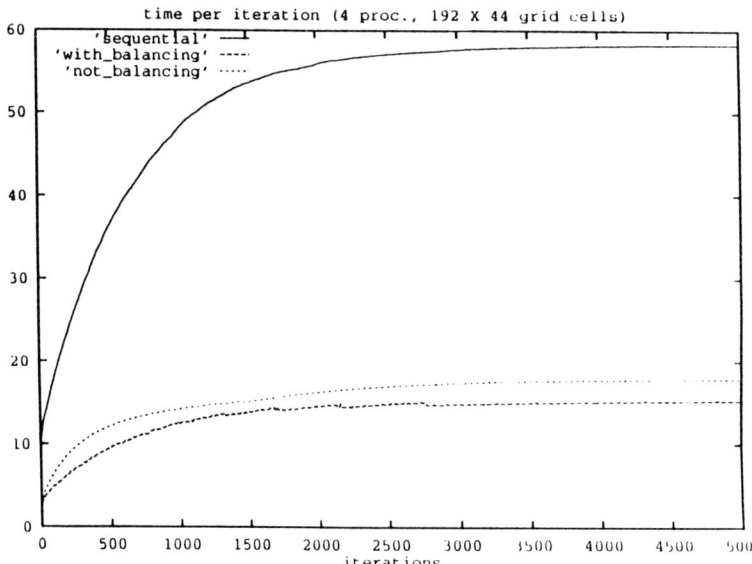

figure 6. Comparison between the times per cycle of the balancing and not balancing versions for a 4 processors case.

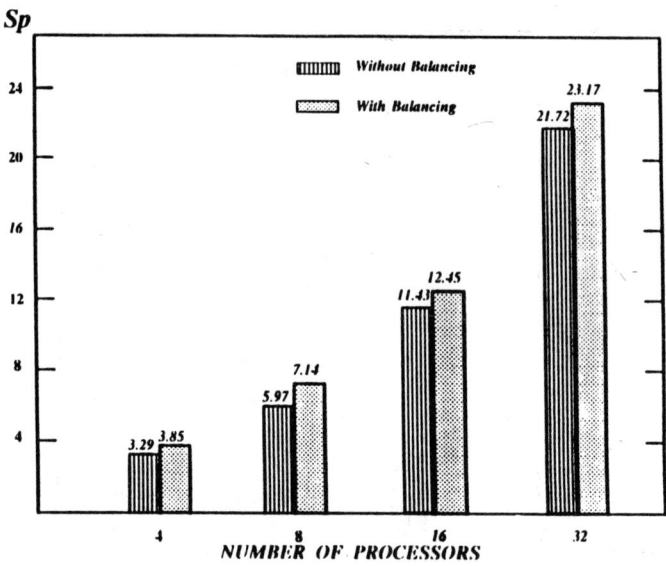

figure 7. Comparison between a fixed and a dynamic partitioning.

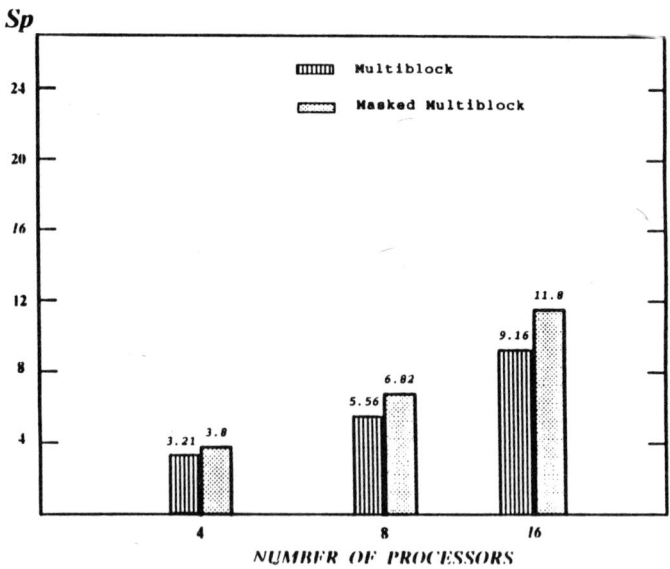

figure 8. Multiblock versus Masked Multiblock.

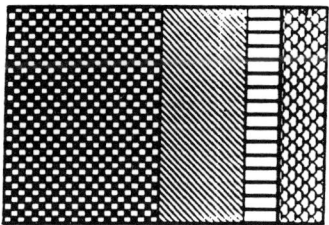

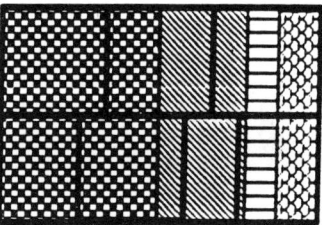

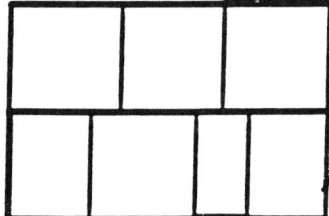

figs. 9-11. Masked blocks are calculated using the load balancing algorithm, independently of the grid blocks.

Applications of a Parallel Pressure–Correction Algorithm to 3D Turbomachinery Flows

Mark E. Braaten

GE Research and Development Center, Schenectady, NY 12031, USA

Abstract

A parallel algorithm for the solution of three-dimensional compressible flows in turbomachinery has been developed and demonstrated on scalable distributed memory multicomputers. The algorithm solves the compressible form of the Euler or Navier–Stokes equations via a compressible pressure correction formulation. To achieve high accuracy for highly turning blade rows, the computational grid is constructed without requiring strict periodicity of the grid points along the periodic boundaries between the blade passages. The impact of this feature on code parallelization and computational efficiency is described. The algorithm has been demonstrated on up to 128 processors of an Intel iPSC/860, and up to 256 processors of the Intel Touchstone Delta prototype. Performance 2.5 times faster than a single Cray Y–MP processor has been achieved for an inviscid turbomachinery calculation on 154,000 grid points with 256 processors of the Delta.

1. INTRODUCTION

The research objective of this work is to develop a parallel turbomachinery code capable of running on massively parallel computers at greater performance levels than are obtainable on conventional vector supercomputers. This capability will allow larger, more accurate simulations of viscous transonic turbomachinery flows than is now possible.

Traditionally, time-marching methods such as Jameson's explicit Runge-Kutta scheme [1] or the Beam-Warming implicit scheme [2] have been used to solve compressible turbomachinery flows. In recent years, pressure correction algorithms, originally developed to handle incompressible flows, have been successfully extended to handle compressible flows. In a series of earlier papers [3-5], this author described the development of a two-dimensional parallel pressure correction algorithm applicable to both incompressible and compressible flow. This algorithm was demonstrated for some simple inviscid and viscous turbomachinery test cases in Reference 5.

In this paper, the compressible pressure correction algorithm described in Reference 5 has been extended to three dimensions, and calculations for actual turbomachinery blade rows used in modern gas turbine engines have been performed on a large scalable distributed memory multicomputer. Complexities in the algorithm required for the proper treatment of highly turning blade rows have been succesfully addressed in the parallel implementation. High parallel efficiency and performance in excess of modern single processor supercomputers have been achieved.

2. BASIC DESCRIPTION OF ALGORITHM

The compressible pressure correction equation developed here solves the three-dimensional Euler equations for inviscid flow, or the fully elliptic form of the Navier-Stokes equations for viscous flows. The equations for conservation of x-, y-, and z-momentum, mass, and enthalpy are solved, along with the ideal gas equation of state. For turbulent flows, the standard k-e turbulence model is used, along with the wall function treatment for the near-wall regions.

The conventional pressure correction formulation [6], originally developed for incompressible flows, derives an equation for the pressure correction by manipulation of the discrete forms of the momentum and continuity equations. The momentum equations are first solved using a guessed pressure field. The pressure correction equation serves both to correct the velocities to enforce continuity and to provide an updated pressure field. During this process, the density field is taken as fixed.

In the compressible pressure correction formulation, both the velocity and density fields are simultaneously updated to enforce continuity. The resulting algorithm has the attractive property of being able to address flows at all Mach numbers, making it very widely applicable.

Upwinding of the densities provides the mechanism for shock capturing in transonic and supersonic flows. The conservative second-order accurate QUICK (quadratic upstream differencing) scheme [7] is used to compute the combined convection-diffusion fluxes. Although the first-order accurate hybrid differencing scheme is still widely used for incompressible flows, it leads to excessive smearing when shocks are present and excessive total pressure errors. Shocks are found to be captured within 3-4 grid cells when QUICK is used, and total pressure conservation in inviscid flows is found to be significantly better.

Highly turning turbine blade rows pose difficulties to the conventional H-grids commonly used for turbomachinery calculations. The requirement of strict grid periodicity at the periodic boundaries causes severe grid shearing, and loss of orthogonality of the grid at the blade surface, as shown in Figure 1. This grid shearing can lead to significant numerical error, and loss of stability of the solution algorithm. Lack of grid orthogonality in the near-wall region also reduces the accuracy of the wall functions used to model the turbulent boundary layer in viscous cases.

A remedy for this problem is to relax the requirement of strict grid periodicity, and allow for a mismatch of the grid points across the periodic boundary. The resulting grid is much less sheared, and grid orthogonality near the wall is much improved, as illustrated in Figure 2. The penalty in this approach is the need to interpolate values communicated across the periodic boundary, and the interpolation errors that consequently arise. These errors can be reduced by taking the periodic boundary to be at the mid-passage line, so

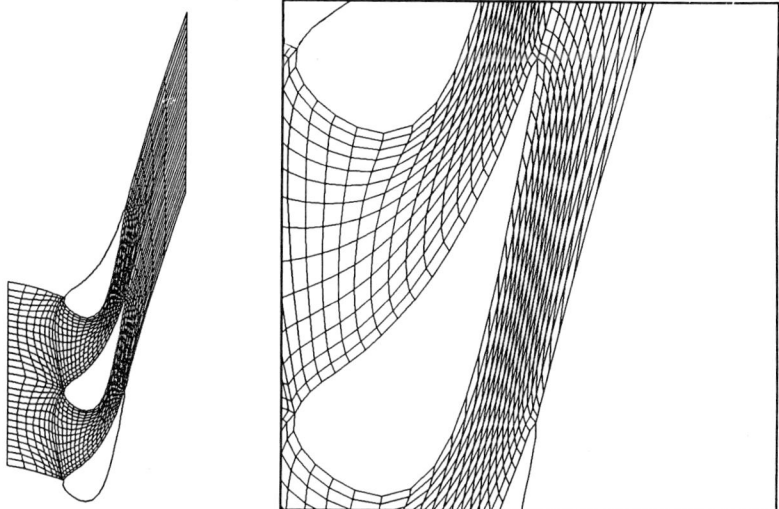

Figure 1. Computational H-grid

Strict grid periodicity causes severe grid shearing near trailing edge

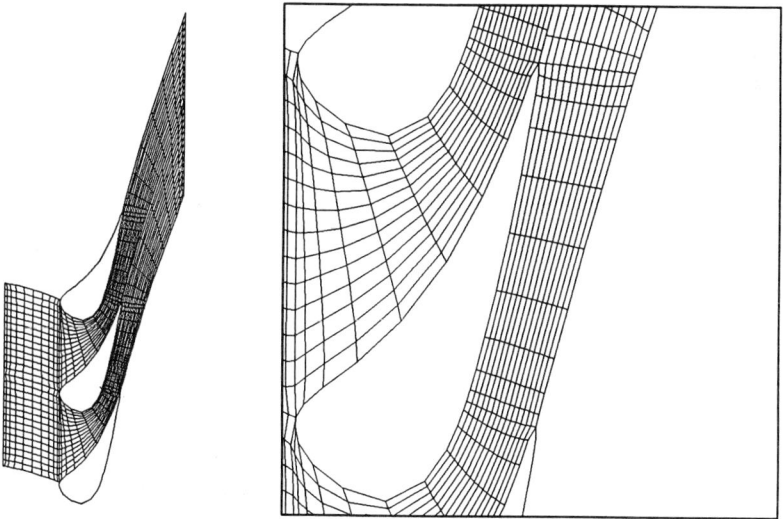

Figure 2. Passage I-grid

Relaxation of strict grid periodicity improves grid orthogonality, but necessitates interpolation across periodic boundary

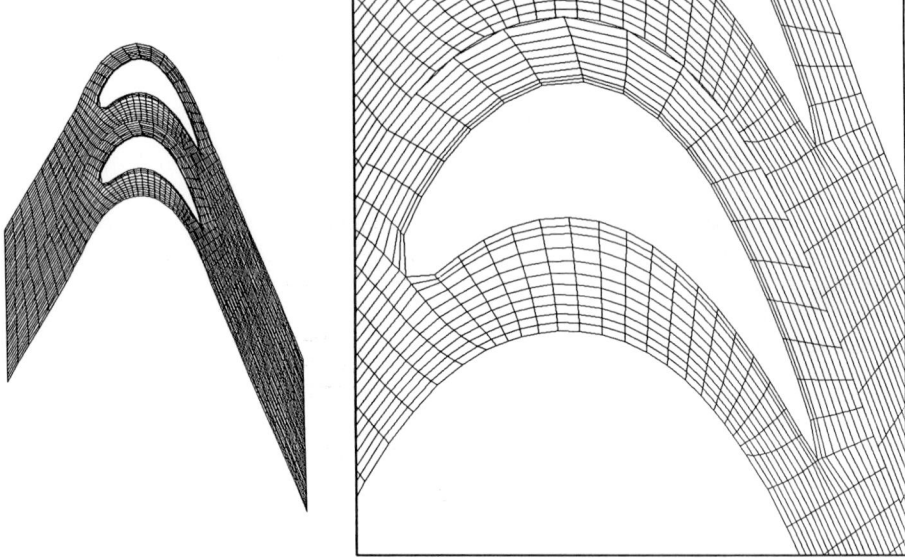

Figure 3. Internal Blade I-grid

Solution domain extends from middle of one flow passage to middle of another; blade is treated as triple grid line

that the interpolations are performed away from the leading and trailing edges of the blade, where the flow gradients are typically highest. This requires the blade to be meshed as part of the computational domain, as shown in Figure 3. Here the blade is treated as a slit in the transformed domain, resulting in a "triple grid line" both upstream and downstream of the trailing edge of the blade, where three grid lines collapse together. The presence of triple grid lines requires special treatment of the zero-volume cells which result, as described in [8].

3. PARALLELIZATION ISSUES

The natural means of parallelizing the basic algorithm is by domain decomposition. A full three-dimensional decomposition is done by dividing the solution domain into overlapping cubical regions (in the transformed space), such that an equal number of internal control volumes are assigned to each processor. This ensures a reasonable load balance since the number of floating point operations per control volume is roughly equal. The desired number of processors in each of the three grid directions can be set by the user at run time. An overlap of one cell in each direction at each face is used to minimize the redundant storage of quantities in the overlapping regions. Six temporary arrays are used to provide storage for the additional planes of values required by the QUICK scheme, rather than by using a two cell overlap. Additional communication is required across the periodic boundaries and across any triple grid lines present, to update the required values. The communication across the periodic boundaries requires special considerations, as described later, but the communication across the triple grid line is straightforward.

The algorithm developed here uses a standard staggered grid formulation for the velocity components to prevent problems with spurious oscillations in the computed pressure field. In the staggered grid scheme, the velocity components are solved on the faces of the control volumes, while the other scalar variables such as pressure correction and temperature are computed at the control volume centers. In a given grid direction, there is always one less staggered velocity component to be solved for than there are corresponding scalar values. Since the solution of the pressure correction equation takes up the majority of the computation time, the best domain decomposition leads to an equal number of scalar grid points (or equivalently, control volumes) on each processor. This ensure a good load balance for the solution of this (and the other scalar) equations. Unfortunately, such a decomposition always leaves one processor with one less velocity value in each line to solve than the others, which results in a computational load imbalance during the solution of the momentum equations. This load imbalance is a major contributor to parallel inefficiency for this algorithm, a fact which suggests that unstaggered grid formulations may ultimately prove superior for parallel machines.

The possible presence of mismatched grid points at the periodic boundary has important implications on the parallel implementation of the algorithm. The resulting communication across the periodic boundary is not necessarily one-to-one between processors on opposite sides of the domain. Rather, it depends on the particular nature of the grid. The situation is analogous to the domain decomposition of an unstructured grid, in which each processor can have different numbers of neighbors, and different amounts of data which need to be communicated with different processors.

The approach taken here was to stay with a regular structured three-dimensional domain decomposition and suffer the inefficiency that results in the transfer of values across the periodic boundary. This approach was chosen because the exchange of values takes very little time relative to the bulk of the calculation for the serial code. At the beginning of the calculation, each processor builds lists of processors that periodic boundary values will be sent to and received from. First, the entire plane of global coordinates on the far side of the periodic boundary is temporarily stored on each processor, and then triangulated. For each point on its side of the periodic boundary, the processor determines the corresponding triangle on the far side, the processor the triangle belongs to, the three vertex locations, and the corresponding area coordinates. These values are assembled into lists, which are subsequently used to pass and interpolate values across the periodic boundary.

Completing the parallelization of these routines proved to be a difficult job because of the extensive bookkeeping involved, and consumed three weeks of effort. Happily, the inefficiency caused by the slight communication imbalance across the mismatched periodic boundary had no significant impact on the parallel efficiency of the overall application for reasonably large grids.

4. COMPUTATIONAL RESULTS

A series of calculations for inviscid and viscous flow in turbomachinery blade rows has been completed. The problems studied included a test case comprised of a cascade of biconvex airfoils, a two-dimensional transonic turbine test rig, a highly turning subsonic turbine nozzle from an industrial steam turbine, and a highly turning transonic turbine nozzle from an industrial gas turbine. Since the geometries and flow conditions for the last two cases are considered proprietary, the paper will focus on the results from the test rig, and show only timings for the other cases.

All parallel computations shown were performed in 32-bit arithmetic on a 16-node Intel iPSC/860 at GE-CRD, a 128-node iPSC/860 at the NASA-Ames Research Center, and the 512-node Intel Touchstone Delta prototype at the California Institute of Technology. Experience with this algorithm indicates that 64-bit arithmetic is not required. However, Cray timings do reflect 64-bit arithmetic since that is the default on that machine.

The parallel algorithm was first tested for the inviscid flow through a cascade of biconvex airfoils, computed on grid with 30,000 grid points. These calculations were done using the Green Hills compiler on the NASA-Ames iPSC/860 machine. With 64 processors, the performance of the iPSC/860 was 40% faster than the vectorized production code running on a single processor of a Cray Y-MP. With the full scalar optimization of the Green Hills FORTRAN compiler, performance 10% better than the Cray was achieved with only 32 processors.

Table 1 shows measured and calculated performance for a turbulent viscous calculation on a highly turning transonic turbine nozzle, computed on a grid with 37,000 grid points. These calculations were done using the pipelining pgf77 compiler. With 16 processors, performance equivalent to 79% of a single Cray Y-MP processor was achieved, along with an estimated parallel efficiency of 80.5%.

Table 1. Performance of parallel algorithm on viscous transonic turbine cascade (37,000 grid points)

# Processors	sec/iter/point/equation	Performance/Cray	Parallel Efficiency
Cray Y-MP (1)	1.1×10^{-5}	1.00	-----
iPSC/860 (8)	2.5×10^{-5}	0.44	89.9 %
iPSC/860 (16)	1.4×10^{-5}	0.79	80.5 %

Table 2 shows measured performance for the calculation of an inviscid transonic turbine cascade on a grid with 130 x 18 x 66 points. An outline of the grid used is illustrated in Figure 4, along with closeup views of the leading and trailing edges of the blade. Since access to the full complement of processors on either the Ames machine or the Delta machine is limited, the author wanted to ensure that the code would run the first time for this problem. For this reason, the grid was generated to be as uniform in spacing as possible, since this helps ensure reliable convergence of the algorithm, at the expense of grid resolution around the leading and trailing edges of the blade. Although this compromises the accuracy of the computed results somewhat, the timings, which are of principal interest, are valid. When parallel machines of this class become more widely available, there will be plenty of opportunity to repeat the calculations on grids with more nonuniform spacing which more accurately resolve the solution.

For reference, Figure 5 shows the computed pressure field. Notice the slight mismatch of the contour lines as they cross the mismatched periodic boundary. Comparision of the results was made to timings on a Cray Y–MP running a highly vectorized production version of the original serial algorithm. Performance 2.4 times faster than a single processor of a Cray Y–MP was achieved with 128 processors of the iPSC/860, and performance 2.5 times faster was achieved with 256 processors of the Delta machine.

Since this test case is too large to run on a single processor, the parallel efficiency must be estimated. Here, this was done using the method described in [4]. The parallel efficiency drops below 50% around the level of 64 processors, suggesting that the given problem is too small for a much greater number of processors. The loss of parallel efficiency is due primarily to load imbalances caused by the use of the staggered grid (as discussed earlier), and from the application of different boundary conditions on different processors, as well as the cost of communications between processors during the solution step.

Table 2. Performance of parallel algorithm on inviscid transonic turbine cascade (154,000 grid points)

# Processors	sec/iter/point/equation	Performance/Cray	Parallel Efficiency
Cray Y-MP (1)	1.10×10^{-5}	1.00	-----
iPSC/860			
4 x 2 x 2 (16)	1.51×10^{-5}	0.73	80.9 %
4 x 2 x 4 (32)	8.85×10^{-6}	1.24	67.1 %
4 x 4 x 4 (64)	6.29×10^{-6}	1.75	50.1 %
8 x 4 x 4 (128)	4.59×10^{-6}	2.40	33.3 %
Touchstone Delta			
4 x 2 x 2 (16)	1.36×10^{-5}	0.81	74.8 %
4 x 2 x 4 (32)	8.09×10^{-5}	1.36	58.9 %
4 x 4 x 4 (64)	5.84×10^{-6}	1.89	41.4 %
8 x 4 x 4 (128)	4.47×10^{-6}	2.46	25.9 %
8 x 4 x 8 (256)	4.30×10^{-6}	2.56	14.9 %

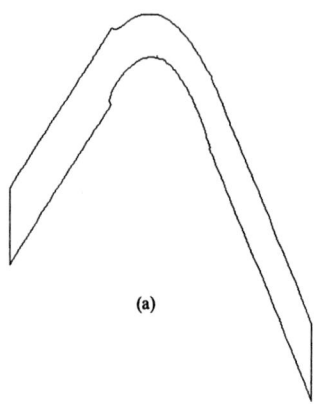

Figure 4. Body-fitted passage I-grid for GEP experimental trubine cascade (154,000 points)

(a) Outline of flow domian
(b) Closeup of grid near leading edge
(c) Closeup of grid near trailing edge

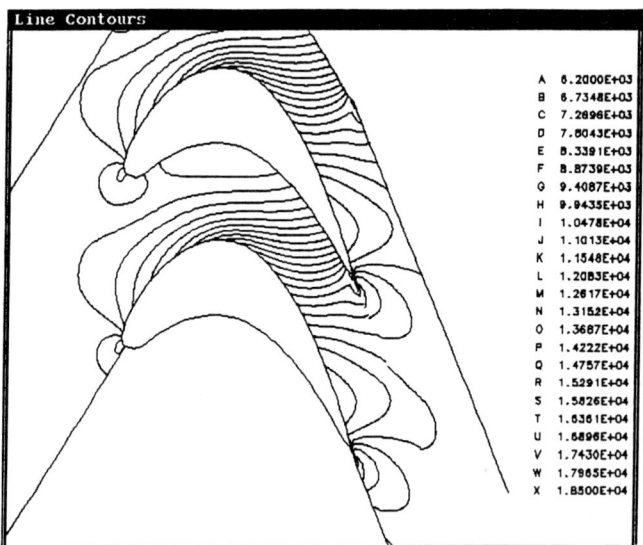

Figure 5. Computed pressure contours for inviscid transonic turbine cascade test case

Parallel efficiency for this problem appears slightly poorer on the Delta machine than on the iPSC/860, for the same number of processors. Although at first this result seems surprising, since the Delta is supposed to have much better communication bandwidth than the iPSC/860, upon closer inspection it can be explained. First, the i860 processors on the Delta machine are a later revision than those on the Ames machine, and are reportedly 10-15% faster. Faster processors lead to a higher communication / computation ratio, if the speed of the underlying communication network remains the same. Most of the local messages passed by this algorithm are of only moderate length (typically 1000 bytes), so latency effects are still important, and increased bandwidth is not that significant.

The global communications required by the code were also found to be somewhat slower on the Delta machine than on the iPSC/860. The global communications routines used in the code were developed prior to Intel's offering of similar library routines such as GSSUM, and are based on the hypercube communications algorithms described in [9]. These algorithms help assure contention-free communication on a hypercube since the messages travel between nearest neighbors along one dimension of the hypercube at a time as the global operation proceeds. The time required to complete the global operation scales as log P, where P is the number of processors in the hypercube. On the iPSC/860, the performance of the hand-written routines is similar to that of the GSSUM library routine, with a slight edge to GSSUM as the number of processors becomes large, as shown in Figure 6.

On the mesh topology of the Delta, however, the situation is very different. The hypercube-based algorithm does not lead to nearest neighbor communication, and the logarithmic scaling does not hold. The effects of contention become very apparent beyond the level of 64 processors. The Intel library routine GSSUM, which implements an appropriate mesh-based algorithm on the Delta, shows comparable performance on the Delta to GSSUM on the iPSC/860, and clearly outperforms the hand-written routine. These results are also shown in Figure 6.

There are some lessons to be learned from this experience. Details of the underlying topology do influence the communication performance, particularly for global communications where all of the processors participate simultaneously, loading up the network with messages. The standard practice in writing parallel codes for earlier hypercube machines was to extensively optimize the code for the hypercube architecture using such devices as gray codes, hypercube-based shuffle and global summation routines, etc. These optimizations, which were specific to the details of the underlying topology of the machine, can be detrimental to performance when the code is ported to a machine with a different underlying topology.

A better practice is to define some standard global communication operations, such as global summation, with standard calls, and let the computer vendors implement these operations on their machines in whatever manner is most efficient for their particular machine architecture. The GSSUM routine provided by Intel has the same syntax on both the iPSC/860 and Delta, so the coding is truly portable, and the performance on both machines is comparable. This is in significant contrast to the optimized hypercube-based routine, which although portable, showed a significant performance loss on the mesh topology of the Delta.

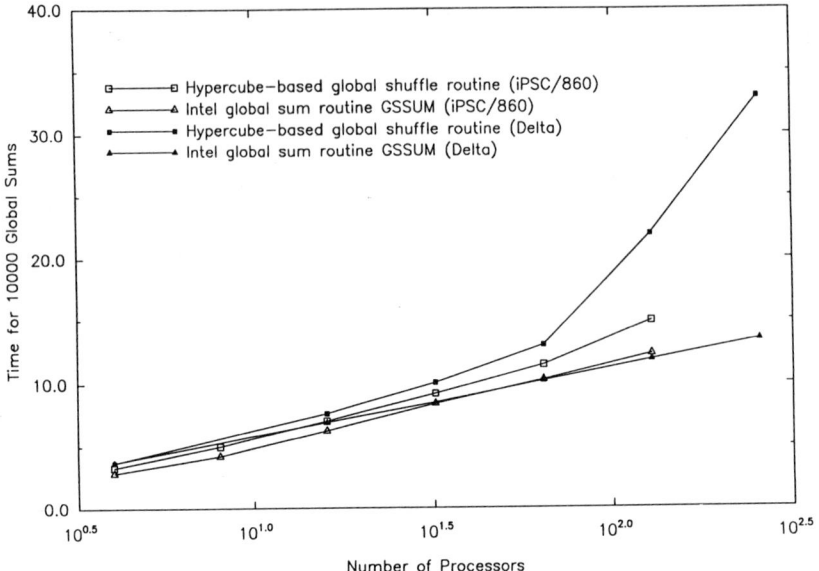

Figure 6. Performance of global Summation Routines on Intel iPSC/860 and Intel Delta

5. ACKNOWLEDGEMENTS

The author would like to thank ICOMP at NASA–Lewis and NAS at NASA–Ames for access to the iPSC/860 at NASA–Ames. Thanks are also due to NASA–Lewis for providing access to the Intel Touchstone Delta System operated by Caltech on behalf of the Concurrent Supercomputing Consortium.

6. REFERENCES

1. Jameson, A., Schmidt, W., and Turkel, E., "Numerical Solution of the Euler Equations by Finite Volume Methods Using Runge–Kutta Time Stepping Schemes, AIAA paper AIAA-81-1259, 1981.
2. Beam, R. W. and Warming, R. F., "An Implicit Finite Difference Algorithm for Hyperbolic System in Conservation Form", **J. Comp. Phys.**, 23, pp. 87–110, 1976.
3. Braaten, M.E.,"Solution of Viscous Fluid Flows on a Distributed Memory Concurrent Computer", **Int. J. Num. Meth. Fluids**, 10, pp. 889–905, (1990).

4. Braaten, M.E.,"Development of a Parallel Computational Fluid Dynamics Algorithm on a Hypercube Computer", **Int. J. Num. Meth. Fluids**, 12, pp. 947–963, (1991).

5. Braaten, M. E., "Parallel Computation of the Compressible Navier–Stokes Equations with a Pressure–Correction Algorithm", in Walker, D.W., and Stout, Q. F. (eds.), Proceedings of the Fifth Distributed Memory Computing Conference, April 8–12, 1990, Charleston, S.C,. pp. 463–469, (1990).

6. Patankar, S. V., **Numerical Heat Transfer and Fluid Flow**, Hemisphere, Washington, D.C., 1980.

7. Leonard, B. P., "A Stable and Accurate Convective Modeling Procedure Based on Quadratic Upstream Interpolation", **Comput. Meths. Appl. Mech. Eng.**, 19, pp. 59–98, 1979.

8. Cedar, R. D., and Holmes, D. G., "The Calculation of Three–dimensional Flow through a Transonic Fan Including the Effects of Blade Surface Boundary Layers, Part–span Shroud, Engine Splitter and Adjacent Blade Rows, AMSE paper ASME 89–GT–325, 1989.

9. Saad, Y. and Schultz, M. H., "Data Communication in Hypercubes", Report YALEU/DCS/RR–248, Yale University, 1985.

Parallel Computational Fluid Dynamics '92
R.B. Pelz, A. Ecer and J. Häuser (Editors)
© 1993 Elsevier Science Publishers B.V. All rights reserved.

The Calculation of Three-Dimensional Compressible Flow Through a Rectangular Nozzle Using a Data Parallel Finite Element Model

Frank P. Brueckner[a], Darrell W. Pepper[a], Thomas H. Sobota[b], and Roger H. Chu[c]

[a]Advanced Projects Research, Inc., 5301 N. Commerce Avenue, Suite A, Moorpark, California 93021, U.S.A.

[b]Advanced Projects Research, Inc., 147 Ward Street, Hightstown, New Jersey 08520, U.S.A.

[c]MasPar Computer Corporation, 749 N. Mary Avenue, Sunnyvale, California 94086, U.S.A.

Abstract

A finite element model (RUBY3D) is developed and used to simulate three-dimensional compressible fluid flow on a massively parallel computer. The algorithm is based on a Petrov-Galerkin weighting of the convective terms in the governing equations. The discretized time-dependent equations are solved explicitly using a second-order Runge-Kutta scheme. A high degree of parallelism has been achieved utilizing a massively parallel MasPar MP-1 computer. The algorithm is applied to flow through an advanced jet engine nozzle configuration. Performance results are presented for various computer platforms.

1. INTRODUCTION

Large computational fluid dynamics (CFD) problems have traditionally been solved using powerful scalar or vector pipelined computers. With the recent developments in software tools and programming environments for massively parallel computers, it is now possible to develop applications that can exploit the computing power of massively parallel architectures with a reasonable program development time. Finite element methods, particularly those that use explicit solution schemes, map very well to massively parallel architectures.

In this paper, a finite element algorithm for the numerical simulation of compressible flow is described. Elements are defined using trilinear velocity, density, and energy with a piecewise constant pressure. An anisotropic balancing diffusion is used via a Petrov-Galerkin weighting. The original vector code is ported onto the MasPar MP-1 family of parallel computers; a simple map of one element per processor in each memory layer is used.

To demonstrate the performance of the algorithm and computing platform, RUBY3D is applied to flow through an advanced jet nozzle configuration. An understanding of the flow physics through this type of nozzle is of current interest for reasons including both thrust vectoring and reduced infrared plume signatures (as compared to a circular outflow geometry). The rectangular nozzle provides a relatively simple means to vector the thrust about a single axis. In addition, the rectangular geometry of the resulting free jet enhances mixing between the jet fluid and the surrounding ambient air. A parametric description of this type of nozzle is presented and used to generate the computational mesh. CPU timing is obtained on the MasPar MP 1216 (16,384 processors) and compared to the Cray Y-MP (single processor) and the Alliant FX/40.

2. FINITE ELEMENT MODEL

Simulations of subsonic and transonic fluid flow generally consist of regions in which compressibility effects are important, often including shock waves, along with regions of low Mach number flow. Physically and mathematically, these two regions exhibit quite different behaviors. When the velocity is large and compressibility effects are significant, the conservation of mass is an equation which describes the evolution and transport of the fluid density. In contrast, as the flow becomes incompressible, the conservation of mass reduces to that of zero velocity divergence. Both types of flows are present in the nozzles under investigation. Included in the following description of the numerical method are modifications implemented for accurate simulations of these types of flowfields.

2.1. Governing Equations

The equations which describe the transport of compressible fluid are written in nondimensional vector form as

Mass (Continuity):

$$\frac{\partial \rho}{\partial t} + \underline{u} \cdot \nabla \rho + \rho \nabla \cdot \underline{u} = 0 \tag{1}$$

Momentum:

$$\rho \left(\frac{\partial \underline{u}}{\partial t} + \underline{u} \cdot \nabla \underline{u} \right) + \frac{1}{\gamma M_\infty^2} \nabla p = \nabla \cdot \underline{\underline{\tau}} \tag{2}$$

Energy:

$$\rho \left(\frac{\partial e}{\partial t} + \underline{u} \cdot \nabla e \right) + (\gamma - 1) p \nabla \cdot \underline{u} = \frac{\gamma}{PrRe_\infty} \nabla \cdot \mu \nabla e + \gamma(\gamma - 1) M_\infty^2 (\underline{\underline{\tau}} : \nabla \underline{u}) \tag{3}$$

where ρ is the density, $\underline{u} = (u,v,w)^T$ is the velocity vector, p is the pressure, e is the specific internal energy, t is the time, and μ is the kinematic viscosity. The dimensionless deviatoric stress tensor $\underline{\underline{\tau}}$ is defined as

$$\underline{\underline{\tau}} = \frac{\mu}{Re_\infty}\left[(\nabla \underline{u} + \nabla \underline{u}^T) - \frac{2}{3}\underline{\underline{I}}(\nabla \cdot \underline{u})\right] \tag{4}$$

where $\underline{\underline{I}}$ is the identity tensor and the kinematic viscosity µ is approximated by Sutherland's formula. To close this system of equations, a thermally and calorically perfect gas is assumed, leading to the equation of state

$$p = \rho T \tag{5}$$

and

$$T = e \;. \tag{6}$$

The nondimensional parameters which appear in these expressions are Re_∞, the freestream Reynolds number; Pr, the Prandtl number; M_∞, the freestream Mach number; and γ, the ratio of specific heats.

Finally, the Euler equations are derived by letting $1/Re_\infty \to 0$, thereby forcing the right-hand-sides of Eqs. (2) and (3) to vanish.

2.2. Weak Formulation and Discretization

The weak forms of the governing equations are derived via the Petrov-Galerkin weighted residual method as

$$\int_\Omega v_\rho \left[\frac{\partial \rho}{\partial t} + \underline{u} \cdot \nabla \rho + \rho \nabla \cdot \underline{u}\right] d\Omega = 0 \tag{7}$$

$$\int_\Omega \left\{\underline{v}_u \cdot \left[\rho\left(\frac{\partial \underline{u}}{\partial t} + \underline{u} \cdot \nabla \underline{u}\right)\right] - \frac{1}{\gamma M_\infty^2} p \nabla \cdot \underline{v}_u + \nabla \underline{v}_u : \underline{\underline{\tau}}\right\} d\Omega$$

$$- \int_\Gamma \underline{v}_u \cdot \left[-\frac{1}{\gamma M_\infty^2} p \hat{n} + \hat{n} \cdot \underline{\underline{\tau}}\right] d\Gamma = 0 \tag{8}$$

$$\int_\Omega \left\{v_e \left[\rho\left(\frac{\partial e}{\partial t} + \underline{u} \cdot \nabla e\right) + (\gamma-1)p\nabla \cdot \underline{u} - \gamma(\gamma-1)M_\infty^2(\underline{\underline{\tau}} : \nabla \underline{u})\right] + \frac{\gamma}{PrRe_\infty}\mu\nabla v_e \cdot \nabla T\right\} d\Omega$$

$$- \int_\Gamma v_e \left[\frac{\gamma}{PrRe_\infty}\mu\hat{n} \cdot \nabla T\right] d\Gamma = 0 \tag{9}$$

Here, v_ρ, $\underline{v}_u=(v_u,v_v,v_w)^T$, and v_e are weighting functions, Ω is the flow domain with boundary Γ, $\hat{n}$ is the unit vector normal to Γ, and the boundary integrals in Eqs. (8) and (9) arise from the application of Green's identity to the respective flux terms.

The density, velocity components, and the internal energy are each expanded in terms of the standard trilinear isoparameteric shape functions $N^i(\underline{x})$, where $\underline{x}=(x,y,z)^T$ is the position vector. To improve the behavior of the algorithm in regions where the local Mach number is very low, a piecewise constant approximation for pressure is used; nodal values are interpolated from these element quantities when required for post-processing [1]. To stabilize the discretized equations, an anisotropic balancing dissipation is introduced into each governing equation via a Petrov-Galerkin weighting function. These weights are obtained by perturbing the shape functions N^i such that term i of the v_j weight ($j=\rho,u,v,w,e$) is given by

$$v_j^i = N^i + \alpha_j \frac{h_e}{2|\underline{u}|}(\underline{u} \cdot \nabla N^i) \tag{10}$$

where h_e is the element mesh length and $|\underline{u}|$ denotes the magnitude of the velocity vector $\underline{u}$. The parameter α_j is defined as

$$\alpha_j = \coth\frac{\gamma_j}{2} - \frac{2}{\gamma_j} \tag{11}$$

in which

$$\gamma_j = \begin{cases} \infty, & \text{for } j=\rho \\ \dfrac{\rho Re_\infty |\underline{u}| h_e}{\mu}, & \text{for } j=u,v,w \\ \dfrac{\rho PrRe_\infty |\underline{u}| h_e}{\mu\gamma}, & \text{for } j=e \end{cases} \tag{12}$$

In this work, the anisotropic balancing diffusion approach of Kelly et al. [2] and Brueckner and Heinrich [3] is adopted. In doing so, the functions v_ρ, $\underline{v}_u$, and v_e are used to weight the advection terms in Eqs. (7), (8), and (9), respectively, while the remaining terms are weighted by the shape functions N^i. The effect of this weighting is to introduce a diffusion along the streamline in each transport equation. In the Euler limit, the Petrov-Galerkin weighting functions become the same for each equation; that is

$$v^i = N^i + \frac{h_e}{2|\underline{u}|}(\underline{u} \cdot \nabla N^i) \tag{13}$$

Finally, the spatial integrations over Ω are computed numerically using a 2×2×2 Gaussian quadrature in the local element coordinate system. Reduced integration (i.e., 1-point quadrature) is used for the pressure terms in the momentum equations and the velocity divergence terms in each equation.

2.3. Temporal Integration

An explicit, second-order Runge-Kutta scheme is used to integrate the semi-discrete equations of the previous section. To produce a fully explicit algorithm, the mass matrix is diagonalized by employing a lumped mass approximation. A stability constraint on each timestep must be imposed; the Courant limits associated with a forward Euler scheme are calculated over each element, and the timestep adjusted to the minimum value within the computational domain.

3. PARALLEL PROCESSING

3.1. MasPar Architecture

MasPar manufactures a family of massively parallel computer systems capable of attaining peak processing speeds up to 26 KMIPS (32-bit integer adds) and 1.2 GFLOPS (32-bit floating point adds and multiplies). The MP-1 family of computers obtains its performance by using an array of processing elements (PE's). The PE array consists of 1,024 to 16,384 processors which operate in a Single-Instruction Multiple-Data (SIMD) fashion. There are three major components to the machine [4]: The PE array, the Array Control Unit (ACU), and the UNIX subsystem. Computational power is attained by using a massively parallel array of PE's. Architecturally, each PE is a RISC processor with a 64-bit wide accumulator, 48 32-bit registers, and either 16 KBytes or 64 KBytes of data memory. All PE's execute instructions in lock step on data stored in its local memory. Each PE can enable or disable itself for part or all of a computation based on a logical expression for conditional execution. To share data with other PE's, there are three communication mechanisms available: the Xnet, the router, and the global-or tree. Xnet, or 8-way nearest neighbor communication, provides a very fast path for moving data between a PE and its eight neighbors. Xnet provides an aggregate bandwidth of up to 18 GBytes/second (16K PE's). In addition to Xnet, the MP-1 has an alternate, multi-stage, circuit-switched network for global or random communication patterns. This network router provides a PE the ability to send or fetch data from any other PE in the array. The aggregate bandwidth of the router communications is 1.3 GBytes/second (16K PE's). A global-or tree is used for moving data from the PE array into the ACU. The ACU performs two functions: execution control and scalar computations, and broadcasting instructions and/or data to the PE array. The ACU is the master and controls all the processing in the MP-1 computer. Programs are written to control the ACU, and hence the PE array. The UNIX subsystem provides application engineers with a programming and run-time environment.

3.2. Porting of RUBY3D onto MP-1

The RUBY3D finite element code consists of 3000 lines of Fortran 77. It is highly optimized for efficient execution on vector machines such as the Cray Y-MP. To execute this code on the massively parallel MasPar MP-1, the existing Fortran 77 had to be converted to Fortran 90 or MasPar Fortran [5]. No algorithmic changes were necessary. The conversion task was divided into three steps:

Step 1: Converting F77 to F90

The majority of the conversion effort was to convert Fortran 77 DO loops into the Fortran 90 array syntax using an automated conversion program, VAST-2 [6]. Figure 1 shows an example of the original Fortran 77 code compared to the equivalent MasPar Fortran.

```
         FORTRAN 77                                    FORTRAN 90

DO 500 I = 1 , NELM
  B1(NODE(I)) = B1(NODE(I)) - WK1(I)         B1(NODE(:NELM)) = B1(NODE(:NELM)) - WK1(:NELM)
  B2(NODE(I)) = B2(NODE(I)) - WK2(I)         B2(NODE(:NELM)) = B2(NODE(:NELM)) - WK2(:NELM)
  B3(NODE(I)) = B3(NODE(I)) - WK3(I)         B3(NODE(:NELM)) = B3(NODE(:NELM)) - WK3(:NELM)
  B4(NODE(I)) = B4(NODE(I)) - WK4(I)         B4(NODE(:NELM)) = B4(NODE(:NELM)) - WK4(:NELM)
  B5(NODE(I)) = B5(NODE(I)) - WK5(I)         B5(NODE(:NELM)) = B5(NODE(:NELM)) - WK5(:NELM)
500 CONTINUE
```

Figure 1. Sample Code for Array Syntax

The colon notation denotes that all elements of the array from 1 (the default lower limit) to NELM are computed in parallel. Most of the computationally intensive DO loops are either the number of elements or number of nodes long. On an MP 1216 (16,384 PE's), an instruction is executed using 16,384 different data simultaneously.

Another significant conversion done by VAST-2 was to translate a condition statement (i.e. IF statement) into a WHERE construct in Fortran 90. The example code segment in Fig. 2 demonstrates this feature.

```
         FORTRAN 77                                    FORTRAN 90

DO 400 I = 1 , NELM
  IF ( INDX(I) .GT. IXYZ) THEN               WHERE ( INDX(:NELM) .GT. IXYZ )
    DR(I) = TR(I) * XR(I)                      DR(:NELM) = TR(:NELM) * XR(:NELM)
    RR(I) = TR(I) + YR(I)                      RR(:NELM) = TR(:NELM) + YR(:NELM)
    RY(I) = TR(I) - ZR(I)                      RY(:NELM) = TR(:NELM) - ZR(:NELM)
  ELSE                                       ELSEWHERE
    DR(I) = TR(I) / XR(I)                      DR(:NELM) = TR(:NELM) / XR(:NELM)
    RR(I) = TR(I) * YR(I)                      RR(:NELM) = TR(:NELM) * YR(:NELM)
    RY(I) = TR(I) + ZR(I)                      RY(:NELM) = TR(:NELM) + ZR(:NELM)
  ENDIF                                      ENDWHERE
400 CONTINUE
```

Figure 2. Sample Code WHERE Construct

Step 2: Storing Arrays in PE's Memory

To maximize the execution efficiency, all arrays must be stored in the PE's memory. The arrays in RUBY3D are declared in COMMON blocks, and are specified in an include file. The example in Fig. 3 illustrates the usage of the MasPar Fortran "ONDPU" directive, which directs COMMON's or variables to be stored in PE memory.

```
         COMMON / ABC / A(NNODE), B(NELM)
         COMMON / XYZ / D(NNODE,8), DEL(NELM,8)
CMPF ONDPU ABC, XYZ
```

Figure 3. MasPar ONDPU Directive

Step 3: Storing 2-Dimensional Array in Desired Order

Two-dimensional arrays such as DEL(NELM,8) could be stored in a Cut-and-Stack fashion using MasPar Fortran. This storage scheme, however, would not produce the most efficient algorithm. To maximize performance, the MasPar Fortran "MAP" directive is inserted in the routines that use 2-D arrays. An example of the use of this directive is listed in Fig. 4. The directive in this example maps the first dimension of the array DEL onto the PE array. The second dimension of DEL is mapped in the PE's memory. In this way, the "MAP" directive effectively converts a 2-D array into a number of 1-D arrays; DEL becomes eight 1-D arrays, each of which is NELM long.

```
         DIMENSION DEL(NELM,8)
CMPF MAP DEL(ALLBITS,MEMORY)
```

Figure 4. MasPar Mapping Directive

The advantage of using an automated conversion program along with the compiler directives is that the source code can be maintained in the Fortran 77 form. The initial conversion task took about one man-week. It is anticipated that future revisions will take only one day with these automated procedures.

4. NOZZLE DEFINITION

A basic nozzle geometry was adopted from sketches of a Pratt and Whitney F100 engine, beginning with an approximate size and area ratio, and incorporating a typical afterburner section and a contraction to a rectangular nozzle or throat. These parameters were adjusted to reflect some of the advancements made in engine technology since this engine was designed, and the trends in engine and airplane design. For example, a lower bypass ratio was assumed as this feature is the trend in higher Mach number fighter aircraft.

4.1. Parametric Definition of Nozzle

The nozzle is defined by a series of cross-sections along the axial (x) direction. The boundaries of each cross-section are specified by the super-elliptic equation

$$\left[\frac{R(x,\theta)\cos\theta}{A(x)}\right]^{e(x)} + \left[\frac{R(x,\theta)\sin\theta}{B(x)}\right]^{e(x)} = 1 \qquad (14)$$

where $A(x)$, $B(x)$, and $e(x)$ represent the major axis, minor axis, and the super-elliptic exponent, respectively. The quantities $A(x)$ and $B(x)$ control the aspect ratio of each cross-section, while the super-elliptic exponent affects the cross-section shape; for $e(x)=2$, an ellipse is defined, while nearly rectangular shapes are generated for larger values. With $A(x)$, $B(x)$, and $e(x)$ specified, the nozzle and centerbody radii R can be calculated for each axial position

4.2. Computational Grid

The computational grid is constructed by creating a series of two-dimensional grids at different axial locations. Each grid is generated by first locating points along the boundary of the nozzle and, if present, along the centerbody wall. Mesh points are then positioned along lines between points on the nozzle boundary and the section origin (or the centerbody wall), thereby forming a quadrilateral grid. To avoid degenerate elements at the origin, the central core region is removed and replaced with 4-sided elements.

5. NUMERICAL RESULTS

The baseline geometry has been established with a defined nozzle outer wall and a blunt centerbody. The inlet to throat area ratio is approximately 2.8; the nozzle outlet aspect ratio is about 4. Figures 5(a) and (b) show the discretized models used for the inviscid and viscous calculations, respectively. Boundary conditions are imposed along the x-y plane to reduce the model to one-half of the nozzle. The inviscid half-model consists of 15,234 nodes and 13,272 elements; the viscous half-model has 77,395 nodes and 71,512 elements. The inflow to the nozzle is annular, with a circular centerbody which extends downstream for a nondimensional distance of 1. The reference length L is taken as the half-width of the model in the x-y plane. The inlet density and velocity components are fixed at their respective freestream values and the pressure ratio p_{inlet}/p_{outlet} is 1.83. Values of the applicable dimensionless parameters are

Freestream Mach Number	0.22
Freestream Reynolds Number	90,000
Prandtl Number	0.72
Ratio of Specific Heats	1.4

Sutherland's formula for air is employed to approximate the kinematic viscosity and thermal conductivity of the fluid, and a temperature of 1644°R is used as the freestream reference.

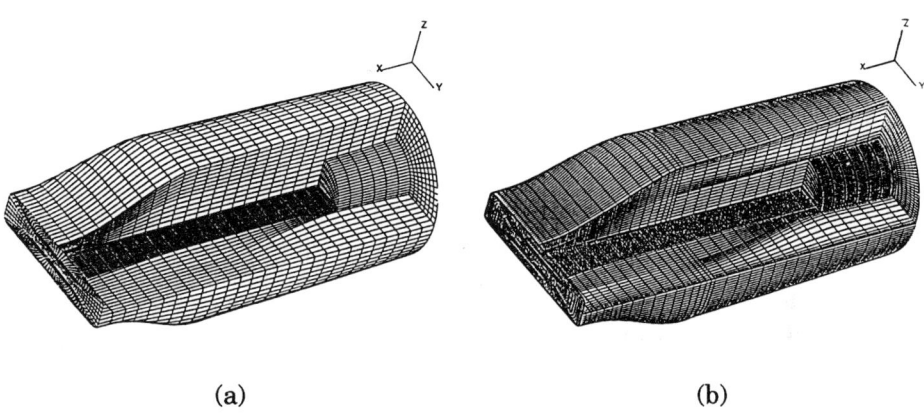

(a)   (b)

Figure 5. Computational meshes for (a) inviscid and (b) viscous simulations.

5.1 Inviscid Simulations

Two solutions are presented; the first with no swirl and a second with a 20° inlet swirl angle. The magnitudes of each inlet velocity vector are equal and each is normalized to the same reference.

A plot of the steady-state velocity field on the upper surface of the nozzle is shown in Fig. 6 for both the no swirl and 20° swirl simulations. The influence of the high pressure region immediately downstream of the centerbody is clearly seen. Figure 7 contains the associated pressure contours on the upper nozzle surface. Apparent in both of these figures is the dramatic effect on the flowfield along the wall due to the introduction of the swirling inflow velocity.

5.2 Viscous Simulation

Contours of the local Mach number in the x-z plane are presented in Fig. 8 for a viscous flow with no swirl, clearly illustrating the development of the boundary layers along the nozzle and centerbody surfaces. As expected, a large low-speed flow region is evident just downstream of the centerbody. The flow separates off of the sharp trailing edge of the centerbody and forms a recirculation zone. The high pressure region downstream of the centerbody also acts to form recirculation zones at the upper and lower nozzle surfaces. Separation occurs near the end of the centerbody, and the flow reattaches at the nozzle contraction. Axial vortices are also formed along the side walls of the nozzle as the fluid progresses through the transition region. A plot of the transverse velocity vectors at the nozzle exit plane is given in Fig. 9, clearly showing a vortex pair at each side wall.

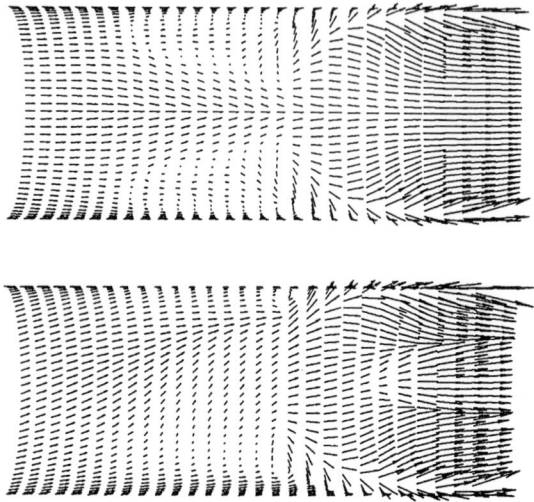

Figure 6. Velocity field on upper nozzle surface for no swirl (top) and 20° swirl (bottom) inviscid simulations.

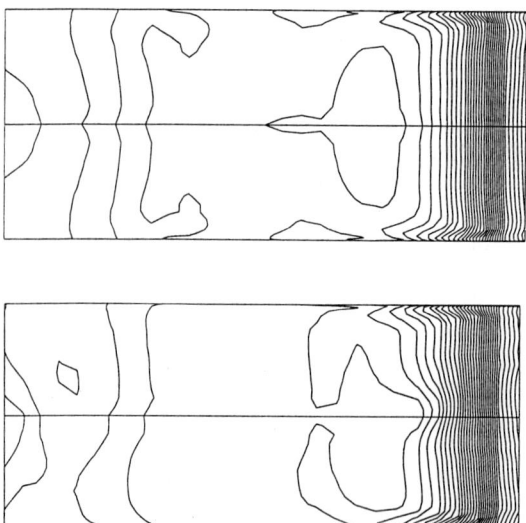

Figure 7. Pressure contours on upper nozzle surface for no swirl (top) and 20° swirl (bottom) inviscid simulations.

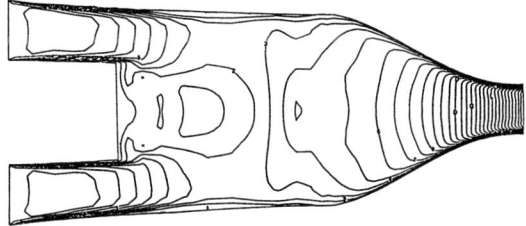

Figure 8. Local Mach number contours in the x-z plane.

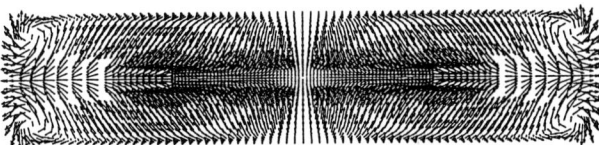

Figure 9. Transverse velocity vector field at the nozzle exit plane.

5.3 Computational Performance

The simulations described above were run on each of three machines: a MasPar MP 1216, a Cray Y-MP (one CPU only), and an Alliant FX/40. Table 1 compares the execution times per time step for both 32-bit and 64-bit word lengths.

Table 1
Execution time comparison (seconds/timestep)

	15,234 nodes/ Euler		77,395 nodes/ Navier-Stokes	
	32-bit	64-bit	32-bit	64-bit
MasPar MP 1216	0.52	0.79	3.4	5.3
Cray Y-MP (one CPU)		0.56		5.2
Alliant FX/40	17.2		147.	

It is seen here that the performance achieved on the MP 1216 was roughly that obtained on a single CPU of the Cray Y-MP. In general, large problems are relatively more efficiently run on the MasPar than smaller problems (in terms of seconds/timestep/node), with maximum efficiency achieved when the number of elements equals the number of PE's. This is nearly the case for the inviscid example, where in excess of 80% of the array of processing elements is utilized. Five memory layers were required for the Navier-Stokes simulations.

6. CONCLUSIONS

The finite element code RUBY3D is used to solve the equations governing compressible fluid flow. The algorithm is based on hexahedral elements with trilinear density, velocity, and internal energy and piecewise constant pressure. This pressure approximation and the selective reduced integration of certain terms in the governing equations improves the behavior of the solution as compared to equal-order, fully integrated (i.e., 8-point), trilinear elements. By converting the Fortran 77 code to Fortran 90, the program is ported onto a MasPar MP 1216 massively parallel computer (16,384 processors). All of the compiler directives that were added to optimize performance were incorporated into the Fortran 77 code; Fortran 90 code is obtained using the conversion program VAST-2. In this way, only the Fortran 77 code needs to be maintained, an important benefit when various computing platforms are utilized. The effort required for future program updates is also significantly reduced.

The performance of the code is demonstrated through simulations of transonic flow through annular to rectangular nozzles. The computed flowfields included large regions of very low local Mach number flow, thereby illustrating the ability of the algorithm to model nearly incompressible flow. Work is currently underway to incorporate mixing enhancing features into the nozzle model, as well as to calculate the plume flowfields. An effort to include finite rate chemistry into RUBY3D is also being conducted.

7. REFERENCES

1 Dyne, B.R., "Finite Element Analysis of Incompressible, Compressible, and Chemically Reacting Flows, With an Emphasis on the Pressure Approximation", Ph.D. Dissertation, University of Arizona, Tucson, AZ, 167 pp., 1992.

2 Kelly, D.W., Nakazawa, S., Zienkiewicz, O.C., and Heinrich, J.C., "A Note on Upwinding and Anisotropic Balancing Dissipation in Finite Element Approximations to Convection Diffusion Problems", *Int. J. Num. Meth. Eng.*, Vol. 15, pp. 1705-1711, 1980.

3 Brueckner, F.P. and Heinrich, J.C., "Petrov-Galerkin Finite Element Model for Compressible Flows", *Int. J. Num. Meth. Eng.*, Vol. 32, pp. 255-274, 1991.

4 Blank, T. "The MasPar MP-1 Architecture", Proceeding of the COMPCON Spring 90, 35th IEEE Computer Society International Conference, Feb. 26 - Mar. 2, San Francisco, CA, 1990.

5 "MasPar Fortran Reference Manual", Software Version 1.1, Aug. 1991.

6 "MasPar VAST-2 User's Guide", Software Version 1.2, Feb. 1992.

Parallel Computational Fluid Dynamics '92
R.B. Pelz, A. Ecer and J. Häuser (Editors)
© 1993 Elsevier Science Publishers B.V. All rights reserved.

A Comparison of Lattice Gas Automata Implementations on the MasPar MP-1[†]

J. Butterworth and J.F. Prins

Department of Computer Science, University of North Carolina, Chapel Hill NC 27599-3175, U.S.A. (butterwo@cs.unc.edu, prins@cs.unc.edu)

Abstract

Implementations of three lattice gas automata (FHP, EHPP, PAIR) and one lattice Boltzmann automaton (EHPP) for the simulation of low-velocity 2-D fluid-flow in porous media are described. Viscosities of the simulated fluids are measured using a Poiseuille flow experiment. Adherence to Darcy's law, describing flow rate through a porous medium as a function of pressure gradient, is assessed for each model. All implementations show good relative agreement with Darcy's law, but differ by a factor of two on the absolute permeability they assign to an artificial sample medium. Performance of the implementations is compared on a 4096-processor MasPar MP-1. The lattice structure of all models can be efficiently mapped to the nearest-neighbor connections of the MP-1 processor mesh. The FHP implementation yields the highest quantititative and qualitative performance metrics.

1. INTRODUCTION

In 1976 [HPP76] introduced a cellular automaton whose time and space averaged behavior exhibited correlation with 2-D fluid-flow. Since that time a variety of *lattice gas automata* (LGA) and related *lattice Boltzmann automata* (LBA) have been developed varying in the internal rules used, the fluid or fluids simulated, and the number of dimensions (2-D or 3-D) in which the simulation is performed. The behavior of several models (e.g. [FHP86]) was shown to be consistent with the Navier-Stokes equations describing fluid-flow in the incompressible limit. The main use to date of the automata models has been in the study of low Reynolds-number complex fluid dynamics, such as multiphase fluid-flow and flow through porous media [RK88][R90][R88][CDD+91][B91].

Lattice gas and lattice Boltzmann automata are particularly well-suited for highly-parallel computation because they possess inherent locality in the form of their evaluation rules, permitting their implementation to be scaled without communication penalty to machines with a large number of processors.

As part of a larger project studying the processes of contaminant transport and fluid flow in mutiphase systems [MM90], we are developing 2-D and 3-D LGA and LBA implementations on highly-parallel computers to simulate some small-scale features of complex flows. In this paper we compare the behavior of several 2-D LG and LB automata and describe their implementation on a MasPar MP-1.

The property of each model that is of principal interest is its conformance with physical fluid flow. It is known that the LGA and LBA techniques only give consistent results within a limited range of physical parameters (such as density and velocity) and automaton-dependent,

---

[†] This research supported in part by ARO-URI contract DAAL 03-91-G0155.

parameters (such as the scale of features in the model). To evaluate a model's physical validity, we performed an experiment described in [RK88] to determine the viscosity of the simulated fluid using an analytic model for Poiseuille flow. For each model we established a range of pressures in which the fluid showed constant viscosity. Subsequently we measured permeability of an artificial porous medium at various pressures within the range established for the model. Darcy's law predicts that flow rate will vary linearly with pressure. In this paper we report on the models' relative and absolute correspondence with Darcy's law.

A second important property of each model is the computational efficiency of its implementation. Although all models scale well to large parallel machines, variations in the rules and lattice connectivity of the various models have implications for the performance on machines like the MP-1 with 2-D mesh interconnection topology. To compare the performance of the various models we examined the simulation size and time required to obtain qualitatively similar results. We also investigated the scalability to 3-D of the basic approaches, since for a number of the models this extension involves intractable memory requirements.

This paper is organized as follows. Sections 2 and 3 describe the LGA and LBA models selected for this paper, and section 4 describes the implementation of these models on the MP-1. Section 5 explores the issues related to extending these models to 3-D. Section 6 describes simulation experiments to measure viscosity and permeability, section 7 presents the simulation and performance results. Section 8 draws some conclusions from the results.

2. LATTICE GAS AUTOMATA

The gas in an LGA consists of particles of identical mass residing at sites connected by edges to form a lattice such as the one shown in Figure 1. The units used in most LGA work are the *lattice unit* for distance, the *time step* for time, and the *particle mass* for mass. All particles have a velocity of one lattice unit per time step along one of the four directions possible in this simple lattice. An exclusion principle limits each site to hold at most one particle moving in a given direction.

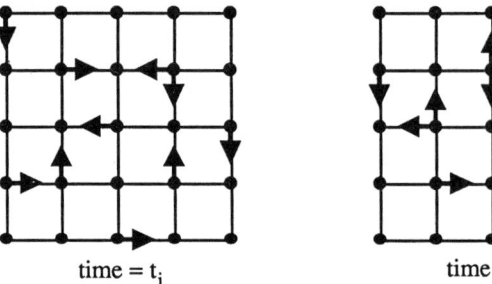

time = t_i time = t_{i+1}

Figure 1. A simple LGA developing through one time step.

The simulation proceeds in discrete time steps. Each time step consists of a movement phase and a collision phase. The movement phase carries each particle to the neighboring site along the edge on which it is traveling. In the collision phase particles arriving at a site collide, potentially causing a change in velocity, subject to conservation of mass, momentum and energy. For example, in the simple lattice of Figure 1, a head-on collision between two particles causes each particle to change its direction of travel by 90°.

If we average the velocities of the particles at each site the resultant vector field defined on the lattice describes fluid flow. The discrete nature of the simulation introduces a large amount of statistical noise at the microscopic level, hence time and space averaging of sites is required to yield a macroscopic vector field that has a more steady behavior. Averaging decreases statistical noise, but requires more sites and/or more time steps in the simulation.

The LGA in Figure 1 is named HPP after its creators, Hardy, Pomeau, and de Pazzis [HPP76]. Although this LGA qualitatively behaves like a fluid, it was shown to be anisotropic and is not a candidate for modeling real fluids [FHP86]. There are a variety of other lattice gas models, characterized by the structure of the lattice, particle velocities and collision rules. We have investigated the LGA models described below.

2.1. FHP

In 1986, Frisch, Hasslacher and Pomeau described a series of 2-D models based on a hexagonal lattice [FHP86]. These LGAs were shown to simulate the incompressible Navier-Stokes equations in the case of low average flow speed. The FHP lattice is shown in Figure 6, and an example FHP collision is shown in Figure 2. The collision rules we use in our implementation are called FHP-I in [FHH+87].

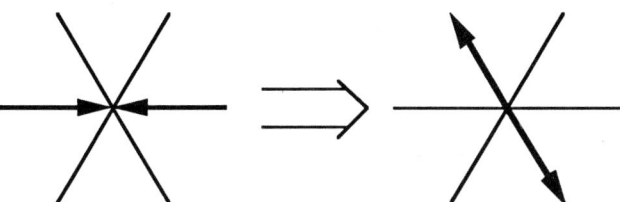

Figure 2. Example of an FHP collision.

2.2. EHPP

A 2-D LGA exists that uses a rectangular mesh like HPP but, unlike HPP, is isotropic [HLF86][CLZ+89]. We will refer to this model as Enhanced HPP (EHPP). On this lattice particles may move not only to nearest orthogonal neighbor sites, but also to nearest neighbors along diagonals. At each site there may be up to 4 particles moving with speed 1 along the x or y axes, up to 4 particles moving with speed $\sqrt{2}$ along diagonals, and possibly 1 particle at rest. By maintaining a balance condition between the concentrations of the different speed particles, this LGA becomes isotropic [HLF86][CLZ+89]. The edges of an EHPP lattice and an example EHPP collision are shown in figure 3.

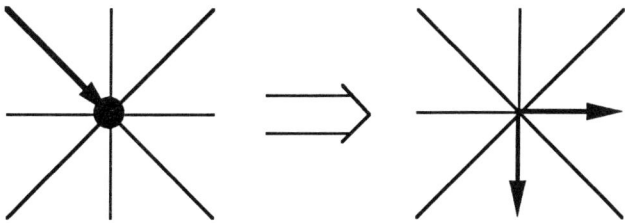

Figure 3. Example of an EHPP collision.

2.3. Pairwise Interaction LGA (PAIR)

Wolf-Gladrow, Nasilowski, and Vogeler have described an LGA that uses a square lattice (rotated by 45°) with particles that have more complex properties than those of other models [WNV91][N91]. Each site can hold up to four particles that move just as in other models. However, a particle carries momentum that is not necessarily equal to its mass times its velocity, so the collision phase becomes more complicated.

The collision phase proceeds by considering incoming particles in a pairwise fashion. First the two particles arriving from NW and NE collide, then the two particles arriving from SW and SE, then two particles from NW and SW, and finally the particles from NE and SE. Figure 4 shows two particles from SW and SE colliding. They do not change velocity but they do exchange some momentum which is indicated by the vectors in the bodies of the particles. We refer to this LGA as PAIR because of its pairwise collision phase.

Unfortunately, the viscosity of this model is anisotropic [N91]. The dynamic viscosity (μ) is 0.125 along the directions of particle movement, but it increases to 0.25 for directions 45° off the movement axes. However, we are still investigating this model because it is expected to scale well to 3-D and may be useful in special situations such as measuring permeability in only one direction.

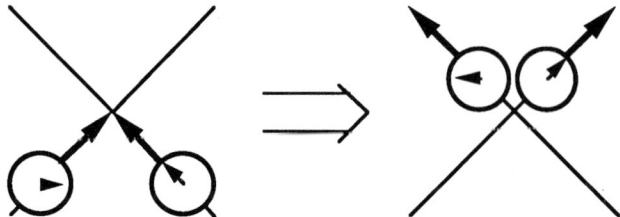

Figure 4. Example of a PAIR collision.

3. LATTICE BOLTZMANN AUTOMATA

Lattice Boltzmann automata are very similar to LG automata but greatly reduce statistical noise, and hence the requirement for large simulations that are time and space averaged [HJ89][MZ88]. The fundamental difference between LB and LG automata is that LBA store particle probabilities at each site rather than discrete particles. However, the LBA still use a discrete time step. The movement phase simply moves particle densities between sites. The collision phase uses rules similar to an LGA except that the conjunctions and disjunctions in the discrete logical description of the collision rules in an LGA are replaced by multiplication and addition of particle probability densities. Thus, the collision phase involves calculating products of the particle densities in various directions and then adding or subtracting these terms from the old particle densities.

4. IMPLEMENTATION

Lattice gas automata are easily implemented on a computer. The state of a site can be encoded using one bit for each possible direction along which a particle may travel. The motion phase then involves an exchange of bits between sites. The collision phase is a function from states to states, which can be implemented by simple table look-up if the number of bits used to encode the state is not too large. This is certainly the case for the 2-D automata described here: the number of entries in the collision tables are, 2^6 for FHP, 2^9 for EHPP, and 2^6 for PAIR.

A lattice Boltzmann automata must represent a particle probability $0 \leq p \leq 1$ for each direction of travel in a state. Thus the EHPP LBA stores 9 floating point numbers per site. Because of the continuous nature of the LBA rules, the collision phase must be calculated explicitly, instead of by look-up table.

To efficiently implement either technique on a parallel computer requires that the lattice be decomposed across processors in such a fashion that neighbors on the lattice are in the same processor or in a neighboring processor in the machine. The LGA or LBA computation is particularly appropriate for a SIMD computer since the operations to be performed at each site are identical, and SIMD-style nearest-neighbor communication can be used. This observation is the basis of many designs of customized parallel cellular automaton machines for LGA simulations [MT90][LF91].

Our interests are in using general-purpose parallel machines to implement LG and LB automata, since these machines can be expected to track rapid improvements in speed and multiplicity of the basic processing elements, as well as accommodating different LG and LB models.

4.1. Mapping the Lattice to the MP-1 Communication Architecture

The particular machine on which we have conducted our experiments is a 4096-processor MasPar MP-1. This SIMD machine organizes processing elements (PEs) into a rectangular array, with each processor directly connected to its eight nearest neighbors as shown in Figure 5. The edges of the processing array are connected to each other to give the processor mesh a toroidal topology. All of the models implemented can be mapped onto the machine so that the movement phase only requires communication between nearest neighbor processors. These mappings thereby ensure a very low communication overhead.

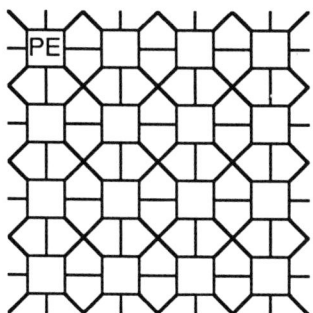

Figure 5. MP-1 communication network.

The EHPP and PAIR models use rectangular lattices, so the mapping into the MP-1 processor array is straightforward. However, FHP uses a hexagonal grid, which requires more effort to map into the MP-1 efficiently. We chose a skewed mapping as shown in Figure 6, which ensures local communication between processors. The toroidal topology in the east-west direction is critical in this mapping, and yields an FHP lattice with periodic boundary conditions in the horizontal direction. In contrast to the rectangular lattices, the lattice top and bottom are not connected in this implementation. By performing extra communication we could have given the FHP lattice periodic boundary conditions in the vertical direction as well, but none of the simulations done so far have required this.

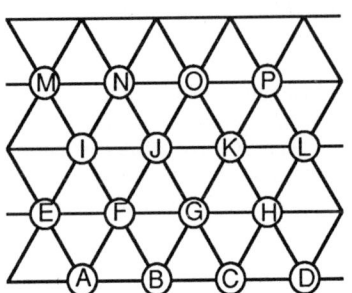 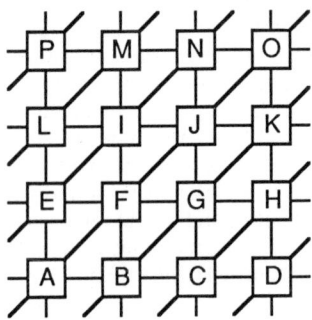

Figure 6. Skewed mapping for the FHP lattice onto the MP-1 processor grid. Unused connections between processors are not shown.

4.2. Controlling the Simulations

4.2.1. Barriers

Barriers are implemented by adding an extra bit to each site indicating the presence or absence of a wall [R88][CDD+90]. The effect of this bit can be factored into the collision rules. Currently we use this bit to simulate no-slip boundary conditions: when a particle collides with a barrier, it reverses direction. It is possible to make arbitrary shaped obstacles by setting barrier bits.

4.2.2. Pressure Gradient

The basic approach to implementing a pressure gradient is to *hit* some particles along an edge of the simulation by adding some momentum to them [CDD+91]. For instance, in an FHP model, changing a particle along the west edge moving west to a particle moving east would impart two momentum units to it and help create a negative pressure gradient toward the east. Averaged over time, these hits cause a pressure gradient that is the momentum imparted per time step divided by the area over which it is applied, divided by the distance over which the pressure drops. The problem with controlling this pressure gradient is knowing how much momentum is imparted per time step on average. For example, if one sets the probability of hitting one particle at a site (the *hit probability*) at 30%, then one might expect to hit 30% of the sites along the edge at each time step. Unfortunately not all sites are candidates for being hit. An FHP site that has no west moving particle or that already has an east moving particle can not be hit. The probability of a site being a candidate for being hit depends on the density of the lattice gas and in general may be difficult to determine analytically.

One improvement of the basic technique is to hit particles throughout the simulation volume rather than just along one edge [R92]. This will avoid pressure oscillations coming from the one edge and will allow lower hit probabilities to be used, which in turn allows higher pressure gradients. Instead of deriving the hit probability, we chose to dynamically determine it. We created a *regulator*, which compares the number of particles hit with the number desired and alters the hit probability for the next simulation step accordingly. The result is that on average just the right number of sites are hit to maintain a desired pressure gradient.

4.3. Software Design for Reusability

Implementations of these lattice gas simulations have many things in common. They all have an initialization phase for setting up obstacles and the initial distribution of particles, a movement phase, a collision phase, and methods for reporting the status of the simulation. We have created a framework for implementing the simulations that takes advantage of these similarities. This framework was written in MPL, a data-parallel dialect of C for the MP-1.

Our simulation system has two conceptual parts called the *front end* and the *simulator* which communicate through a standardized interface. This interface takes the form of MPL function prototypes that specify how the front end can give information to the simulator (such as the initial state of the simulation) or request information from the simulator (such as the average flow rate). The interface allows us to develop front ends that have no dependence on the exact model being used for the low level simulation. It also allows a simulator to be written without any knowledge of whether it will be used for a viscosity experiment or a graphical display.

Currently we have three front ends: one for graphical display of flow, one for measuring viscosity of the fluid, and one for measuring permeability of a medium. Each of these simulators can use any of the lattice gas models that we have implemented by simply setting a command line switch. This approach will be particularly useful as we create more front ends and implement more lattice gas models.

5. EXTENSION OF MODELS TO 3-D

In practice, it is difficult to extend EHPP and FHP to 3-D because of the collision phase. In 2-D, the state at each site is small with a correspondingly small collision-phase lookup table. In 3-D, the lattice connectivity increases dramatically with a commensurate increase in the maximum number of particles per site: a 3-D EHPP site has 19 particles and a 3-D extension of FHP, called FCHC, has 24 particles per site [HLF86]. The resultant lookup tables for the collision function become too large to replicate across the limited-size memories of the processing elements in highly parallel machines like the MP-1.

An advantage of PAIR is that it scales well to 3-D. In the 2-D collision phase, particles interact in pairs, and a collision table of only 2^6 entries can be used. In a 3-D version of PAIR particles still interact in pairs, and the collision table would need 2^8 entries. Such a table could easily fit on each processor.

Extending the LBA to 3D is computationally very demanding due to the increase in the number of products to calculate during the collision phase and the increase in number of terms in each product. Each product represents a specific collision between particles that conserves both mass, momentum and energy. In 2-D, the number of possible collisions for EHPP is small, but in 3-D the number of collisions becomes huge.

It is possible to reduce the computation in 3-D LBA models by expanding the collision operator around equilibrium densities [HJ89][SBH91]. Then the non-equilibrium portion of the densities can be operated on by a linearized form of the collision operator. This technique reduces the collision calculation from many large products to a vector-matrix multiplication. For 3-D models the vectors contain at least 19 densities and the linearized collision operator is represented by at least a 19×19 matrix.

6. EXPERIMENTS

In order to verify the physical validity of these models, some simple experiments were performed relating to the applicability of lattice gas simulations to porous media. One of the most basic properties to measure of a porous medium is its permeability, a measure of the resistance to forced flow. However, in order to measure permeability one must know the viscosity of the fluid being forced through the medium [R88][CDD+91].

We measured dynamic viscosity by simulating plane Poiseuille flow in 2-D and then comparing the results to the known analytic solution for the flow. The top and bottom sides of the chamber shown in Figure 7, separated by a distance d, have no-slip boundary conditions while the left and right sides connect to model an infinitely long tube using periodic boundary conditions. A uniform pressure gradient dp/dx is applied forcing the fluid in the positive x direction. By measuring the volumetric rate of flow per unit area q in the x direction in the steady state, one can calculate dynamic viscosity from the analytic description of Poiseuille flow:

$$q = -\frac{d^2}{12\mu}\frac{dp}{dx} \quad (1)$$

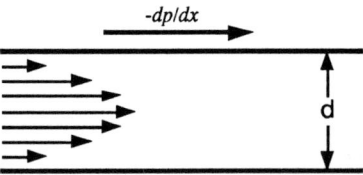

Figure 7. Plane Poiseuille flow Vectors indicate velocity of flow.

Once viscosity is known, the permeability k of a nearly arbitrary medium is described by Darcy's Law

$$q = -\frac{k}{\mu}\frac{dp}{dx} \quad (2)$$

which establishes the linear relationship usually found in porous media between pressure gradient and volumetric flow rate. Permeability measurement can be performed in a fashion very similar to viscosity measurement. A pressure gradient is applied to the porous medium and q is measured, from which permeability is calculated [R88][CDD+91].

7. RESULTS

We have implemented 2-D versions of HPP, EHPP, FHP, and PAIR as LGAs and EHPP as an LBA on a 4096-processor MasPar MP-1. These simulations incorporate arbitrary boundary geometries and forced flow.

7.1. Viscosity

Table 1 records the dynamic viscosities of the FHP-LGA, PAIR-LGA, EHPP-LGA, and EHPP-LBA models we measured using the viscometer front end. Large pressure gradients result in fluid speeds close to the maximum particle speed which causes unphysical effects. As a result, we measured viscosities at various pressure gradients in order to find where these speed limiting effects became important. We also found lower bounds for the pressure gradients because at very low speeds, statistical noise dominates the flow rate measurements in LG automata. Channel width can also limit the range of pressure gradients, since wide channels lead to higher speed flow in the center of the channel. The table includes the range of pressure gradients for which the viscosity measurements were essentially constant over a range of channel widths from $32\sqrt{3}/2$ to 64 lattice units wide since this is the range of channel widths that occur in the permeability experiment in section 7.2.

Densities are listed in Table 1 because they affect the viscosities. All densities are normalized by the number of possible particles per site. Thus, a density of 0.4 for FHP means an average of 2.4 particles per site. For the low speed flows we simulated, density is essentially constant. Note that PAIR viscosity is anisotropic and the measurement in Table 1 is along the direction leading to the minimum value.

Model	Density	linear range of dp/dx	Measured Viscosity	Published Viscosity
FHP	0.4	$10^{-4} \leq dp/dx \leq 10^{-3}$	$2.17 \leq \mu \leq 2.45$	2.31 [R88]
PAIR	0.5	$10^{-5} \leq dp/dx \leq 10^{-4}$	$0.12 \leq \mu \leq 0.14$	0.125 [WNV91]
EHPP (LGA)	0.106	$5 \times 10^{-5} \leq dp/dx \leq 2 \times 10^{-4}$	$0.81 \leq \mu \leq 0.86$	(none)
EHPP (LBA)	0.106	$10^{-8} \leq dp/dx \leq 10^{-4}$	$0.49 \leq \mu \leq 0.51$	(none)

Table 1. Dynamic viscosity of LG and LB automata fluids.

7.2. Permeability

We measured the permeability of a simple porous medium similar to one used in [R88]. The medium is shown in Figure 8. The steady state flow rate was measured at various pressure gradients to determine consistency of behavior with that predicted by (2). Figure 9 shows the steady-state flow rate observed for each model at various pressure gradients. The error bars indicate statistical variation even after steady state was reached. For all three LGA, the shaded region is the range of pressure gradients for which viscosity is constant as determined by the Poiseuille flow experiment. However, at very low pressure gradients EHPP-LBA diverged from Darcy's law, so permeability was only calculated using dp/dx in the range 10^{-6} to 10^{-4}.

Figure 8. Forced flow in steady state through a simulated porous medium. Gray rectangular regions are obstacles. In the channels, brightness indicates flow rate.

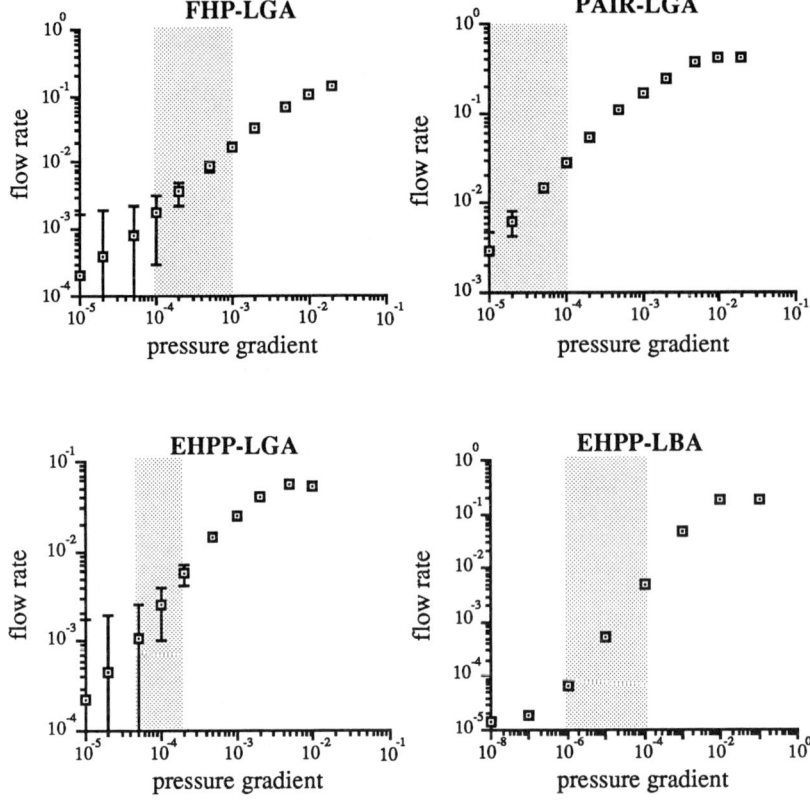

Figure 9. Relationship between pressure gradient and flow rate.

For all three LGA, the shaded regions of Figure 9 were used to calculate the permeability of the medium according to each model. The results are summarized in Table 2. Each model shows good relative consistency with Darcy's law over the constant-viscosity range of dp/dx, but the models are not in absolute agreement on the permeability of the medium. To investigate the dependence of the results on feature size in the medium, we doubled the number of sites used to represent each dimension of the artificial medium. The observed permeabilities were essentially unchanged. Therefore we believe that the absolute variation in permeability observed is an inherent feature of the simulated fluid.

	FHP-LGA	PAIR-LGA	EHPP-LGA	EHPP-LBA
Permeability	41.1 ± 1.5	24.6 ± 0.97	21.9 ± 1.6	27.0 ± 1.4

Table 2. Measured Permeability of artificial medium.

7.3. Comparative Performance of the Models

Table 3 gives the basic memory requirements and performance of each of the models on a 4096-processor MP-1. Since the implementation is scalable, the performance on a larger or smaller machine can be extrapolated directly. Slightly better performance is achieved in larger simulations because of the decreased communication to computation ratio. To compare the performance of the models on a realistic experiment we have included the simulation size, number of steps to steady-state and total time required to perform the permeability calculation of the previous section at a fixed dp/dx.

Model	Space $\frac{bytes}{site}$	Update rate $\frac{sites}{sec} \times 10^6$	PERMEABILITY EXPERIMENT		
			Sites	simulation steps	total time (secs)
FHP (LGA)	1	31	512 × 512	3610 ± 507	17
PAIR (LGA)	2	9.8	512 × 512	4863 ± 1337	108
EHPP (LGA)	2	19	512 × 512	1760 ± 278	56
EHPP (LBA)	72	0.2	64 × 64	2180 ± 605	44

Table 3. Performance of the implementations on a 4096-processor MP-1

8. CONCLUSIONS

The objective of our work is to simulate fluid flow through porous media. Since all four of the models presented here give behavior consistent with Darcy's law within a range of pressure gradients, each could serve as the basis for more complex fluid flow simulations. Of the three lattice gas automata, FHP and PAIR yield constant viscosity measurements over the largest range of pressure gradients. However, the EHPP lattice Boltzmann automaton yields constant viscosity over a larger range of pressure gradients than any of the LGA. This suggests that the LBA approach will be superior over a larger variety of porous media and pressure gradients.

The FHP implementation yields the highest update rate per site, due to the smaller number of lattice connections per site compared to EHPP and a simpler evaluation rule compared to PAIR. Although LBA models have dramatically lower update rates per site compared with LGA models, the smaller number of sites per simulation (because of the lack of statistical noise) and the smaller number of simulation steps to steady-state give the EHPP-LBA a performance advantage over the EHPP-LGA in the permeability experiment. We expect that an FHP-LBA will give a commensurate performance advantage over the FHP-LGA.

Thus we conclude that an FHP-LBA will maximize flexibility in porous flow experiments and will maximize performance on the MP-1. However, we will continue to investigate alternatives to FHP as we add more complexity to our simulations, such as multiphase and 3-D fluid flow. It is possible that these features may prove to be more efficiently added to models other than FHP.

9. REFERENCES

[B91]   B. Boghosian, "Lattice Gases Illustrate the Power of Cellular Automata in Physics", *Computers in Physics*, Nov/Dec (1991) pp. 585.

[CDD+91]   S. Chen, K. Diemer, G. Doolen, K. Eggert, C. Fu, S. Gutman, and B. Travis, "Lattice Gas Automata for Flow Through Porous Media", *Physica D*, **47** (1991) pp 72.

[CLZ+89]   S. Chen, M. Lee, K. Zhao, and G. Doolen, "A Lattice Gas Model with Temperature", *Physica D*, **37** (1989) pp. 42.

[FHH+87]   U. Frisch, D. d'Humières, B. Hasslacher, P. Lallemand, Y. Pomeau, and J. Rivet, "Lattice Gas Hydrodynamics in Two and Three Dimensions", *Complex Systems*, **1** (1987) pp. 649.

[FHP86]   U. Frisch, B. Hasslacher, and Y. Pomeau, "Lattice-Gas Automata for the Navier-Stokes Equation", *Phys. Rev. Lett.*, **56** (1986) pp. 1505.

[HJ89]   F. Higuera and J. Jimenez, "Boltzmann Approach to Lattice Gas Simulations", *Europhysics Letters*, **9** (1989) pp. 663.

[HLF86]   D. d'Humières, P. Lallemand, and U. Frisch, "Lattice Gas Models for 3D Hydrodynamics", *Europhys. Lett.*, **2** (1986) pp. 291.

[HPP76]   J. Hardy, O. de Pazzis, and Y. Pomeau, "Molecular Dynamics of a Classical Lattice Gas: Transport Properties and Time Correlation Functions", *Phys. Rev. A*, **13** (1976) pp. 1949.

[LF91]   F. Lee and M. Flynn, "Architectural Mechanisms to Support Three-Dimensional Lattice Gas Simulations", Proceedings of the 3rd Symposium on Parallel Algorithms and Architecture, ACM, (1991) pp. 115.

[MM90]   A. Mayer and C. Miller, "A Compositional Model for Simulating Multiphase Flow, Transport and Mass Transfer in Groundwater Systems", *Proceedings of the 8th Conference on Computational Methods in Water Resources*, Springer-Verlag, (1990) pp. 217.

[MT90]   N. Margolus and T. Toffoli, "Cellular Automata Machines", *Lattice Gas Methods for Partial Differential Equations*, SFI SISOC, Eds. Doolen et al., Addison-Wesley Publishing Co., (1990) pp. 219.

[MZ88]   G. McNamara and G. Zanetti, "Use of the Boltzmann Equation to Simulate Lattice-Gas Automata", *Phys. Rev. Lett.*, **61** (1988) pp. 2332.

[N91]   R. Nasilowski, "A Cellular-Automaton Fluid Model with Simple Rules in Arbitrarily Many Dimensions", *Journal of Stat. Phys.*, **65** (1991) pp. 97.

[R88]   D. Rothman, "Cellular-Automaton Fluids: A Model for Flow in Porous Media", *Geophysics*, **53** (1988) pp. 509.

[R90]   D. Rothman, "Macroscopic Laws for Immiscible Two-Phase Flow in Porous Media: Results From Numerical Experiments", *Journal of Geophysical Research*, **95** (1990) pp. 8663.

[R92]   D. Rothman, personal communication.

[RK88]   D. Rothman and J. Keller, "Immiscible Cellular-Automaton Fluids", *Journal of Stat. Phys.*, **52** (1988) pp. 1119.

[SBH91]   S. Succi, R. Benzi, and F. Higuera, "The Lattice Boltzmann Equation: A New Tool for Computational Fluid-Dynamics", *Physica D*, **47** (1991) pp. 219.

[WNV91]   D. Wolf-Gladrow, R. Nasilowski, and A. Vogeler, "Numerical Simulations of Fluid Dynamics with a Pair Interaction Automaton in Two Dimensions", *Complex Systems*, **5** (1991) pp. 89.

Solving CFD methods involving global communication on distributed memory MIMD multiprocessors

Alan G. Chalmers[a] Steven P. Fiddes[b]

[a]Dept. of Computer Science, University of Bristol, Bristol, United Kingdom

[b]Dept. of Aerospace Engineering, University of Bristol, Bristol, United Kingdom

Abstract

A panel method is frequently used in aerodynamics to calculate the incompressible potential flow over a three dimensional body. The global data dependencies and variations in computational complexity makes the parallelisation of this method on large numbers of MIMD processors a formidable task, unless effective methods can be found to minimise the inherent message densities. In this paper minimum path configurations of processors are used to achieve this. The performance superiority of these topologies over torus and ring configurations is demonstrated.

1 Introduction

Computational fluid dynamics has been identified as an application area that can benefit greatly from the application of parallel processing. This is due to a large part due on the "near-neighbour" communication requirements of most time marching finite-difference CFD methods, for example finite-volume methods where fluxes between adjacent cells are responsible for "driving" most of the calculation. However, there is a wide range of CFD techniques that do not involve only near-neighbour communication, but require *global* communication between various components of the calculation method. For example, free Lagrange methods use a system of particles that interact with neighbours, but the neighbours change during the calculation. Spectral methods (such as those used in many direct solution procedures for Navier-Stokes equations) recast the governing equations as a global interaction of modes in the frequency domain. Indeed, one of the earliest CFD techniques - panel methods for subsonic flows, builds on the elliptic nature of the flow and each panel influences the flow conditions at every other panel immediately - again a global interaction problem.

Global communication manifests itself as a high communication requirement when implementing panel methods on distributed memory MIMD multiprocessors. Furthermore, the global data dependencies and variations in computational complexity makes their parallel implementation on large numbers of MIMD processors a formidable task, unless effective methods can be found to minimise the inherent message densities. In this paper we describe a new family of minimum path processor configurations to achieve this and show the performance superiority of these topologies over torus and ring configurations which are traditionally used to solve problems involving global communication.

Some solutions of the panel methods have been achieved on large SIMD architectures, such as the Connection Machine. This type of architecture lacks the flexibility for exploiting asymptotic expansions of the exact solution for panels that are "far" apart. In this paper we will show that the flexibility of MIMD multiprocessors connected in minimum path configurations leads to the most efficient parallel implementation of panel methods.

Furthermore, panel methods can be regarded as a special case of boundary element methods, a technique that has a wide range of applications outside CFD, such as electromagnetic scattering, structural analysis and acoustics, and thus the computational approach that we follow can give an efficient parallel implementation of not only other CFD methodologies requiring global communication, but also a wide range of applications from other disciplines.

2 Panel Methods

Panel Methods are applicable to any problem that is governed by Laplace's equation. These methods were originally known as surface singularity methods, however, the technique of covering the domain of the problem with small quadrilaterals led to the method being named the panel method.

For some years, panel methods have been extensively used on serial computers for the solution of a range of potential flow problems [8]. The routine application of this method to very large problems in aerodynamics has been restricted by the high computational cost. The availability of inexpensive MIMD systems of processors has provided the opportunity to develop methods which allow cost effective solutions to these problems. Full details of the method formulation and numerical approximation may be found in [2, 9].

Panel methods of the type described in [2, 9], or based on computationally similar formulations, have found a wide range of applications in aerospace and other industries, for example the design of racing cars, sailing vessels, locomotives etc., where a need arises to predict the low speed flow past complex configurations. Panel methods have become established as a viable industrial design tool and a substantial amount of commercial supercomputer time is spent on running them. In a recent review, Hess [8] stated that his company (Douglas) computes the flow about a complete aircraft using panel methods about ten times a day. The number of panels used is typically in the range of 1000 − 7000. Hess quotes a 5000 panel solution as requiring some 40 minutes on an IBM 3090. On a historical note, Hess and Smith originally developed their method using an IBM 7090. A 5000 panel problem would have taken twelve and a half hours on this machine!

The actual method used to calculate the velocity influence coefficients depends on the relative distance between the collocation point (when the boundary condition is applied - normally the centroid of the panel) and the panel being considered. If this distance is small then a full, exact expression for the influence coefficients must be used [6]. This exact expression involves time consuming logarithm and arctangent evaluations. If the collocation point is some distance from the panel then an asymptotic expansion of the exact expression may be used to produce much simpler expressions that are faster to evaluate with no substantial loss in accuracy. In our calculations two expansions are used, one for *far-field* points and a more complex expansion for *mid-field* points. For *near-field* points the full, exact expression is used.

A very large number of influence coefficients have to be calculated in a panel method. If we consider calculating the flow past a single body whose surface is defined by a mesh of $m \times n$ points, this produces $(m - 1) \times (n - 1)$ panels which therefore requires $(m - 1)^2 \times (n - 1)^2$ influence coefficients per velocity component be calculated. Because of this size of computation posed by a panel method and its frequency of use as a design tool, there is a strong incentive to develop a parallelised panel method.

3 A Parallel Solution

The calculation of flow past a body by means of a panel method comprises four stages:

1. Calculation of panel geometries
2. Influence function calculation and assembly of the matrix
3. Solution of the influence matrix
4. Calculation of flow quantities on or off the surface of the body

Each stage can be solved as a separate parallel problem, but using different strategies to achieve an efficient solution. However, there is an interaction between each stage which cannot be solved without regard for the other stages, because an independently constructed efficient solution for a particular stage of the problem may result in data being placed on processors which is inappropriate for the next stage of the problem, as described below. To reposition this data may require a considerable communication overhead with a corresponding large impact on the performance of the overall solution. The approach for each stage is described below.

3.1 Calculating the Panel Geometries

The first stage of the panel method computation involves calculating the geometries, collocation points and normals for each panel. Each panel's geometry calculations are totally independent of other panels. A demand driven model of computation is appropriate to minimise processor idle time and avoid an initial data load penalty [2]. As we wish to address very large problems there will, in general, be insufficient space available in each processors local memory to hold the information of the geometries of all panels. A copy of the code for calculating the panel geometry is available on each processor and each processor is thus provided with the panel geometry on a panel-by-panel basis from the system controller as and when the processor requires more work. Buffering these tasks ensures that panel data is always available at the processor when it is ready to accept it.

Rather than some arbitrary allocation of tasks a *preferred bias* allocation schema is adopted. In this schema the panels to be processed are *conceptually* divided equally amongst the processors. When a panel is to be allocated to a processor, rather than some arbitrary allocation, the processor task is matched to the conceptual section. Of course, variations in computational complexity make it very difficult to accurately decide the preferred bias, therefore, we must expect the computation of some conceptual regions to be completed while others continue. This load balancing problem is easily overcome by allocating a panel from the portion belonging to the processor that has processed the least number of panels so far. The processor performing the work then communicates the result for storage at the processor from whose conceptual region the panel was allocated. This ensures that a balance is attained in the number of panel geometries stored at each processor at the end of the first stage of the solution sequence.

3.2 Setting up of the Influence Matrix

In the second stage of the solution, the velocity induced by the panel at the collocation points of all panels (including itself) is calculated. As a result of the first stage the panel data has been distributed evenly across the processors. For the influence coefficient calculation some of this data must be exchanged between processors. Two communication options are available for evaluating the influence functions:

1. Pass the collocation point co-ordinates and normals between processors and have each processor evaluate the influence coefficients for the panels it holds on all the collocation points it holds and receives. This is an attractive procedure as the amount of data that needs to be communicated between processors for each panel is simply the collocation point co-ordinates and normals for that panel, ie. six real numbers.

2. Pass the complete panel geometry between processors and have each processor evaluate the influence of all panels on the collocation points that it holds. This involves the communication of some thirty real numbers per panel between processors, ie. a five-fold increase on the communication required for the first strategy.

Although the first strategy appears the most desirable, the final choice is dictated by another consideration. If the first possibility is chosen then the influence matrix is distributed over the system with each processor holding *columns* of the influence matrix. The second option results in the *rows* of the matrix being distributed over the processors. Whether the influence matrix is distributed column-wise or row-wise across the processors has an important effect on the strategy that may be adopted to solve the influence matrix in the next stage. Only row storage results will be discussed in this paper, but "column-based" solvers are possible.

At the end of the second stage, the influence matrix has been calculated and is stored in a distributed manner across the processors along with the relevant portion of the right hand side.

3.3 Solving the Influence Matrix

The influence matrix is, in general, full, non-symmetric and usually not strictly diagonally dominant. The matrix will not always meet the requirements for guaranteed convergence of an iterative scheme, such as Jacobi or Gauss-Seidel. However, early work with the method considered test cases consisting of a single component that produced diagonally dominant matrices, so some time was spent examining the effectiveness and efficiency of iterative solvers. Parallel row and column based solvers were developed using a mixture of Gauss-Seidel and Jacobi techniques [6].

A global Gauss-Seidel iteration procedure can only be performed sequentially, as each update of the solution vector, depends on the latest updates of the elements above it. However, we have used a hybrid scheme to produce the results given below. In this hybrid the update of the solution vector on each processor is performed using the Gauss-Seidel iteration on the part of the solution vector corresponding to the rows held by that processor, and then this is communicated globally before the next update. Each global iteration is synchronised. Thus each processor updates a block of the matrix using a Gauss-Seidel procedure while a Jacobi-like scheme iterates over the blocks. This method we have termed *block Gauss-Seidel global Jacobi*.

The iterative method worked well for the test cases studied. However, to ensure that the solution method is robust and can handle any non-singular matrix, a direct solver is essential, and work is currently in hand to develop an efficient parallel direct solver.

3.4 Tangential Velocity Calculations

This final stage of the calculation requires simple matrix multiplication, using the already calculated velocity influence coefficients and may easily be performed in parallel. However, the storage requirements associated with the velocity influence coefficient matrices is very large, and they have not been retained in the results discussed below. However, because of the very effective parallelisation of the influence matrix calculations it is feasible to consider re-computing the influence coefficients for this final stage and thus avoid the large storage overheads.

4 System Configurations

The principle reason for the panel method being described as difficult to effectively parallelise is the requirement for the interaction of all panels with each other, ie. a global data dependency. The implication being that a processor will need access to all panel information on every other processor, with the consequent high cost in communicating this information. Although the panel method gives rise to a fixed number of messages initiated at any processor, the system configuration will determine the total number of messages that permeate the system.

The performance of a system of distributed memory MIMD processors depends in large part on the efficiency of the message passing system that provides the interface between the co-operating processors [5]. The suitability of the underlying processor interconnection topology is an important factor affecting the performance of this message transfer system. In a fully interconnected network every processor is adjacent to every other processor. For an p processor system this requires that each processor has $(p-1)$ physical links available for interconnection. For large values of p this may not be possible, in which case processors will have to communicate with each other not directly but via intermediate processors. For a constant number of links per processor, as the number of processors in the network increases, so to does the number of intermediate processors through which a message must be routed. This increase in maximum distance between any two processors is a function of the underlying interconnection topology. For the class of problems that require every processor to communicate frequently with all other processors it is desirable that the maximum distance (in terms of the number of links to be traversed) between any two processors be small if the system performance is not to be too severely compromised.

The topology usually recommended for dealing with a global data requirement is a ring, see for example [7]. However, in the ring configuration the maximum distance between any two processors, and therefore the maximum message path length, is large and only a simple linear arrangement of processors would give a higher total of messages on the system. Algorithms that exhibit global dependencies have been used on hypercubes of MIMD processors, for example [1], and on hypercubes of SIMD processors (the connection machine) [4, 14]. However, for a fixed number, k, of links per processor, a maximum of 2^k processors can be included in the hypercube with a maximum distance of k between processors. Torus configurations are another possibility which provide, for a system of k processors a maximum distance between any two processors of order($\sqrt{k}$). A ternary tree of height h, consists of a single processor at the top level, the *root processor*, connected to a maximum of three other processors, each of which is a root processor of a ternary subtree of height $(h-1)$. Examples of these configurations are shown in Figure 1.

We have adopted minimum path systems for dealing with problems that exhibit global data dependencies. The philosophy underlying the construction of a minimum path (AMP) configuration is to minimise the distance a message has to travel between any source processor and any other destination processor within the configuration. This principle is maintained even at the expense of the loss of symmetry in a system [3]. The processor currently used in AMP configurations is the T800 transputer [11] and so the following results for AMPs are obtained with each processor only having *four* links available for connecting to other processors. (AMP configurations are, of course, not limited to only four links per processor, but are possible for

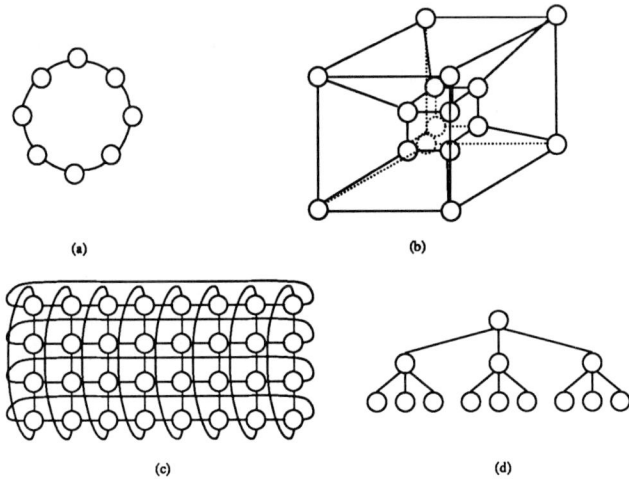

Figure 1: (a) 8-processor ring (b) 16-processor hypercube (c) 32-processor torus (d) 13-processor ternary tree

any number of links per processor).

The diameter of a configuration may be defined as the maximum number of links a message has to traverse between any source processor and any other destination processor within the configuration. Table 1 gives the diameters of a number of processors arranged in each of the configurations. As can be seen from the table, for a given number of processors, the diameter of AMP configurations are less than any of the other configurations, and indeed the 64-processor AMP configuration has the same diameter as a 8-processor ring.

	Processors							
	8	13	16	23	32	40	53	64	128
AMP	2	2	3	3	3	4	4	4	5
Hypercube	3	-	4	-	5	-	-	6	7
Torus	3	-	4	-	6	7	-	7	12
Ternary Tree	4	4	5	6	6	6	8	8	10
Ring	4	6	8	11	16	20	26	32	64

Table 1: Comparison of configuration diameters

Table 2 gives the average interprocessor distance values, for the different configurations. This average interprocessor distance is equivalent to the average number of links a message has to cross from any source processor to its desired destination processor. So, for example, a message from a source processor in a 64-processor AMP would have to cross 2.92 links on average while, for a processor on a 64-processor ring, the same message would have to traverse on average 16 links. The 53-processor AMP is shown in figure 2.

The average interprocessor distances for the AMP configurations are lower than all the other configurations shown except for the 128-processor hypercube where it is marginally higher.

	Processors								
	8	13	16	23	32	40	53	64	128
AMP	1.28	1.55	1.73	2.05	2.31	2.53	2.76	2.92	3.58
Hypercube	1.50	-	2.00	-	2.50	-	-	3.00	3.50
Torus	1.50	-	2.00	-	3.00	3.50	-	4.00	5.64
Ternary Tree	1.97	2.56	2.91	3.39	3.93	4.25	4.77	5.01	6.25
Ring	2.00	3.23	4.00	5.74	8.00	10.00	13.25	16.00	32.00

Table 2: Comparison of average interprocessor distances

However, it must be remembered that the 128-processor hypercube requires *seven* links per processor as opposed to the *four* links that are used in the AMP configurations, and despite this, the diameter of the 128-processor AMP is less than that of the hypercube with the same number of processors.

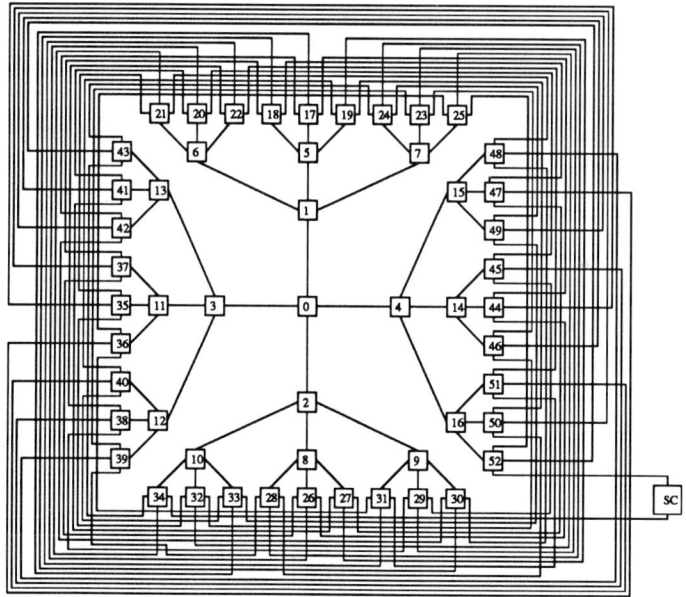

Figure 2: 53-processor AMP

5 Other Parallel Panel Method Implementations

Chriske and Bogucz [4] and Tseng and Egolf [14] have implemented versions of panel methods on Connection Machines [10]. Chriske and Bogucz consider the source panel method of Hess and Smith. They restrict their work to considering the prediction of only two-dimensional external flow. The SIMD nature of the Connection Machine is used by allowing each processing element to calculate an influence matrix coefficient. If the number of processing elements is less than the total number of matrix elements then the concept of the virtual processing element is utilised. While this concept does not actually allow all the calculations to be performed in parallel, it does, however, act like a pipeline. If the number of influence matrix elements is less than the number of processing elements then some of these processing elements may not be used. Chriske and Bogucz use a Gauss-Jordan reduction scheme to solve the influence matrix. No pivoting of rows was required as the influence matrix for the test case they chose was well conditioned [4].

Tseng and Egolf implement a Greens Identity formulation [12] of the panel method on a Connection Machine. Their work examines a three-dimensional steady-state subsonic compressible potential flow. Once again a number of coefficients are calculated in parallel, utilising the concept of the virtual processing element when necessary. To perform the calculations, each virtual processing element receives the co-ordinates of the four vertices of a panel and then performs the required algorithm serially to obtain the source and doublet coefficients. In their paper, Tseng and Egolf explore several different algorithms to solve the resultant matrix [14].

Due to the nature of the selected architecture, both the above papers are unable to exploit an important feature of panel methods. These papers both utilise the exact expression for the influence coefficients. While this exact expression must be used if the distances between the panels are "small", if the collocation point is some distance from a panel then an asymptotic expansion of the exact solution should be used. These expansions are much simpler expressions and thus may be computed in a fraction of the time of the exact solution, as shown for a single source method coefficient in table 3. Obviously if a large number of coefficients may be solved by these expansions then the total time saved will be significant. Indeed, these asymptotic expansions may produce a *more accurate* solution for panels that are "far" apart than the exact solution [13]. This is because if the panels are "far" apart then the exact expression involves the difference of nearly equal quantities which are then subject to "round-off" errors on computers.

Using the rigid SIMD architecture of the Connection Machine it is not possible to perform the near-, mid- and far-field calculations at different processing elements at the same time, but rather these different calculations have to be performed separately. This means that those processing elements not involved in the appropriate calculation are ignored and, therefore, for that particular computation, do not contribute to the overall system performance. Multiprocessor MIMD architectures are more flexible and thus better suited for implementing problems which require different calculations depending on the individual data items, that is, where the type of computation is data dependent.

	near	mid	far
Influence calculation	1.000	0.221	0.151

Table 3: Normalised times required for *near-*, *mid-* and *far-field* calculations

6 Results

The results shown are for a parallel panel method run for a variety of configurations on a Parsys Supernode system which contains a maximum of 64 T800 transputers [11]. It should be noted that the parallelisation strategy is language and even processor independent.

Results were obtained using ring, torus and AMP system configurations for a test case of a 10% thick ellipsoid at incidence above the ground plane. The body was represented by an increasing number of panels from 196 (corresponding to a 15 × 15 mesh) to 3481 panels (corresponding to a 60 × 60 mesh). Due to memory limitations it was only possible to solve up to a 576 panel problem on a single processor.

To investigate the performance characteristics of the panel method on the system configurations, test runs were carried out using systems ranging from 1 to 63 processors. Execution times were recorded for the geometry calculation stage, setup stage and the matrix solution stage of the method. (Labelled *Geom*, *Set up* and *Solve* in the following tables. The total time to solve the first three stages of the panel method is labelled *Total*). The setup stage used the method described above, for calculating the rows of the influence matrix matrix directly on the processors where the rows are required for the matrix solution stage.

Processors	Geom	Set up	Solve	Total
1	1.78	94.09	68.0	163.87
2	1.43	46.8	63.6	111.83

Table 4: Single processor results for a 576 panel problem

Processors	Geom	Set up	Solve	Total
4	1.14	24.9	35.8	61.84
8	1.32	13.9	21.1	36.32
16	1.32	8.3	14.1	23.72
23	1.32	7.5	17.0	25.82
32	1.32	6.7	17.2	25.22
53	1.33	5.9	31.9	39.13
63	1.34	5.5	35.3	42.14

Table 5: AMP topology results for a 576 panel problem

Processors	Geom	Set up	Solve	Total
4	1.32	25.2	35.9	62.42
8	1.57	14.2	22.0	37.77
16	1.31	8.9	17.3	27.51
25	1.31	8.2	20.3	29.81
32	1.40	7.2	32.8	41.40
52	1.40	7.3	45.8	54.50
63	1.42	6.8	40.2	48.42

Table 6: Torus topology results for a 576 panel problem

Table 4 give the time required for the first three stages of the panel method for a problem with 576 panels (25 × 25 mesh) when run on one and two transputers. Tables 5, 6 and 7 give corresponding results for increasing number of processors for the AMP, torus and ring

Processors	Geom	Set up	Solve	Total
4	1.42	24.9	36.1	62.42
8	1.32	14.6	23.1	39.02
16	1.32	12.8	21.5	35.62
23	1.32	13.4	27.0	41.02
32	1.32	13.3	26.4	41.02
53	1.34	15.5	74.8	91.64
63	1.34	17.6	88.23	107.17

Table 7: Ring topology results for a 576 panel problem

configurations respectively. The iterative solver used in the single processor version is Gauss-Seidel while that used on the multiprocessors is the block Gauss-Seidel global Jacobi. Of note is the rapid reduction in time required for calculating the influence matrix as the number of processors increases, and how the fraction of total execution time spent in this calculation reduces. For a single processor version of this code, some 60% of the execution time is spent in generating the influence matrix. This drops to some 30% of the time for 32 processors connected in an AMP.

The speedup of the matrix solution is not as marked as the influence matrix calculations. In fact, for all three topologies there is a marked **increase** in the time for this stage of the calculation in going from 32 to 63 processors. This is due to increasing communication overhead that is introduced as the number of processors is increased. The impact of this overhead can be seen to be lower on the AMP than either the ring or torus.

Finally, figure 3 shows the time taken in seconds for the first three stages of the panel method for problem sizes increasing from 576 to 2916 panels, on 63-processor AMP, torus and ring configurations. Results were not possible for the problem sizes greater than 1936 panels on 63-processor rings as the resultant message densities were too high and buffer overflow occurred. The solution of the 576 panel problem on the AMP configuration is 18% faster than the same problem on the torus and 200% faster than the same problem on the ring. For a problem size of 1936 panels, the AMP is still approximately 18% faster than the torus and 165% faster than the ring. When the problem size is increased to 2916 panels, the AMP configuration is still 15% faster than the torus.

7 Discussion

The results above show the advantage of the AMP topology over rings and tori for parallel panel methods. The advantage is twofold. Firstly, for a given number of processors the AMP is faster, and secondly, the fall off in efficiency with decreasing grain size (eg increasing the number of processors for a fixed size problem), is slower for the AMP than the other two configurations. This means that an AMP configuration can make efficient use of more processors than either a ring or torus is able to.

Work is currently in hand to implement a so-called Dirichlet boundary condition method similar to that proposed by Morino [12]. This will permit the calculation of flow past general lifting configurations and produce matrices that will be a more stringent test of the solution procedures, with the advantage of only requiring the generation and storage of one influence matrix.

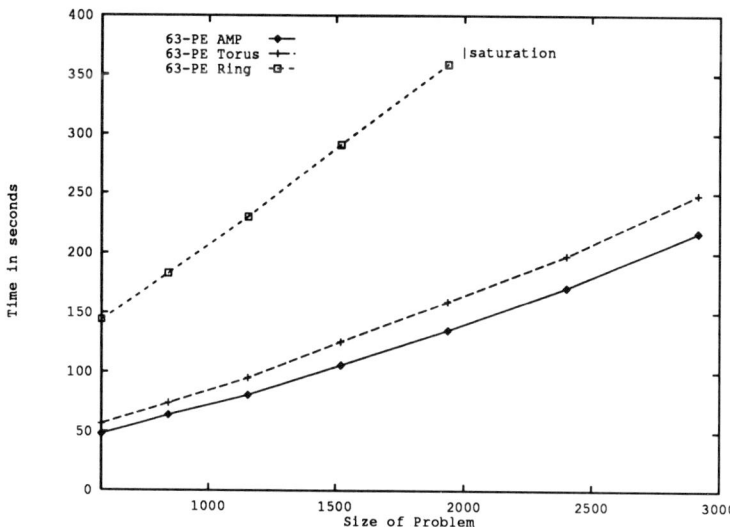

Figure 3: Total time taken in seconds for increasing problem size on 63-processor configurations

8 Conclusions

We have demonstrated that the panel methods can be effectively parallelised, and particularly so when the message complexity is minimised by using an appropriate system architecture such as the AMP. The methodology described here for the efficient implementation of an aerodynamic panel methods on distributed memory MIMD processors can be immediately applied to other boundary element methods by replacing the particular influence function computations, and similarly to other CFD methodologies with inherent global communication requirements. This offers the opportunity for large engineering applications to be attempted without recourse to expensive vectorising architectures.

9 Acknowledgements

We would like to thank Dr Roger Miles and the Bristol Transputer Centre for the use of their equipment.

References

1 C. J. Catherasoo. Separated flow simulations using the vortex method on a hypercube. In *AIAA 8th Computational Fluid Dynamics Conference*, Honolulu, 1987.

2 A. G. Chalmers, S. Fiddes, and D. J. Paddon. Parallel panel methods. In H. S. M. Zedan, editor, *13th Occam Users Group conference*, pages 313–321, IOS Press, York, 1990.

3. A. G. Chalmers and S. Gregory. *Constructing Minimum Path Configurations for Multiprocessor Systems*. Technical Report CSTR-92-12, Department of Computer Science, University of Bristol, Bristol, Apr. 1992.

4. D. M. Chriske and E. Bogucz. Performance evaluation of the Connection Machine for panel method calculations. In *29th Aerospace Sciences Meeting*, Reno, USA, Jan. 1991.

5. T. Feng. A survey of interconnection networks. *IEEE Computer*, 12–27, Dec. 1981.

6. S. P. Fiddes and A. G. Chalmers. Aerodynamic panel methods on transputers. In *Transputers for Industrial Applications*, Antwerp, Belgium, 1990.

7. G. Fox et al. *Solving problems on concurrent processors - Volume 1*. Volume 1, Prentice Hall International, 1988.

8. J. L. Hess. Panel methods in computational fluid dynamics. *Annual Review of Fluid Mechanics*, 22, 1990.

9. J. L. Hess and A. M. O. Smith. *Calculation of Non-Lifting Potential Flow about Arbitrary Three-Dimensional Bodies*. Technical Report ES 40622, Douglas Aircraft Company Inc., March 1962.

10. D. W. Hillis. *The Connection Machine*. The MIT Press, 1985.

11. Inmos. *IMS T800 Architecture*. Inmos Technical Note 6, Inmos Ltd., Bristol, 1988.

12. L. Morino and C. Kuo. Subsonic potential aerodynamics for complex configurations: A general theory. *AIAA Journal*, 12(2), Feb. 1974.

13. J. N. Newman. Distribution of sources and normal dipoles over a quadrilateral panel. *Journal of Engineering Mathematics*, 20:113–126, 1986.

14. K. Tseng and T. A. Egolf. Parallel implementation of a boundary integral formulation for potential flow aerodynamics. In *29th Aerospace Sciences Meeting*, Reno, USA, Jan. 1991.

A Fast Vortex Method for the Simulation of Three-Dimensional Flows on Parallel Computers

K. Chua and T.R. Quackenbush

Continuum Dynamics, Inc., P.O. Box 3073, Princeton, NJ 08543, U.S.A.

Abstract
Previous flow simulations using vortex methods on parallel computers have suffered from large computation and inter-processor communication time. In this paper, a parallel fast vortex method which addresses these difficulties is presented. The new method has a reduced asymptotic time complexity and is computationally efficient. Implementation on parallel-processing computers is carried out using a new communication strategy with low inter-processor communication requirements. Sample calculations are presented which highlight the enhancements in both computation and communication time.

1. INTRODUCTION

Recently, there has been much interest in the use of a vortex method for practical flow simulations [1-3]. In this method, compact regions of vortical fluid in a flow are discretized using a collection of three-dimensional vortex elements. The elements convect at the local fluid velocity and automatically track the vortical regions as they evolve in the flow. This allows the computational grid to continuously adapt to regions of steep flow gradients where high numerical resolution is needed.

The Lagrangian nature of the vortex method makes it well suited to the treatment of complex flow configurations. However, because of the high computational cost of the method, application to very large scale computation has not been feasible. The bulk of the computational work load is attributed to the Biot-Savart velocity calculation. Given N vortex elements, the velocity calculation at each time step requires N^2 machine operations. The quadratic nature of the method confines its use to calculations involving fewer than several thousand elements and thus limits the flow regimes that the method can analyze with confidence.

The computational intensiveness of the vortex method makes its implementation on a parallel-processing computer particularly attractive. As discussed by Fox and Otto [4] and Seitz [5], the long-range pair-wise interactions between the vortex elements define a problem format which is easily implemented on the hypercube concurrent processor. The elements are partitioned among the set of processors and a ring topology is used for the communication between the processors. Although the strategy requires exchange of large array of data between the processors at every time step, the associated communication time is insignificant compared to the time required by the N^2 calculation and a relatively high parallel efficiency has been reported [6,7].

Recently, a robust fast vortex method has been developed [8-10]. The method has been successfully applied to a very large scale computation of a complex flow around a representative helicopter rotor/body configuration. As many as 10,000 vortex elements were used and significant improvements in the computation time were reported. The reduced computational intensiveness of the fast vortex method implies that an implementation of the new method on a parallel-processing computer using the classical communication strategy [4,5], would result in significant communication overhead. A new communication strategy with reduced inter-processor communication requirements is therefore needed.

In this paper, a new communication strategy for the efficient parallel implementation of the fast vortex method is presented. The new strategy is designed to require minimum inter-processor communication, in synergy with the reduced computational requirement of the fast vortex method. The paper is organized as follows: Section 2 describes the mathematical formulation of the fast vortex method; Section 3 describes the new communication strategy; Section 4 presents demonstration calculations on the Intel iPSC/860 parallel computer and Section 5 gives a summary of this work.

2. MATHEMATICAL FORMULATIONS

In a Lagrangian vortex method, the vorticity field $\underline{\omega}$ is discretized into a collection of discrete vortex vector element:

$$\underline{\omega}(\underline{x}, t) = \sum_{i=1}^{N} \underline{\omega}_i(t) \, \delta v \, \zeta_\sigma(\underline{x} - \underline{x}_i(t)) . \tag{1}$$

Each element is represented by a vector strength $\underline{\omega}_i$, a position $\underline{x}_i$, a volume δv, and a spherically symmetric smoothing function ζ_σ. The elements convect at the local velocity:

$$\frac{d\underline{x}_i}{dt} = \underline{u}(\underline{x}_i) , \tag{2}$$

and stretch and rotate according to the local velocity gradient:

$$\frac{d\underline{\omega}_i}{dt} = \underline{\omega}_i \cdot \nabla \underline{u}(\underline{x}_i) . \tag{3}$$

Given N vortex elements, the local velocity and velocity gradients are needed at each vortex location for the integration of Equations (2) and (3) at every time steps. Classically, this is computed by summing the Biot-Savart inductions due to all the vortex elements and thus exacts a high computational workload.

2.1 Far-field approximation

In the present formulation, the velocity computation is decomposed into a near-field vortex-to-vortex velocity interaction and a far-field group-to-group velocity interaction. The use of a far-field group-to-group approximation reduces the asymptotic time complexity of the method and enhances its computational efficiency.

To initiate a calculation using the far-field approximation, a grid of three-dimensional cubic boxes is first superimposed on the flow domain. The grid of boxes is chosen to contain all the vortex elements in the flow domain and the elements are clustered into groups according to the boxes. Near-field velocity interactions between vortices which are in the same box or immediate neighboring boxes are evaluated by a direct summation of the Biot-Savart induction. For the evaluation of the velocity interactions between vortices which are in well-separated boxes, a far-field group-to-group approximation is used. The far-field approximation is formulated based on multipole expansions, geometric influence coefficients and Taylor series expansions.

2.2 Multipole expansion

The vortex-induced velocity is computed by the Biot-Savart law for the appropriate vortex element representation. For evaluation points which are in the far-field, because of the quadratic drop off of the Biot-Savart kernel with distance, the induced velocity can be computed using a point vortex representation:

$$\underline{u}(\underline{x}, t) = -\frac{1}{4\pi} \sum_{i=1}^{N} \frac{(\underline{x} - \underline{x}_i(t)) \times \underline{\omega}_i(t) \, \delta v}{|\underline{x} - \underline{x}_i(t)|^3} . \qquad (4)$$

A multipole expansion of Equation (4) can be written as:

$$\underline{u}(\underline{x}) = -\frac{1}{4\pi r^3} \sum_{k=0}^{\infty} \underline{E}_k \frac{1}{r^k} , \qquad (5)$$

where $\underline{E}_k$ are the moment coefficients of the expansion, $r = |\underline{x} - \underline{x}_{cm}|$, and $\underline{x}_{cm}$ is the point of expansion. The fixed center of each group is used as the point of expansion. Given any observation point which is well separated from the group, the multipole expansion given by Equation (5) converges. In particular, the truncation error of the expansion is bounded by:

$$e < c\left(\frac{D}{r}\right)^{p+1} , \qquad (6)$$

where e is the truncation error, p is the number of terms in the expansion, D is the diameter of the vortex group, r is the separation distance of the evaluation point from the center of the group, and c is a constant [11].

2.3 Geometric influence coefficients

For the evaluation of the far-field group-to-group velocity interactions, the velocity due to each inducting vortex group is evaluated at the center of all distant vortex groups. The multipole expansion of the group induced velocity, given by Equation (5), is used and is rewritten in the following form:

$$\underline{u}(\underline{x}) = C_0 \underline{E}_0 + C_1 \underline{E}_1 + C_2 \underline{E}_2 + \ldots , \qquad (7)$$

where

$$C_0 = \frac{-1}{4\pi r^3}, \qquad (8a)$$

$$C_1 = \frac{-1}{4\pi r^4}, \qquad (8b)$$

$$C_2 = \frac{-1}{4\pi r^5}. \qquad (8c)$$

For each pair of groups, the separation distance r between the centers of the groups is fixed at all time. Therefore, the set of influence coefficients C_0, C_1, C_2, etc. given by Equation (8), can be computed and stored at the outset of the calculation. At each subsequent time step in the calculation, the moment coefficients E_0, E_1, E_2, etc. are computed, and the group induced velocity due to one vortex group on another vortex group can be immediately obtained via simple multiplication.

2.4 Taylor series expansion

Using Equation (7), the velocity due to an inducting vortex group is evaluated at the center of a distant vortex group, and a Taylor series expansion is subsequently used to compute the velocity at an individual vortex location within that group:

$$\underline{u}(\underline{x}) = \underline{u}(\underline{x}_{cm}) + \delta\underline{x} \cdot \nabla \underline{u}(\underline{x}_{cm}) + \frac{1}{2!}(\delta\underline{x} \cdot \nabla)^2 \underline{u}(\underline{x}_{cm}) + \ldots , \qquad (9)$$

where $\delta\underline{x} = \underline{x} - \underline{x}_{cm}$ and $\nabla = \partial/\partial\underline{x}$. In Equation (9), the induced velocity $\underline{u}$ and its gradients $\partial\underline{u}/\partial x_i$, $\partial^2\underline{u}/\partial x_i \partial x_j$ and so on (here, x_i and x_j denote the ith and j^{th} components, respectively, of the vector $\underline{x}$) are computed using multipole expansions.

The far-field group-to-group approximation based on the multipole expansions, geometric influence coefficients and Taylor series expansions as described above forms the foundation of the present fast vortex method. The method has an asymptotic time complexity of $O(N^{4/3})$ but can be further improved to $O(N\log N)$.

3. PARALLEL IMPLEMENTATION

In our previous effort in the implementation of vortex methods on parallel computers [7], the ring topology proposed by Fox and Otto [4] and Seitz [5] is used. The set of multiple processors is linked into a ring structure (Figure 1), and the vortex elements partitioned among the processors on the ring. The N^2 velocity calculation workload is divided equally among the processors, and the inter-processor communication is effected by message passing between neighboring processors on the ring. The following is an outline of the calculation steps in each processor:

(i) Compute the velocity interactions between all resident vortices of the processor.

(ii) Make two copies of the information of the resident vortices - host copy and guest copy.

(iii) Send guest copy to front neighbor processor and receive guest copy from rear neighbor processor.

(iv) Compute velocity interactions between vortices in resident host copy and vortices in visiting guest copy.

(v) Repeat steps (iii) and (iv) until all the velocity interactions have been computed, i.e. $N_p/2$ times for N_p processors.

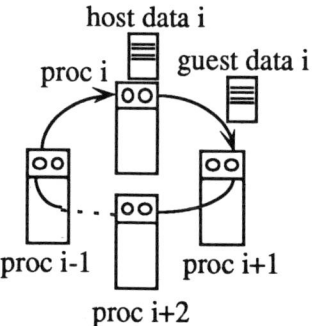

Figure 1. Schematic showing the communication strategy (ring topology) used for long-range data transmission.

In the above, all the information of the resident vortices in each processor needs to be transmitted. This results in a large communication message size and a corresponding large communication time. However, in the classical vortex method, the communication overhead is insignificant compared to the computation time.

In the present fast vortex method, the computational load is considerably reduced. It is therefore expected that the communication time would become a significant fraction of the total parallel computation time. In the following, we describe a new communication strategy with reduced communication requirement for the efficient parallel implementation of the fast vortex method.

3.1 A new communication strategy

The new strategy builds on the structure of the fast vortex method and takes advantage of the near-field/far-field velocity decomposition. The flow domain is divided into a grid of three-dimensional cells and the cells are partitioned onto a set of multiple processors (Figure 2). Each cell may be further divided into a set of smaller boxes. A set of geometric influence coefficients between all non-immediate neighboring boxes (including boxes from different processors) is computed and stored in each processor.

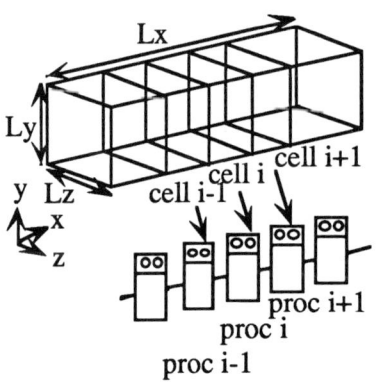

Figure 2. Schematic showing the mapping of cells onto processors.

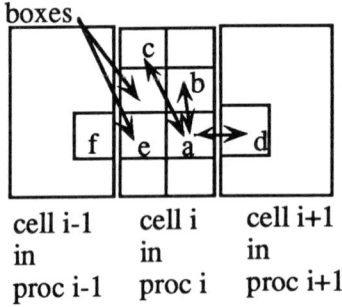

Figure 3. Schematic showing communication steps for transmission of data used in near-field/far-field velocity calculation.

The vortex elements are partitioned onto the set of processors according to the cell in which they reside. In each processor, the vortices are clustered into groups according to the boxes that lie within each cell, and a multipole expansion is computed about the center of each group. The global velocity interaction between all vortices is then processed according to the following sequence, as illustrated in Figure 3:

(i)   Near-field velocity interactions between vortices from the same processor.

(ii) Far-field velocity interactions between vortices from the same processor.

(iii) Near-field velocity interactions between vortices from different but immediate neighboring processors.

(iv) Far-field velocity interactions between vortices from different processors.

Steps (i) and (ii) involve calculations of the velocity interactions between vortices residing on the same processor and thus do not require interprocessor communication. Step (i) involves the near-field velocity interactions between vortices which are from the same box or immediate-neighboring boxes (i.e. interaction between boxes a and b in Figure 3). This is computed using a direct summation of the Biot-Savart law. Step (ii) involves the far-field velocity interactions between vortices which are from non-neighboring boxes (i.e. interaction between boxes a and c). This is computed using the far-field group-to-group approximation.

Steps (iii) and (iv) involve the calculation of the velocity interactions between vortices which reside on different processors and require the transmission of information between processors. Step (iii) involves the near-field velocity interaction between vortices from neighboring boxes at the cell interface (e.g., interaction between boxes a and d). This is computed using a direct summation of the Biot-Savart law and requires the transfer of the information of all the vortices in the respective boxes. This is a one-step data exchange between immediate neighboring processors and can be carried out very efficiently. Step (iv) involves the far-field group-to-group velocity interactions between all pairs of non-neighboring boxes which are from different processors (e.g., interaction between boxes d and e). This requires the long-range communication of information between all processors. For this purpose, the ring topology described earlier is employed (Figure 1). In the present application, because of the use of a vortex group velocity induction, only the moment coefficients of the group expansion need to be communicated. The set of moment coefficients involves only a few floating point numbers, and can be transmitted between the processors very efficiently.

4. RESULTS

The present parallel algorithm for the fast vortex method has been implemented on the Intel iPSC/860 parallel computer at the Department of Energy's Oak Ridge National Laboratory. Sample test calculations involving single and multiple vortex rings are presented and the results are compared with those of the classical N^2 method.

In the present application of the fast vortex method to the vortex ring calculations, a one-dimensional grid of cells is chosen. The cells are evenly distributed along the circumference of a circle with the same radius as the initial condition of the vortex ring. The azimuthal dimension of the cell is determined by the number of processors, while the radial and axial dimensions are chosen to be sufficiently large to contain all the vortex elements. A schematic of the discretization strategy is given in Figure 4. The cells are partitioned onto the set of processors and within each processor, the cell may be divided into smaller boxes. The use of an azimuthal array of cells avoids the allocation of regions of flow with no vortex elements to any processor and ensures the load-balancing of the computation.

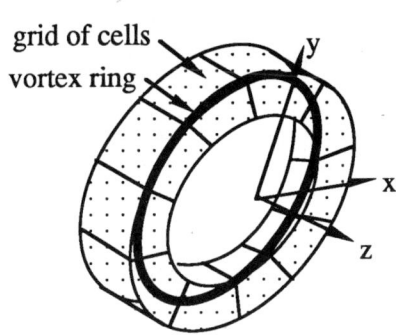

Figure 4. Schematic showing computational cells for the vortex ring problem.

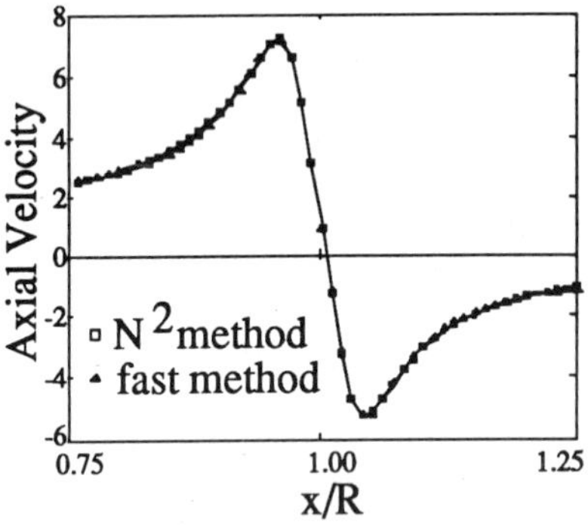

Figure 5. Plot of axial velocity computed at points along radial line on the plane of the vortex ring; fast vortex method △; N^2 vortex method □.

4.1 Isolated vortex ring

In this test case, the calculation of the isolated vortex ring is carried out. The vortex ring has radius R=1, circulation $\Gamma=4\pi$, and core thickness $\sigma_t=0.05$. The cross-sectional core has constant vorticity distribution. In the discretization, two layers of vortex filaments are used, i.e. one filament in the center of the core, and 8 around the circumference of the core. Each filament is further discretized into 256 vortex elements, each of which has a smoothing parameter $\sigma_s=0.025$.

Figure 5 shows a plot of the vortex induced axial velocity on the plane of the ring along a radial cut that passes through the core of the vortex. The squares represent data points computed using the N^2 method and the triangles denote results computed using the fast vortex method. The agreement between the two calculations is very good.

Figure 6 shows a plot of the computational speed for the calculation versus the number of vortex elements. 32 nodes on the iPSC/860 are used in the calculation. Three different sets of calculations, using the fast vortex method with 32, 64 and 128 vortex groups, respectively, are performed. These correspond to 1, 2, 4 groups per cell with each cell assigned to a processor. The enormous reduction in computation time using the fast vortex method is clearly evident. Also, for a fixed number of processors, the computational savings is increased as the number of groups is increased. This results from the reduced number of near-field velocity calculations that each processor has to carry out.

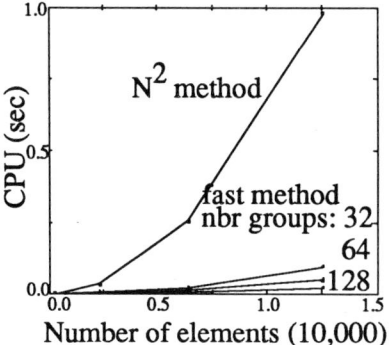
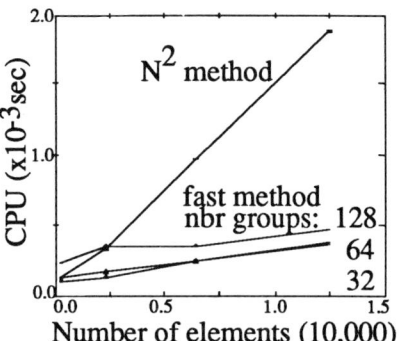

Figure 6. Comparison of CPU time on the iPSC/860: fast vortex method △; N^2 vortex method ▫.

Figure 7. Comparison of communication time on the iPSC/860: new communication strategy △; communication strategy of Ref. [4,5] ▫.

Figure 7 shows a plot of the inter-processor communication time in these calculations. In the N^2 method, all the information of the vortices in a processor is communicated using the classical long-range communication strategy [4,5]. This results in a large communication message size and a corresponding large communication time. In the fast method, the communication process involves two components. The first component is a one-step communication between near-neighboring processors and incurs negligible clock time. The second component is a long-range inter-processor communication process and involves information of the vortex groups within each processor, i.e. the moment coefficients of the multipole expansions. This also exacts negligible clock time because of the small communication message size. A comparison of the communication time taken by the present fast method to that by the classical method shows significant improvement. In particular, for the classical calculation, a steep linear increase in the communication time with the number of vortices is observed, whereas in the case of the fast calculation, this increase is much slower. Also, with a fixed number of processors, as the number of vortex groups per cell (processor) is increased, there is a slight increase in the communication time. This reflects the increased number of moment coefficients that are transmitted.

The above comparisons for a fixed number of processors (Figure 6 and 7) show that a large number of vortex groups per processor is desirable for improved computational speed. However, this can also increase the interprocessor communication time. This conflict can be eliminated with the use of a hierarchical clustering approach [12] which will be implemented in a future effort.

Figure 8 shows a plot of the computational speed of the fast vortex calculation versus the number of vortex elements. Three sets of calculations, using 8, 16, and 32 nodes on the iPSC/860, are presented. In these calculations, the number of vortex groups per processor is held fixed at one. It is observed that as the number of processors is increased, the computational time is dramatically reduced. In particular, a doubling of the number of processors more than doubles the computational speed. This enormous speed increase with the number of processor can be accounted for in two parts:

(i) Speed increase with number of processors: in the new parallel algorithm, the long-range inter-processor communication time is reduced to a minimum and an increase in the number of processor results in a corresponding increase in computational power with little loss due to communication overhead.

(ii) Fewer near-field interactions: with a fixed number of vortex groups per processor, an increase in the number of processors results in an increase in the number of vortex groups; this reduces the number of vortices per group and the corresponding number of near-field vortex-to-vortex interactions that need to be computed.

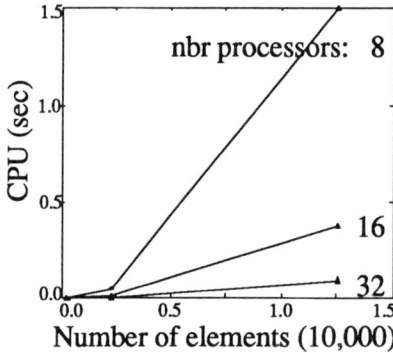
Figure 8. Comparison of CPU time of the fast vortex method for different numbers of processors (one vortex group per processor).

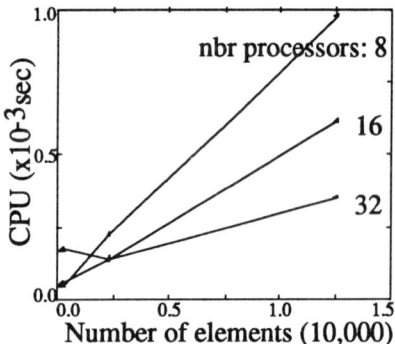
Figure 9. Comparison of communication time of the fast vortex method with different numbers of processors (one vortex group per processor).

Figure 9 shows a plot of the communication time in these calculations. It is observed that as the number of processors is increased, the communication time is reduced. This is contrary to typical observations where communication time is increased with the number of processors. This may be explained as follows: in the present implementation, the far-field inter-processor communication message size involves only a few floating point numbers (independent of the number of vortices) and therefore requires negligible communication time. In contrast, the one-step near-neighbor communication may require significant time since it involves the information of all the resident vortices in the processor. For example, when using 8 processors, the calculation with 12,000 vortices involves 1,500 vortices per processor and the communication message array containing the information of all the vortices is large. As the number of processor is increased, the number of vortices per processor is reduced, resulting in lower communication time.

4.2 Multiple interacting vortex rings

To demonstrate the accuracy of the present fast method for long-time integration, we have applied the method to the simulation of the leap-frogging of two vortex rings. Figure 10 shows an oblique view of the results of two calculations of the model problem using the fast vortex method and the classical vortex method. Each vortex ring is discretized using 256 elements and advanced in time using a 4th-order Runge-Kutta scheme. 100 integration steps, each of size $\Gamma \Delta t/R^2 = 0.003$, are used. Good qualitative agreement between the two sets of results is observed.

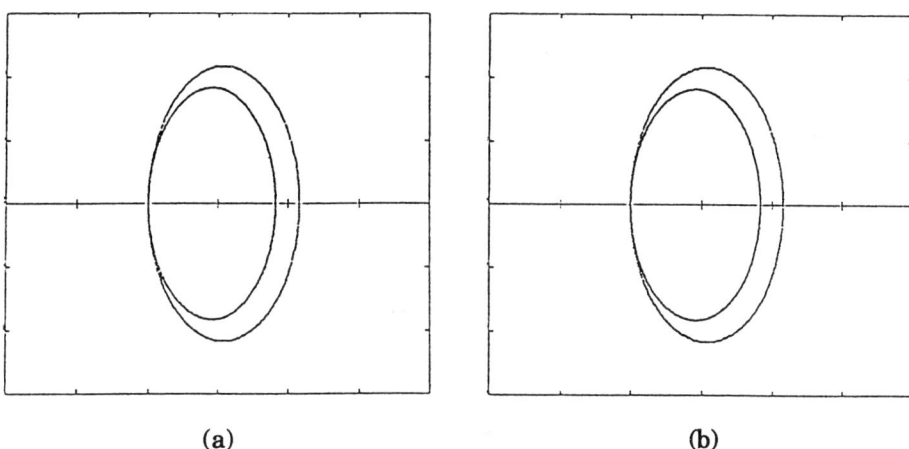

(a)   (b)

Figure 10. The leap-frogging of two vortex rings simulated on the iPSC/860: (a) fast vortex method and (b) classical vortex method.

5. SUMMARY

In the present effort, a fast vortex method for unsteady flow calculations has been successfully formulated and implemented. Robustness and accuracy of the method has been demonstrated and a computational speedup in excess of a factor of 20, compared to the classical N^2 approach, has been achieved for calculations using $O(10^4)$ vortex elements. More importantly, the asymptotic time complexity of the method is dramatically reduced. Because of this, the speedup factor will grow enormously for very large scale computations involving $O(10^5\text{-}10^6)$ elements.

The fast vortex method has been implemented on the Intel iPSC/860 parallel computer to assess the additional speedup possible relative to serial processing machines. In addition to the inherent speedup from the use of multiple processors, this implementation has been enhanced by the application of a novel, low-overhead parallel algorithm that greatly reduces the communication overhead for large-scale vortex calculations. The present implementation not only involves greatly reduced computational load relative to classical methods but also achieves high parallel efficiency. Thus, we have successfully addressed the two key issues of computational and communicational intensiveness of a parallel vortex computation.

6. ACKNOWLEDGMENTS

This work was supported by the U.S. Department of Energy under contract DE-FG05-91ER81207. The Technical Monitor was Dr. Thomas Kitchen. Use of the Intel iPSC/860 parallel computer was provided by the Oak Ridge National Laboratories.

7. REFERENCES

1. Leonard, A., Jour. Comp. Phys., 37 (1980) 289.
2. Leonard, A., Ann. Rev. Fluid Mech. 17 (1985) 523.
3. Nicolaides, R.A. and Gunzburger, M.D. (eds.), Incompressible Computational Fluid Dynamics - Trends and Advances, Cambridge Univ. Pr., Cambridge, 1991.
4. Fox, G.C. and Otto, S.W., Physics Today, 37 (1984) 50.
5. Seitz, C.L., Comm. ACM, 28 (1985) 22.
6. Catherasoo, C.J., AIAA Paper No. 87-1189, 1987.
7. Pepin, F., Chua, K. and Leonard, A., Proc. 3rd Conf. Hypercube Concurrent Computers and Applications, Pasadena, CA, Jan. 1988.
8. Chua, K., Leonard, A., Pepin, F. and Winckelmans, G., Proc. ASME Ann. Winter Meet., Chicago, IL, Nov. 1988.
9. Chua, K. and Quackenbush, T.R., AIAA Paper No. 92-2624, 1992.
10. Chua, K., Quackenbush, T.R. and Leonard, A., Continuum Dynamics, Inc. Report No. 92-03, final report to DoE under grant DE-FG05-91ER81207, June 1992.
11. Greengard, L., ACM Disti. Dissertation Ser., MIT Press, Cambridge, MA, 1988.
12. Pepin, F., Ph.D. Thesis, Caltech, May 1990.

CFD on the 1-K node nCUBE/2

D. D. Cline

Massively Parallel Computing Research Laboratory, Division 1421, Sandia National Laboratories, Albuquerque, New Mexico, USA, 87185

Abstract

The numerical simulation of fluid motion demands the utmost in high performance computing hardware. At Sandia's Massively Parallel Computing Research Laboratory, we have been developing CFD applications to execute on distributed memory massively parallel, Multiple-Instruction, Multiple-Data (MIMD) computers. The 1024 node nCUBE/2 provides a unique capability for exploring the scalability of algorithms, as well as performing CFD simulations at an unprecedented degree of fidelity. In the present paper we discuss fundamental communication issues which are of relevance to CFD implementations on large-scale parallel architectures.

1. INTRODUCTION

Massively parallel computers offer the potential for unprecedented performance in carrying out numerical simulations of fluid motion. *Hardware scalability* enables machines of increasingly higher performance to be constructed without significant new breakthroughs in computer technology. In addition, researchers may access only the necessary computer resources to carry out a particular simulation and thereby scale their application to the degree of resolution required.

The primary limiting factor which prohibits these machines from entering mainstream computing is the development of *scalable software*. As hardware performance grows, so does the degree of complexity of the applications researchers are willing to undertake. As such, there are two competing effects in the present arena of CFD research. On one hand, massively parallel performance enables more complex simulations to be carried out in a cost-effective manner, while at the same time, increased performance through parallelism introduces programming difficulties which may significantly increase software development time.

In the present paper we will review some fundamental aspects of CFD code development which specifically targets large-scale massively parallel multiple instruction, multiple data (MIMD) computers. We will describe some of the basic concepts in high performance message passing and discuss some of the lessons learned in developing data structures and domain decomposition techniques for Sandia's 1024 processor nCUBE/2.

2. NCUBE/2 OVERVIEW

The nCUBE/2 is a massively parallel MIMD computer consisting of 1024 processors, each with their own self-contained operating system. The nCUBE chip is a proprietary *VLSI* design which supports both 32-bit and 64-bit IEEE floating point arithmetic with a clock speed of 20 MHz. Each processor or "node" of the machine has access to 4 MBytes of local memory for a total *in-core* machine capacity of 4 GBytes. An option in the current design allows for each node to be configured with up to 64 MBytes of memory using a *double-wide* module. In addition, the processors can access a 16 GByte parallel disk system for mass storage or performing *out-of-core* computations. The processors have 14 *Direct Memory Access* (DMA) channels which provide communication for up to 8192 processors under the current design (2^{13} nodes). The 14th DMA channel provides for interfacing the nodes with external I/O boards which in turn access devices such as graphic screens or mass disk storage. For a 4096 node configuration, four cabinets, each containing 1024 processors are tied together at the systems' backplane. A second cluster of four cabinets may be stacked on a lower set to produce an 8192 processor system. The nCUBE system supports FORTRAN-77, C, and C^{++} languages. The nCUBE/2 represents the second generation of massively parallel computers from nCUBE corporation and is approximately 5-6 times faster than the first generation nCUBE/10 while performing 64-bit arithmetic.

3. HIGH PERFORMANCE MESSAGE PASSING

The development of an efficient parallel algorithm requires careful consideration of the ratio of computation time to the communication time for a particular granularity. In general, communication time is composed of two parts: a message start-up time associated with software initiation of the message, and a message transmittal time which is controlled by the hardware. The start-up time is fixed for a given message, whereas the transmittal time is proportional to the number of bytes transmitted over the DMA channels. For the current generation nCUBE, message start-up is approximately 100 microseconds, while the network bandwidth is approximately 4 MB/sec bandwidth, duplex. To reduce message start-up, data structures can be organized to combine messages so that the total number of messages to be passed is reduced. Although transmittal time cannot be reduced directly, its overhead can be considerably reduced by *overlapping communication*. A simplified diagram of the overlap concept is depicted in Figure 1. As an example, we will consider a processor exchanging messages with two of its neighbors in the upward and downward directions. We assume for the sake of simplicity, that to write and read the message requires a unit length of time. Consider the situation where messages are sent in a *non-overlapped* manner. In this case, the mes-

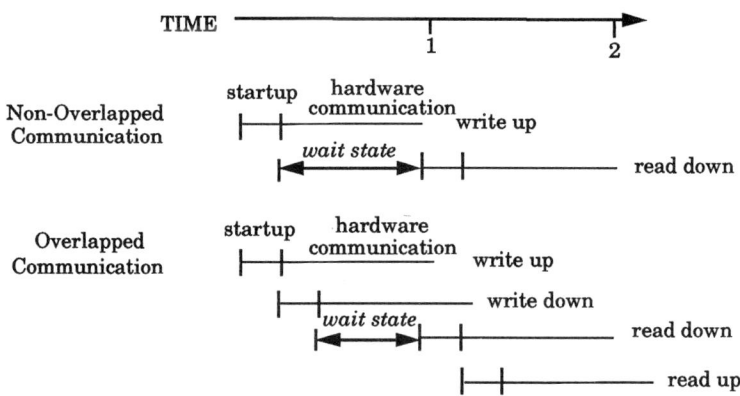

Figure 1. Efficient Communication with Overlap.

sage start-up time is incurred in the write operation, after which control is returned to the execution stream. Since the messages are non-overlapped, the controlling processor must *wait* until the message has arrived before it can begin reading in the downward direction. This is referred to as a *blocking* read operation - control to the execution stream is delayed until the message has been placed into the processor's message queue. This wait state constitutes idle time for the processor and reduces the parallel efficiency of the algorithm. If communication is *overlapped*, we see that we can perform two writes and two reads in approximately the same amount of time it takes to perform a single write/read pair in the non-overlapped case. On the nCUBE, the write operation is *non-blocking* and after the initial message start-up, control is returned to the program execution stream, thus allowing additional writes to be performed. The overlapped write/read ordering shown in Figure 1 is enabled through simultaneous operation of the DMA channels. The wait state encountered in the overlapped case is much smaller and therefore the communication is more efficient. If additional messages are written to other processors, this wait state may be further reduced or eliminated altogether [1].

Proper ordering of write/read pairs is essential to achieve high performance in message passing. To reduce the frequency of wait states, each processor executes all of its write statements, followed by the corresponding read statements in identical order. Since messages must be contiguous in memory, it is necessary to perform gather/scatter operations when passing non-unit stride arrays. The time required to perform the gather/scatter operations can be overlapped with the transmittal time, so that messages written over the DMA channels will arrive before the corresponding reads are executed. In all nCUBE/2 implementations, the gather/scatter operations are performed using very fast assembly code. Tests indicate the assembly-coded gather/scatter operation is as much as 2-4 times faster than FORTRAN, depending on the granularity.

Efficient overlapping of messages assumes that the messages are of sufficient length that they can be effectively overlapped. For shorter messages, this may not be the case. This point is illustrated in the individual communication timings presented in Table 1. These timings were measured as the total time required for a given processor to exchange 6 faces of its 3-D subdomain with adjacent processors. The message length was varied by changing the decomposition granularity (mesh size) assigned to each processor.

Table 1
Comparison of Communication Times (msec)

granularity	unoptimized	optimized
4^3	2.99	2.44
8^3	6.93	3.94
16^3	20.81	8.70
32^3	72.89	25.97

The unoptimized routine was written entirely in FORTRAN and write/read pairs were organized so as to maximize the possibility of wait states. The optimized version of the same routine was written with assembly-coded gather/scatter operations and overlapped message transmittal. Note the increasing performance gain with increased granularity. For coarse-grain decompositions, the optimized communication is 2.8 times faster than the unoptimized routine. The finest decomposition, however, shows only a 1.2 factor of improvement in the optimized code. The fine-grain decompositions do not take full advantage of the overlapped message transmittal, which diminishes the advantage of the overlapped routine. In this case, message start-up time is a significant fraction of the overall communication time and parallel efficiency is subsequently reduced. A parallel solver for the Navier-Stokes equations may require several hundred messages to be passed by each processor every timestep. Coarse-grain decomposition with overlapped communication is essential to achieving high efficiency in a parallel Navier-Stokes application.

4. SCALED AND FIXED-SIZE SPEEDUP

Decomposition granularity can have a significant effect on the overall performance of a parallel algorithm due to the message passing issues raised in the previous section. This point is further illustrated in a series of simple numerical experiments conducted with a parallel, explicit 3-D *Flux-Corrected Transport* (FCT) algorithm [2]. *Scaled* and *fixed-size* solution times are presented in Table 2 as a function of the global domain resolution. For each solution, two timings are presented. The number below the diagonal line indicates the average total solution time to advect a simple scalar field 25 timesteps. The number above the diagonal represents the average communication time, including the gather/scatter operations and wait states. Tabular entries with dashes indicate the decomposition required a grain size too large to fit on the 4 MByte memory of the nCUBE processor. Notice that for fixed domain size, the communication time becomes a significant fraction of the overall solution time as more processors are added to the decomposition. For the 512 node solution on the 32^3 grid, the distributed grain size is only 4^3 cells per processor, with communication time accounting for over 40 percent of the total solution time. The scaled decompositions are represented in the table as diagonal entries where the grain size per processor is fixed. Examining diagonal entries for both communication and total solution time show them to remain essentially constant as the problem is scaled to larger hypercubes. This trend is an indication of high parallel efficiency. For these decompositions, the communication time accounts for less than 5 percent of the total solution time.

Table 2
Communication and Total Solution Times - 3D FCT (sec)

domain \ nodes	1	2^3	4^3	8^3
256^3	—	—	—	55.4 / 1188
128^3	—	—	55.3 / 1187	18.3 / 173.9
64^3	—	56.3 / 1190	18.1 / 173.6	7.5 / 32.5
32^3	1137	18.1 / 173.7	7.4 / 30.8	4.1 / 9.5

From the timings collected in the performance study, the parallel speedup may be determined by the following relations [1],

$$\text{Fixed Speedup} = T(1) / T(p) \qquad (1)$$

$$\text{Scaled Speedup} = pT(1) / T(p) \qquad (2)$$

where p is the number of processors and $T(1)$ and $T(p)$ are the solution times for the single node and multiple node problems, respectively. Fixed and scaled speedups are provided in Table 3, where fixed speedups appear along the lower horizontal entries and scaled speedups appear along the diagonal entries. The parallel efficiency for the fixed size decompositions can be seen to drop off as the grain size per node is reduced. On 512 processors a fixed speedup of 120 corresponds to a parallel efficiency of 23 percent. The poor efficiency of the fine grain result is not surprising considering the lower communication performance for fine-grain decompositions. The degradation in fixed-size efficiency is a consequence of what is commonly known as Amdahl's Law [3]. For the scaled speedups, the parallel efficiency is quite high. The speedup of 490 on 512 processors corresponds to a parallel efficiency of 95 percent. The parallel implementation of the FCT algorithm is identical to the serial code other than for modifications to introduce communication. With tuning and optimization of the flux correction procedure, the fine-grain efficiency could be improved. Finally, it is interesting to note that the 256^3 problem, if solved by a single processor of the nCUBE/2, would require over 160 hours of execution time compared to the 20 minute parallel execution time on 512 processors.

Table 3
Fixed-Size and Scaled Speedup - 3D FCT

nodes \ domain	1	2^3	4^3	8^3
256^3	—	—	—	490
128^3	—	—	61	—
64^3	—	7.6	—	—
32^3	1	6.5	37	120

5. MESSAGE PASSING ON HIGHER-ORDER STENCILS

Message passing on MIMD computers is frequently handled by the introduction of *ghost cells* which surround each computational subdomain. The ghost cells reserve memory locations for storing messages, and provide a natural region of overlap between the subdomains. A typical second-order central difference expression is shown in Figure 2a below. In order to compute the second derivative of u at point i to second-order accuracy, information from point i+1 is required. This value is passed to the ghost cell location on the processor containing point i from the adjacent processor containing u at i+1. The information required to compute the derivative then resides locally on the processor.

$$\Delta x^2 \frac{\partial^2 u}{\partial x^2} = u_{i-1} - 2u_i + u_{i+1}$$

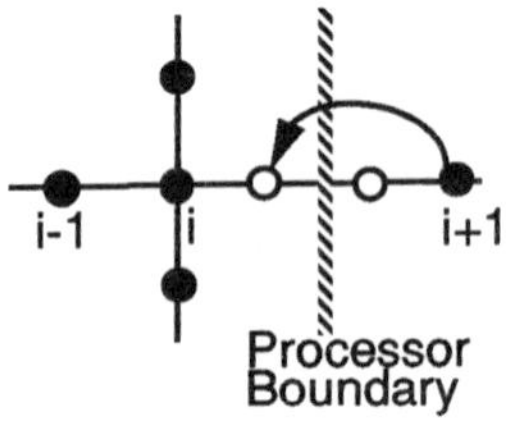

Figure 2a. Message Passing for Second-Order Difference Stencil.

For higher-order approximations at point i, the difference stencil requires information at points i+1 and i+2. At first, one may be tempted to enlarge the data structure to incorporate 2 layers of ghost cells, and pass both data points onto the local processor. In practice this is very inefficient from the standpoint of memory utilization and requires twice as much communication as the second-order approximation. As shown in Figure 2b, it is possible to preserve the single layer of ghost cells by calculating the incremental part of the difference approximation which resides on the adjacent processor. This increment may then be passed to the local processor containing point i, where it is included as an additional term in the overall higher-order difference expression.

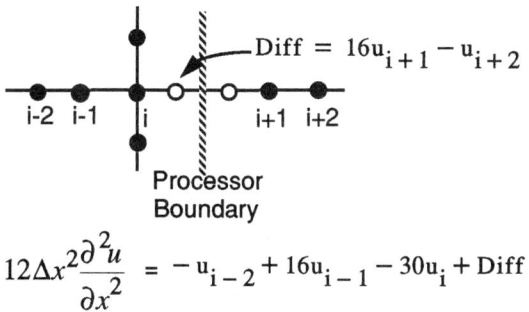

Figure 2b. Message Passing for Fourth-Order Difference Stencil.

6. MIMD DATA STRUCTURES FOR MESSAGE PASSING

As with serial application codes, data structures play an important role in numerical algorithm design for message passing architectures. First it must be understood that messages are gathered into a contiguous array (unit stride) in memory prior to transmittal to a neighboring processor along the interconnect. Similarly, when the processor receives the message, the unit stride array must be scattered back into the original data structure in order to complete the message passing process. A relevant example is illustrated in the 2-D data representation shown in Figure 3a below. The data is assumed to be local to a single processor, with ghost points denoted by dashed cell boundary lines. In conventional FORTRAN ordering, the 2-D numbering scheme begins at the lower left ghost point and proceeds sequentially from left to right across the subdomain. From the viewpoint of message passing, this data structure has several interesting features. When passing messages between processors in the up and down directions, the numbering scheme is unit stride along those two sides so that no gather/scatter operations are necessary. However, when messages are passed to the left and right directions, the data layout is non-unit stride and so messages passed in these directions must be buffered into and out of a unit stride array on the write and read operations, respectively. The gather-scatter operations are typically overlapped with the actual communication time using the techniques described in Section 3.

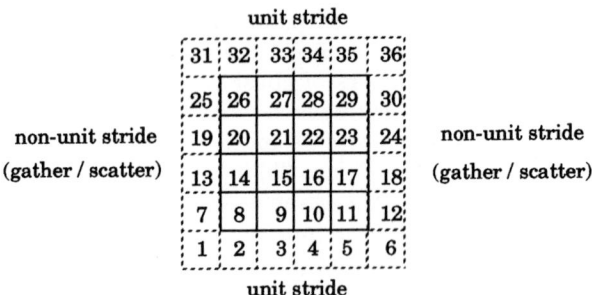

Figure 3a. Data Organization with Conventional FORTRAN Ordering.

As an alternative to the conventional numbering scheme, the data layout shown in Figure 3b is better designed for explicit message passing. In this case particular attention is focused on the numbering of the ghost cells. Interior points are numbered using conventional ordering techniques. Ghost points however, are numbered sequentially along the perimeter of the subdomain, beginning along the lower right boundary, followed by the sequential ordering of the upper left boundary. This ordering scheme preserves a unit stride along the ghost boundary cells. In this case, left and right message passing can be handled more efficiently since the buffer scatter on the read operation is eliminated. Because of the non-unit stride on the left and right interior points, messages must still be gathered for the write operation. Using this data organization, adjacent processors with unit stride perimeters are aligned for efficient message passing as illustrated in Figure 3c. Application of this type of data structure to parallel iterative solvers for sparse matrices may be found in [4].

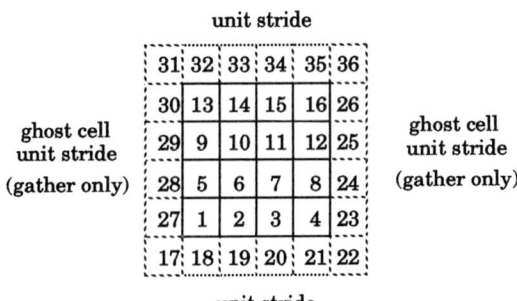

Figure 3b. Data Organization for Eliminating Scatter Operation.

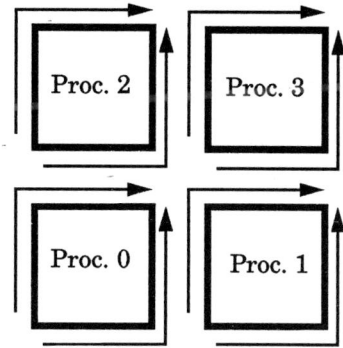

Figure 3c. Alignment of Unit Stride Arrays Along Processor Boundaries.

7. DOMAIN DECOMPOSITION STRATEGIES

A critical issue with CFD applications on massively parallel computers is the way in which the computational domain is partitioned among the ensemble of processors. Poor implementation of the domain decomposition can result in poor parallel performance in what may otherwise be a highly parallel algorithm. To illustrate this point we will consider two decomposition strategies for a 3-D domain which is distributed over 64 processors. Each decomposition is illustrated in Figure 4. Although the number of subdomains in each decomposition is identical, the number of messages to be passed in the 3-D case is higher owing to the greater number of interfaces between processors. If message latency is small, the primary issue related to efficiency of the two decompositions is the volume to sur-

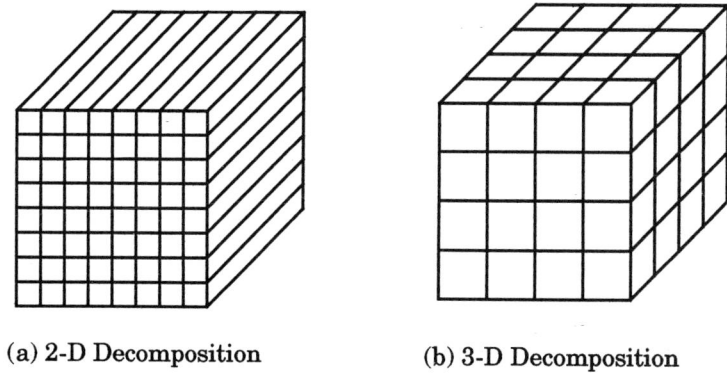

(a) 2-D Decomposition   (b) 3-D Decomposition

Figure 4. Decomposition Strategies for 3-D Domain on 64 Processors.

face area ratio of the subdomains. Generally, it is desirable to maximize the volume of the subdomain with respect to the surface area in order maximize the computation time to communication time. A numerical experiment was performed on the Intel Delta machine at Caltech using an explicit advection code to simulate the propagation of a blast wave across a Cartesian domain [5]. The domain was distributed using both methods shown in Figure 4. In the first case, the domain was subdivided using a 2-D decomposition in which long slabs were mapped to each processor. The Delta's 2-D mesh interconnect was utilized to determine neighbor processors to insure nearest neighbor communication. In a second case, the domain was subdivided into 3-D subdomains and neighboring processors were determined using a gray-code scheme designed for a hypercube network. In this decomposition strategy, nearest neighbor communication is no longer guaranteed and intermediate message hops between processors is incurred. Timings for these two cases are shown in Table 4 for the solutions on 64 and 512 Delta processors. The average communication time refers to the average time spent by a processor in the ensemble performing communication operations, including gather/scatter and wait states. In addition to the timings, we also show the measured speedup and corresponding parallel efficiency. Notice the superior performance of the gray-coded decomposition, which is indicative of the higher volume to surface area ratio for the 3-D subdomains. Scaling from 64 to 512 processors leads to further degradation in performance and parallel efficiency in the 2-D decomposition in spite of the nearest-neighbor communication for the 2-D mesh. We may conclude that 3-D applications targeted for mesh-based architectures must be decomposed into 3-D subdomains if high parallel efficiency is to be achieved.

Table 4
Performance Comparison for 2-D and 3-D Decompositions

No. of Processors	Decomposition Strategy	Grid Size per Processor	Avg. Comm. Time per Timestep(sec)	Scaled Speedup	Parallel Efficiency
64	2-D mesh	8 x 8 x 64	0.54	46.5	0.73
64	3-D gray code	16 x 16 x 16	0.47	53.4	0.83
512	2-D mesh	4 x 8 x 128	0.78	298.8	0.58
512	3-D gray code	16 x 16 x 16	0.49	401.6	0.78

8. SUMMARY

We have examined some of the fundamental concepts which must be understood in order to develop highly parallel and efficient CFD applications for MIMD computers. CFD code developers must focus their attention on writing truly scalable software. That is, software must be written in a general purpose manner so that it will be equally functional when distributed over anywhere from one to one-thousand processors. Serial bottlenecks must be eliminated in all aspects of the application. These bottlenecks include user input, mesh generation for complex geometries, linear equation solvers, and parallel output for scientific visualization. Clearly, as this technology evolves, there will additional issues which must be addressed in order to bring these new machines into mainstream computing. Parallel I/O, parallel visualization, and the manner in which researchers interface with distributed memory machines will be of considerable importance to the CFD community in the future.

9. Acknowledgments

The author would like to acknowledge the Caltech MIMD Consortium for use of the Delta Machine to conduct some of the performance studies discussed in this paper. The author would also like to acknowledge the numerous technical discussions with the MPCRL staff which contributed to the preparation of this paper.

10. REFERENCES

1 J. L. Gustafson, G. R. Montry, and R. E. Benner, Siam J. Sci. Stat. Comp., **9**, (1988), 609.

2 D. L. Book, J. P. Boris, and K. Hain, J. Comput. Phys., **18**, (1975), 248.

3 G. Amdahl, AFIPS Conf. Proc., **30**, (1967), 483.

4 J. N. Shadid and R. S. Tuminaro, "Sparse Iterative Algorithm Software for Large-Scale MIMD Machines: An Initial Discussion and Implementation," MPCRL Report, Sandia National Laboratories, Albuquerque, NM, (1991).

5 D. R. Gardner, D. D. Cline, C. T. Vaughan, "Implementation of a Single-Material Version of PAGOSA on MIMD Hypercubes," SAND92-0640, Sandia National Laboratories, Albuquerque, NM (1992).

Multidomain Computations of Compressible Flows in a Parallel Scheduling Environment

F. Dellagiacoma, S. Paoletti, F. Poggi and M. Vitaletti

IBM European Center for Scientific and Engineering Computing (ECSEC)
Viale Oceano Pacifico 171-173, 00147 Rome (Italy)

Abstract

PARAGRID is a parallel multi-block environment in which it is possible to integrate a large class of codes for the computation of fields in three-dimensional space. A code solving the discretized field equations on a *single-block structured grid* can be integrated in the proposed environment as a *subdomain solver*, so that *multiple copies* of the code can work concurrently on different grid blocks and exploit the computing nodes of parallel computers. The environment has been experimented with an implicit code for steady-state computations of compressible flows.

1. MULTI-BLOCK METHODS

Domain decomposition methods can be very useful for the computation of fields in three space dimensions. The basic idea is to divide the region of interest into smaller *subdomains* and to build the global solution for the whole region by computing intermediate data related to the subdomains. This approach may help to reduce the amount of memory required to run a computer simulation. Moreover, the region of interest has often a complex shape and it is difficult to generate a single-block *structured grid* covering the whole domain. Algorithms based on *unstructured grids* are seen as an attractive solution to solve field equations on regions with a complex geometry, but they are generally less efficient and their implementation is more difficult. An intermediate level of flexibility is offered by *multi-block methods*, where multiple, structured grid blocks are used to cover a set of non-overlapping subdomains. The grid blocks can be generated in two steps:

- the first one requires to create structured two-dimensional grids on the subdomain delimiting surfaces. A major constraint, at this stage, is the use of the same distribution of nodes for a surface shared by two adjacent blocks.

- In the second step, a three-dimensional structured grid is generated, for each block, starting from the distribution of nodes at the boundaries.

Multi-block codes for solving the Euler and the Navier-Stokes equations have been successfully applied in the last decade to the computation of compressible flows of aeronautical interest. Emphasis was on the reduction of the core memory requirements and the gain of flexibility with respect to single-block grids, while most implementations were targeted to serial computers. Recently, with the advent of parallel computers and the availability of large networks of powerful workstations, parallel implementations of multi-block algorithms have begun to appear.

The flow solution can be computed by a spatial discretization of the governing partial differential equations (PDEs) in each sub-domain, e.g. by *finite difference* or *finite volume* methods, followed by a numerical integration in time of the ordinary differential equations obtained for the discretized variables. A numerical algorithm for single-domain computations can be converted into a *block-solver* with a small effort, by including an appropriate treatment of the points falling on *internal boundaries*, which are shared by two or more adjacent grid blocks.

This procedure is well suited to solve *initial value* problems for fields governed by *hyperbolic* PDEs, which describe the propagation at finite speed of both linear and non-linear waves. Indeed, the field variables associated to a given grid point at time $(n+1)\Delta t$ can be computed from the knowledge of the field at time $n\Delta t$ over the *domain of dependence* of the point [4]. The domain of dependence is itself contained in a sphere centered at the point and whose radius is given by the product $c\Delta t$, where c is the speed at which information is propagated. The values of the time-step size Δt adopted in the numerical integration are usually small enough to ensure that only few neighboring points fall in the domain of dependence of a grid node. As a result, one step of advancement in time can be computed, for each subdomain, from the knowledge of the initial status of the field over a region which is the union of the domains of dependence of all points. This region contains the subdomain itself, plus a "halo" of surrounding points, hereafter called the *interaction region*. Because of these features, field problems described by hyperbolic PDEs are solvable through *spatially localized computations*.

The numerical analysis of mixed initial and boundary value problems for *parabolic* equations shows that these also can be solved by spatially localized computations. In general the solution at some point of space depends, at any time $t > 0$, on its initial value over the whole domain. However, it is still possible to solve diffusion-type problems, e.g. the heat equation, by applying *time-explicit* schemes to compute the evolution in time of the field variables associated to a grid discretization, e.g. by finite-differences [4]. The domain of dependence of a grid point, *as defined by the numerical scheme*, thus includes only a finite number of its closest neighbours. By the unions of such domains of dependence it is possible to arrive at the definition of the *interaction region* of a subdomain already introduced for hyperbolic problems.

Time dependent compressible flows are described by systems of PDEs of hyperbolic type — the Euler equations — when the flow is treated as *inviscid*, or of mixed type, hyperbolic and parabolic — the Navier-Stokes equations —, when *viscous* and *thermal diffusion* terms are taken into account. As we have seen, the spatial decomposition of the computational domain is a natural method, due to the character of the governing equations, to divide the computational work among several concurrent processes. Each process may be in charge of one or more subdomains. It requires to exchange only a limited amount of information, at each step, with the other processes, in order to perform the update of the field. The large experience which has been developed on multi-block time-marching algorithms on serial computers suggests that parallel multi-block codes could be successfully exploited in a parallel production environment.

There are several problems in CFD which are not easily solvable through spatially localized computations. One of them is the computation of *incompressible flows*, where the equation for the pressure field has a non-local character. Another is the computation of stationary flows by a direct solution of the steady-state equations. On the other hand, most of the theoretical work on domain-decomposition methods has been traditionally concerned (see [6], [2], and [3]) with *boundary-value* problems for *elliptic* partial differential equations (PDEs), a class which includes the "difficult" non-local problems encountered in CFD. The reader is referred to [3] for a review of domain-decomposition methods. Here we confine the discussion to problems where the evolution in time of field quantities defined over a certain region of space, are governed by laws possessing a local character, like in the case of hyperbolic and parabolic PDEs.

2. PARAGRID: A PARALLEL MULTI-BLOCK ENVIRONMENT

The existence of a well defined *region of interaction* between adjacent subdomains has been introduced in the previous section as a characteristic feature of local time-dependent PDEs, and their finite-difference discretizations. The same concept may be easily extended to field equations casted in *integral* form, which are considered in connection with finite-volume or finite-element discretizations. On a structured grid, the field variables may be associated to different *grid sites*, depending on the scheme adopted for space discretization. Most algorithms use data defined on the *cell vertices*, the *cell centers*, and the cell *face centers*. The major functions of a multi-block code can be grouped into the following categories:

- the **topology management functions**, which are responsible to identify the adjacency relationships between different subdomains,

- the **communication management functions**, by which computations related to a given subdomain have access to field data from adjacent domains which fall in the interaction region, and

- the **scheduling functions**, which prepare the environment for the block-solver, i.e. the program implementing the time-update procedure at the subdomain level.

The parallel implementation of a multi-block code depends primarily on the design of the communication and the scheduling functions, while it seldom requires to modify the topology management or the block-solver. The main motivation for developing the PARAGRID multi-block environment was the fact that the multi-block functions, not including the block-solver itself, are independent of the physical meaning of the field variables related to any specific problem and they can be designed to support an arbitrary allocation of the physical variables on different grid sites. Following this approach, the user is left with the task of providing only the block-solver, which is integrated in PARAGRID through a well defined API (application programming interface). The PARAGRID API defines, among other things, the storage arrays holding the field variables associated with the subdomain and with the interaction region. These arrays are passed as arguments to the block-solver program, which is called directly by the environment. The block solver is responsible to compute one step of advancement in time, for a given subdomain, on the basis of the input initial data provided by PARAGRID. The details of the parallel implementation are hidden to the user, because the same API may be supported on a wide range of parallel architectures.

2.1. The Application Programming Interface

The field variables, are classified by PARAGRID according to the grid site to which they are associated, which may be a vertex, a cell center or a cell face center. The data are also classified by the different type of management into three classes:

- **GEO**-type variables, are initialized at the beginning of the simulation, and their values do not change along the whole sequence of update steps. These type of data are useful to keep, for example, *static* geometric data like cell volumes, cell face areas, jacobians of the grid mapping, etc...

- **FLO**-type variables, are also initialized at the beginning of the simulation, but their values are updated at each step by the block-solver. All the *dynamic* information defining the status of the field at any instant of time must be hold in FLO-type variables. Note that moving grids are supported by PARAGRID, provided that the adjacency relations between subdomains do not change. In this case, geometric data like cell volumes and areas do change in time, so they must be hold in FLO-type variables and be updated by the block-solver.

- **AUX**-type variables, are not initialized and their content, on enter to the block-solver, are *unpredictable*. The storage reserved for AUX-type arrays is meant to hold temporary data which are used by the block-solver during the update process.

Three input files are processed by PARAGRID. The first one contains the coordinates of the grid nodes for each block. The coordinates are relative to the same cartesian reference frame. The cartesian coordinates of corresponding nodes in the two adjacent blocks must be *identical* and the coincidence of boundary nodes is used to identify the internal boundaries between adjacent blocks.

PARAGRID creates and manages the *region of interaction* between adjacent subdomains in two stages:

- in the first stage PARAGRID performs an *automatic detection of the internal boundaries*. In the input phase the grid file is read and the complete set of grid blocks is analyzed to detect all pairs of adjacent faces. Two coupled faces may belong either to different blocks or to the same one. The latter situation occurs, for example, for an annular region of space.

- in the second stage the domains are modified by an *enlargement procedure*, in which the original grid is placed in the *core* of a larger block. Each domain is extended beyond its internal boundaries by a certain number of *cell-layers*, whose precise value is specified by the input parameter *NEXLAY*. Thus all domains are enlarged into their neighboring regions and as a consequence any internal boundary is surrounded by an *overlap zone* .

This procedure is followed by PARAGRID to initialize the array, denoted by XYZ, containing the grid coordinates. Because of the previous argument it should be clear that the input mesh coordinates are first copied into the innermost part of the array XYZ. To complete the initialization procedure the coordinates of the grid nodes belonging to the overlap regions are added by fetching their values from the block arrays relative to the contiguous domains. PARAGRID computes the size of the arrays associated with the enlarged domains and allocates storage space in a dynamic fashion, as needed by the particular application. Let us illustrate the previous ideas with a specific example. Consider an internal boundary separating two domains D_1 and D_2. The region of overlap consists of a total number of $2 \times NEXLAY$ layers. An interaction region formed by $NEXLAY$ layers from the adjacent domain D_2 is added to the core region of domain D_1; likewise domain D_2 is extended by $NEXLAY$ layers into domain D_1. As we have seen when introducing the enlargement process, the original grid block is identified with the innermost portion of a larger three-dimensional region. For example, the grid coordinates are stored in arrays whose dimension bounds are increased by $2 \times NEXLAY$ to account for the points belonging to the outer shell. The latter is made up at most of 26 items:

- *6* volumes sharing a boundary *face* with the core block

- *12* volumes sharing a boundary *edge* with the core block and
- *8* volumes sharing a boundary *corner* with the core.

PARAGRID identifies the source of the additional sets of points among the neighboring subdomains. Due to the constraints imposed on the input grids it is easy to find the source block to be added to an internal boundary face. Of course such a block does not exist if the face corresponds to a *physical* boundary, e.g. a wall delimiting the region of interest. Things may become more complex when dealing with edges and corners, because the corresponding blocks might not exist. In general PARAGRID attempts to fill all the 26 enlargement chunks when there is no ambiguity on the source domain from which the enlargement data must be obtained. When this is not possible, the lack of enlargement data is signalled to the block solver through a set of flags related to the different types of enlargement areas, i.e. faces, edges and corners.

The same filling procedure followed to create enlarged grid blocks is used by PARAGRID, at each step, to manage the field information relative to the interaction region, which is stored in FLO-type variables. In this case, the "halo" of initial data associated to the region of overlap with the adjacent blocks is *replaced* with the values computed in the previous time-update of such blocks.

A generic update step, for each individual subdomain, is thus accomplished by the following phases, most of which are *transparent* to the user:

1. The enlargement region of the subdomain FLO-type arrays are filled with replacement data from the core region of its neighboring blocks. Some of these data may be stored on remote computing nodes and are received through a communication network (see below, phase 3).

2. PARAGRID passes control to the block-solver program provided by the user. The block-solver updates the FLO-type variables corresponding to the core region of the subdomain. The consistency of computations performed on different subdomains must be ensured by the proper use of the initial field values in the interaction region, which have been provided by PARAGRID in the previous phase.

3. PARAGRID prepares blocks of data from the core region of the subdomain which must be used to replace the initial values relative to the interaction region of its neighbours. On a distributed memory system these data are sent to the computing nodes holding the data of the target domains (see above, phase 1).

The PARAGRID API includes the format of additional input files, which contain information about the multi-block structure of the grid, the type of variables used by the block-solver, the depth of the interaction region, the identification of physical boundary faces, and several others. The main output information is stored in the *restart* file, which can be used to resume the simulation from an intermediate step and for post-processing. Another entry point, in addition to the block-solver, is provided for the *global analysis program*, which must be implemented by the user. This facility has access to the field data stored in the restart file and can be used to interface a post-processor for the dynamic display of the whole field, to monitor the performance of the simulation, to evaluate global residuals or to perform any other operations which need informations from all the subdomains.

2.2. Implementation on a Cluster of Workstations

One of the major goals, in the design of PARAGRID, was to define a general purpose multi-block environment which could be efficiently implemented on a large variety of parallel systems, including both shared-memory multi-processors and distributed-memory systems. The user is responsible to implement block-solvers suited for specific applications, either *ex novo* or adapting existing single-block codes. Block-solvers conforming to the PARAGRID API can run without change on any platform where PARAGRID is available. The present version of PARAGRID, has been implemented at IBM ECSEC on a cluster of IBM RISC/6000 workstations running the AIX operating system, which can be interconnected by various communication networks. The modularity of the code and the use of standard software interfaces for memory and task management and for interprocessor communication will ease the porting to other parallel platforms.

The efficiency of a parallel program running on multiple processors is measured by the average level of *processor utilization*. The two major factors which tend to reduce the parallel efficiency are the *communication delays*, and the *workload unbalancing*. In PARAGRID one copy of the block-solver is scheduled for execution on each of the available computing nodes, and the distribution of the grid-blocks among the nodes can be dynamically adjusted by PARAGRID to increase the level of *load balancing*. The PARAGRID API helps to reduce the communication delays, because the block-solver computes an entire update step working on subdomain data. The delay involved in the replacement of FLO-type variables in the interaction region usually represents a small fraction of the computing time. For a given algorithm, the influence of such delay may be decreased by increasing the size of the grid-blocks or by increasing the speed of the network. High level of efficiency on several workstations are already achievable today, with PARAGRID, on the available local area networks. The technological progress in this area is expected to allow parallel multi-block computations on several tens of interconnected nodes in the next future. Load unbalancing, on the other hand, may be the major limiting factor on the number of processor which can be efficiently applied to a given problem. In order to use a large number of processors it may be necessary to refine a given multi-block grid into smaller blocks of comparable size.

3. A BLOCK-SOLVER FOR INVISCID TRANSONIC FLOWS

In general curvilinear coordinates, in three space dimensions, the Euler equations, describing the flow of an inviscid gas in thermodynamic equilibrium, may be written in conservation form as:

$$\frac{\partial \mathbf{q}}{\partial t} + \frac{\partial \mathbf{F}}{\partial \xi} + \frac{\partial \mathbf{G}}{\partial \eta} + \frac{\partial \mathbf{H}}{\partial \zeta} = 0 \qquad (1)$$

where $\xi(x,y,z;t)$, $\eta(x,y,z;t)$, and $\zeta(x,y,z;t)$ are body-fitted curvilinear coordinates, and $\mathbf{q}$ is a vector whose five components are proportional to the conserved physical quantities of the flow. The flux-vectors $\mathbf{F}$, $\mathbf{G}$, $\mathbf{H}$ are nonlinear functions of the flow variables $\mathbf{q}$ and depend also on the mapping derivatives of cartesian coordinates with respect to the curvilinear coordinates.

Using the Euler backward first order implicit scheme for time discretization, the solution incremental change $\Delta \mathbf{q}^n = (\mathbf{q}^{n+1} - \mathbf{q}^n)$ at time $n\Delta t$ ($\mathbf{q}^n = \mathbf{q}(n\Delta t)$) is expressed as a function of the nonlinear flux vectors at time $(n+1)\Delta t$. A system of linear algebraic equations is then obtained by a linear expansion of the flux vectors $\mathbf{F}^{n+1}$, $\mathbf{G}^{n+1}$ and $\mathbf{H}^{n+1}$ about the solution vector $\mathbf{q}^n$ and by a finite-difference spatial discretization of the linearized PDEs. Spurious oscillations are dumped by the adaptive artificial dissipation method proposed by Jameson. The linear algebraic system generated by the unfactored scheme has the following form:

$$\{1 + \Delta t [\delta_i \mathbf{A}^n + \delta_j \mathbf{B}^n + \delta_k \mathbf{C}^n]\} \Delta \mathbf{q}^n = -\Delta t [\delta_i \mathbf{F}^n + \delta_j \mathbf{G}^n + \delta_k \mathbf{H}^n] \qquad (2)$$

where δ_i, δ_j, and δ_k are symbols to denote centered finite differences along the grid coordinate lines, and $\mathbf{A}^n = \frac{\partial \mathbf{F}^n}{\partial \mathbf{q}}$, $\mathbf{B}^n = \frac{\partial \mathbf{G}^n}{\partial \mathbf{q}}$ and $\mathbf{C}^n = \frac{\partial \mathbf{H}^n}{\partial \mathbf{q}}$ are the jacobians of the flux vectors (second order and fourth order artificial dissipation corrections are not shown in the above scheme). The sparse matrix appearing on the left hand side of (2), which has seven block diagonals, is computed at each step and it may be efficiently solved by an iterative algorithm [5].

An alternative method is to follow the Beam and Warming approach [1], where the original matrix is approximated by the product of three *block-tridiagonal* factors. This produces the following *alternating direction implicit* (ADI) scheme:

$$[1 + \Delta t \delta_i \mathbf{A}^n][1 + \Delta t \delta_j \mathbf{B}^n][1 + \Delta t \delta_k \mathbf{C}^n] \Delta \mathbf{q}^n = -\Delta t [\delta_i \mathbf{F}^n + \delta_j \mathbf{G}^n + \delta_k \mathbf{H}^n] \qquad (3)$$

The computational cost of the ADI implicit scheme is higher compared to that of multi-stage explicit schemes, but this cost is often compensated by a more favourable stability bound. The same code can be used for the simulation of both *unsteady* and *steady* flows. A steady flow solution is obtained by computing a sequence of update-steps starting with a convenient initial field, e.g. a homogeneous flow, until the transient solution is eliminated. It is important to observe that the *pseudo-unsteady* method reduces the problem of solving a complex system of mixed elliptic-hyperbolic equations (the steady-state Euler equations) to that of finding the asymptotic solution of a much simpler initial value problem.

The multi-block partition of the grid does not create serious problems for unsteady computations. However, the acceleration methods usually employed in pseudo-unsteady computations of a steady state solution need a special care when implemented in a block-solver. A very simple and effective approach, to speed-up the convergence of the time-marching scheme to a steady-state solution, is to use a spatially variable time step-size $\Delta t_{ijk} = \Delta t_0 / \Psi_{ijk}$, where Ψ_{ijk} is an estimate of the cell Courant number for $\Delta t_0 = 1$. Therefore Δt_0 is the value of the Courant number applied to every grid-node, and the domains of dependence of nodes on a block boundary face include approximately Δt_0 cell layers belonging to an adjacent block. In principle one could set the PARAGRID $NEXLAY$ parameter equal to Δt_0, in order to provide enough initial data to the block solver. This is hardly feasible when using cell Courant numbers of about 10 or greater, because of the communication and the computational costs associated to the management of the interaction region. On the other hand, the experience with multi-block algorithms tells that reducing the depth of the interaction region to a few layers of cells has only a small impact on the rate of convergence of an implicit algorithm to a steady solution. Some numerical experiments have been performed by using PARAGRID with a block-solver implementing the ADI factored scheme to assess the impact of grid-blocking in the computation of steady transonic flows.

The PARAGRID block-solver was implemented by adapting an existing single-block code developed at IBM ECSEC. On a single-block structured grid boundary conditions are applied to the six faces of the block. For example, *slip-flow* conditions are applied to solid walls and *non-reflective* conditions are applied to the open boundaries. The field values on the boundary nodes are computed *explicitely* by the program, at the beginning of a time-step update, to enforce the appropriate conditions. The update of the *internal nodes* is computed with *frozen* boundary values (i.e. $\Delta \mathbf{q} = 0$ for a boundary node) by solving the linear system (3). Note that the boundary data affect only the explicit evaluation of the right hand side. The simple strategy followed for adapting the code to the multi-block environment was to apply the block solver directly to the *enlarged* grid blocks provided by PARAGRID. The treatment of internal boundaries did not require a special care, because the initial field values on those boundaries are automatically set by the replacement procedure, before entering the block-solver. The depth of the *overlap* zone by which one block is extended into the adjacent neighbours can be chosen at run time by setting the

PARAGRID input parameter $NEXLAY$. Experiments were performed by using either *one* or *two* layers of overlap. The first observation which can be made is that nodes on an internal boundary, which are shared by two adjacent blocks, are computed twice and the associated values are *duplicated* (this problem does not occur with cell-centered finite volume schemes). On the other hand, the computation of the field on internal boundaries uses in an essential way the initial values associated to the "halo" of grid cells outside the core region, which were computed in the previous update of an adjacent block. With explicit schemes, the identity of duplicated values can be a direct consequence of the numerical implementation. Implicit schemes will generally produce non-identical values for the duplicated nodes, and may require some additional care. In the case of steady-state calculations it is important to ensure that data computed in different subdomains for the same duplicated node will converge to identical values.

3.1. Numerical experiments

A standard test for inviscid transonic codes is the computation of the flow around a swept wing. A C-type single-block grid of $95 \times 17 \times 25$ ($= 40,375$) nodes was used to create two multi-block grids with 15 and 30 blocks, respectively. The first grid was obtained by creating 5 intervals in the streamwise direction of the original grid and 3 intervals in the spanwise direction. The second grid was created by adding a two-interval splitting along the normal direction.

The reported data refer to an angle of attack $\alpha = 0.$ and to a free-stream Mach number $M_\infty = 0.92$. Figure 1 illustrates the convergence history obtained on the three grids. The single-block grid was used with the original code, while the multi-block grids were used with the PARAGRID version of the block-solver by setting the input parameter $NEXLAY$ to 1. The implicit code run with a constant cell Courant number of 10. The decrease in convergence rate between the single-block and the multi-block algorithm is evident. On the other hand, a similar convergence rate is obtained on the two blocked grids, in spite of the difference in the number of blocks.

The parallel efficiency measured for the parallel run on 3 IBM RISC/6000 mod. 530H workstations was about 95%, and the parallel speed-up was about 2.85 compared to a serial run of the multi-block code. A smaller figure for the speed-up — about 2 — is obtained comparing the execution times of the single-block code with that of the parallel multi-block code.

The faster convergence of the single-block algorithm was expected and it must be evaluated against the many advantages offered by a multi-block code. After all, it is simply impossible to generate a single-block structured grid for more complex geometries.

A careful tuning of the boundary conditions applied on internal boundaries may help to improve the convergence rate. A simple way to achieve this goal is to increase the depth of the interaction region. Figure 2 shows the convergence history obtained for the single-block grid and those obtained for the 15−block grid using different values — 1 and 2 — of the *NEXLAY* parameter. The increase of the depth of the interaction region by an additional layer of nodes improves significantly the convergence rate. Similar results are obtained under different flow conditions.

4. CONCLUSIONS

Parallel multi-block codes can be implemented by integrating single-domain solvers in the PARAGRID environment. Clusters of many interconnected workstations can be efficiently exploited in a transparent fashion, and offer a cost-effective platform to run time-critical simulations. The application programming interface offered by PARAGRID is also very helpful for the development and testing of parallel multi-block algorithms which are portable across different platforms. The experience with an implicit block-solver demonstrated that the multi-block approach can be successfully applied to solve both unsteady and stationary flow problems.

References

[1] R.M. Beam and R.F. Warming, *An Implicit Finite-Difference Algorithm for Hyperbolic Systems in Conservation-Law Form*, J.Comp.Phys., Vol. 22, No. 1 (1976), 87-110.

[2] P.E. Bjørstad and O.B. Wildlund, *Iterative methods for the solution of elliptic problems on regions partitioned into substructures*, SIAM J. on Numer. Anal. 23/6 (1986), pp. 1097-1120.

[3] T.F. Chan, *Domain Decomposition Algorithms and Computational Fluid Dynamics*, in "Vector and Parallel Computing Issues in Applied Research and Development", J.Dongarra, I. Duff, P. Gaffney and S. McKee (Editors), E. Horwood Ltd., Chichester, UK, 1989, pp. 65-82.

[4] P.R. Garabedian, *Partial Differential Equations*, John Wiley & Sons, Inc., New York·London·Sydney, 1964.

[5] S. Paoletti, M. Vitaletti and P. Stow, *An Unfactored Implicit Scheme for 3D Inviscid Transonic Flows*, AIAA paper 92-2668-CP, 10th AIAA Applied Aerodynamics Conference, Palo Alto, CA, 1992, pp. 518-527.

[6] H.A. Schwarz, *Über einige Abbildungsaufgaben*, Ges. Math, Abh., 11 (1869) pp.65-83

Figure 1: Convergence history - overlap = 1.

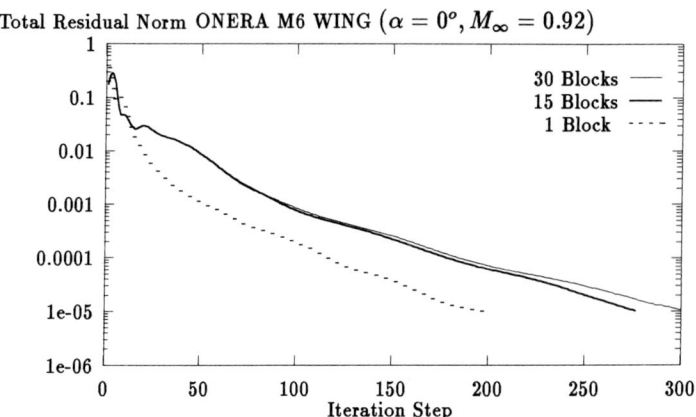

Figure 2: Effect of different overlap sizes

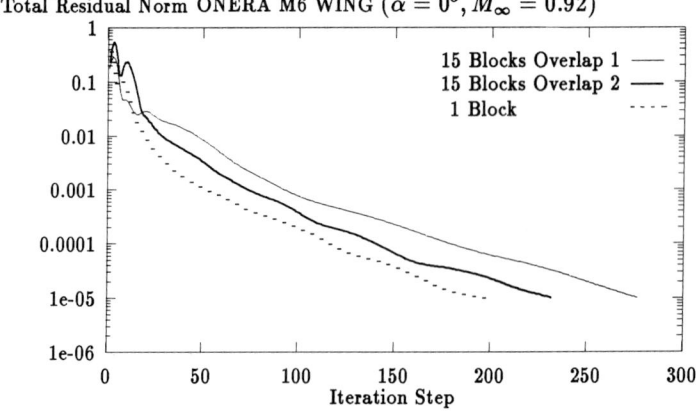

CFD Experiences on a Range of Novel Architecture Systems

D. R. Emerson, R. J. Blake and R. J. Allan

S.E.R.C., Daresbury Laboratory, Warrington WA4 4AD, U.K.

Abstract

An explicit Total Variation Diminishing (TVD) MacCormack scheme is employed to solve the unsteady Navier-Stokes equations. The production code has been developed as a model CFD benchmark and embedded in Fortnet, a portable message-passing harness, to allow machine-independence when porting the code to new or existing systems without modifications to the solver decomposition or calling routines. An assessment of the architecture under investigation can therefore be obtained. Results are presented for three novel-architecture systems: a quasi-heterogeneous system of Silicon Graphics workstations; a Parsytec transputer cluster; and a Kendall Square 32-node virtual shared memory system.

1. INTRODUCTION

Computational Fluid Dynamics (CFD) is now one of the key subjects to exploit parallel and novel architecture computer systems. However, one of the disadvantages that is frequently encountered when a new system becomes available is that, in general, the existing code cannot run on the new system without some re-writing of the code. This re-write can sometimes be extensive. A desirable feature of any CFD production code would, therefore, be machine-independence. This can be achieved for a wide range of computer architectures, using both shared memory and distributed memory, by employing one of several message-passing harnesses currently available e.g. PICL [1] from Oak Ridge National Laboratory (ORNL), USA; Express [2] from CalTech, USA (now marketed by ParaSoft Inc.); PARMACS [3] from the Gesellschaft für Mathematische Datenverarbeitung (GMD), Sankt Augustin, Germany and Argonne National Laboratory (ANL), U.S.A.; PVM [4] from ORNL; and tcgmsg [5] from ANL. At Daresbury Laboratory we employ Fortnet, a portable message-passing environment developed by Allan et al. [6], to achieve machine independence. Details of Fortnet and the machines on which it is available were given at the Parallel CFD '91 meeting [7].

2. THE NUMERICAL METHOD

The Navier-Stokes equations are solved using a finite-volume, operator-split approach, as described by MacCormack and Paullay [8]. The full details of the implementation have been given in Emerson and Poll [9]. Only brief details of the method will, therefore, be presented here. The 2-D equations of motion are integrated over a control volume as follows:

$$\frac{\partial}{\partial t} \iint_\Omega Q dA + \int_\Gamma \underline{H} \cdot \hat{n}\, dl = 0 \tag{1}$$

where

$$Q = (\rho, \rho u, \rho v, E)^T \tag{2}$$

and $\underline{H}$ is a column vector containing the inviscid fluxes and viscous stress and heat transfer terms. The symbols ρ and E represent the density and total energy per unit volume, respectively, and the pressure is related via the ideal gas equation of state. The operator-split scheme can be written:

$$Q^{n+1} = L_y(\Delta t/2) L_x(\Delta t) L_y(\Delta t/2) Q^n \tag{3}$$

and is second order accurate in time and space.

3 PROBLEM DESCRIPTION

The problem addressed is high-speed air flow over a rectangular cavity. A cavity is a fundamental aeronautical feature which can be used to house internal stores, optical sensing equipment, etc. The flow that subsequently develops in the cavity exhibits many features that are of interest to the aeronautical engineer e.g. fluctuating pressure levels, localised heating and recirculating flow regions. The computation of such flows requires a time-accurate solution. The time step imposed by physical constraints also means that explicit time-marching schemes are ideally suited to this type of problem. However, the time constraint also means that large amounts of CPU time are required for the oscillatory flowfield to become fully established.

The explicit time-marching schemes provide an ideal model for parallelisation because there is only a local coupling between the computational cells. Therefore, to test the code on a variety of architectures, a benchmark version of the cavity code was developed. The original code was reduced to a kernel application by removing all unnecessary computation and I/O and allowing the problem size to scale with one parameter, NMAX: i.e. a rectangular computational domain is specified.

3.1 INITIAL AND BOUNDARY CONDITIONS

To initialise the flowfield a growing, laminar boundary layer was specified. The temperature can be obtained from Crocco's relation and Sutherlands law is used to determine the viscosity. The freestream Mach number was 1.5 and the Reynolds number, based on cavity depth, was 4.5×10^5. Within the cavity, all parameters were set to their wall values.

At the inflow boundary, all parameters were held constant, whilst at the downstream boundary, zeroth order extrapolation of conserved quantities was employed. Outflow parameters were set after checking the flow direction to avoid reflections at the upper computational boundary. When the flow was outgoing, zeroth order extrapolation was used, and for incoming flow, the pressure was extrapolated from the computational domain. Along all solid boundaries, the no-slip condition was specified and the wall pressure obtained from the cell-centre normal to the boundary. An isothermal wall was specified and the density was determined from the equation of state.

4. GRID PARTITIONING AND LOAD BALANCING

In order to port the benchmark code to parallel computers the computational grid is partitioned into a number of subdomains equal to the number of processors available. Some difficulty does arise, however, when deciding how best to allocate the computational domain to each processor to achieve a well balanced problem. The simplest geometric approach is to divide the computational region into either vertical or horizontal sections, or a combination of both, as indicated in figure 1. However, this leads to load balancing difficulties. A second approach is to allocate an equal number of computational cells to each processor. Whilst this leads to an evenly balanced load (ignoring any imbalance caused by evaluating boundary conditions), it does add extra complexity to the communications, as illustrated schematically in figure 2. Care is also needed in the grid partitioning strategy to prevent disconnected regions occurring. For simplicity, we have opted to use either vertical or horizontal sections or a combination of both, more complicated schemes being the subject of further research.

5. RESULTS AND DISCUSSION

The following sections detail results obtained on a variety of computer architectures. Where possible, the results were obtained by employing the machine-independent message-passing harness, Fortnet. However, it was not possible to efficiently implement Fortnet on the KSR-1 in the available time. A parallel implementation was, however, achieved by employing compiler directives to achieve fine-grain parallelism at the DO loop level. Further details are given in section 5.3.

5.1 RESULTS FOR A QUASI-HETEROGENEOUS SYSTEM

A heterogeneous system would consist of any number of computers connected in a network and in general, if this system consisted of a group of identical workstations, it would be referred to as a cluster. The system that we have investigated consisted of 2 Silicon Graphics workstations: (i) an Iris 4D/220GTX, and (ii) an Iris Crimson. This combination has a mismatch in processor capacity, data/instruction cache size, and memory, and results in an unbalanced system. Hence we have termed this a "quasi-heterogeneous" system. The machine-specific details are as follows:

1. Silicon Graphics Iris 4D/220GTX
- 2 × 25 MHz processors
- MIPS R3000 CPU RISC Architecture
- 64 MBytes main memory
- 64 kBytes data/instruction cache

2. Silicon Graphics Iris Crimson
- 1 × 50 MHz processor
- MIPS R4000 CPU RISC Architecture
- 32 MBytes main memory
- 16 kBytes data/instruction cache

Table 1
Single-Processor Results For Sequential Code On Silicon Graphics Workstations

NMAX	Machine	Time (seconds)
256	Iris 4D/220	25.2
256	Crimson	10.4

From Table 1 it is evident that the single processor performance of the Iris Crimson is approximately 2.5 times that of the Iris 4D/220. To obtain the optimum performance from this combination it is clear that a dynamic load-balancing strategy should be employed or that the decomposition be statically "tuned" for the specific implementation. In general, the latter choice is not an acceptable alternative because if the system were to change the decomposition would have to be tuned again. The former choice of dynamic balancing is an acceptable approach and can have added benefits in a multi-user environment. However, if the network load is variable, as in most systems, then a significant amount of time could be wasted trying to obtain an ideal balance between computation and communication. For illustrative purposes, we have used only a static grid decomposition, dynamic strategies being the subject of future work.

A wide range of results was obtained from this system for both horizontal and vertical decompositions. However, it was found that the best balance was obtained when 3 processes were running on the Crimson (with time sharing) and a single process was running on the Iris 4D/220. Table 2 illustrates the results with a horizontal decomposition being employed.

Table 2
Results For Parallel Code On Silicon Graphics Workstations

NMAX	Machine	Elapsed CPU	Elapsed Time
256	Iris 4D/220	9.26	13.13
256	Crimson	8.24	13.13
256	Crimson	8.83	13.13
256	Crimson	8.73	13.13

As indicated in Table 2, a significant proportion of the time is spent communicating over the network. Some time is also lost when the Crimson is idle. It is also difficult to assess the efficiency of such a combination. However, if we selected the fastest machine to solve the problem on, then the efficiency is only 20% and the cost of system idle time and communications is very apparent.

5.2 RESULTS ON THE PARSYTEC MULTICLUSTER

The following set of results were obtained on a small Parsytec MultiCluster system located in Aachen, Germany, which utilises the T800 transputer in a 2–D topology. The maximum number of processors available was 16. Each T800 has a peak performance of 1.5 Mflop/s and 4 Mbytes of memory. The small amount of memory, coupled with the limited number of processors, restricted the problem size. A further constraint of the Parix software was that a rectangular mesh had to be used. As the current version of Fortnet requires one processor to act as a server process, only 12 (plus 1 for the server process) of the available 16 processors could effectively be utilised. This issue is currently being addressed.

Only a limited number of changes to Fortnet were required to implement it on the Parix system. The Parix Makelink() and Select() libraries were used to connect processors dynamically and the timing routine was modified to call GetTimeNowLow(). The T800 communications are done via software emulating the proposed T9000 virtual channel hardware, and have a peak speed of 1.2 Mbytes/s. The T9000 will have a peak communications speed of 20 Mbytes/s. The current Parsytec system has limited compiler optimisation, as indicated in Table 3, and it is anticipated that the next compiler release will see a factor of 2 increase in performance.

Table 3
Results For Sequential Code On Parsytec MultiCluster

NMAX	Compiler Option	Time (seconds)	Improvement
32	- - - -	4.65	- - - -
32	-OI	4.29	7.7%
64	-OI	16.65	- - - -

As can be seen in Table 3, the optimised version of the code shows less than 8% improvement in performance. However, for the following results the optimised version of the code was employed. One of the good features of the transputer system, however, is the scaleablity. As the problem scales as $NMAX^2$, doubling the number of grid points should increase the computational time per iteration by a factor of 4. However, as Table 3 shows, the performance of the T800 has improved for this particular problem size.

Table 4 shows the results for a variety of grid topologies where, for example, a 2 × 2 decomposition indicates that 2 horizontal and 2 vertical domains have been allocated to the 4 processors. The most efficient implementation occurs for the 4 × 1 configuration i.e. only a horizontal decomposition is employed.

Table 4
Results For Parallel Code On Parsytec MultiCluster

NMAX	Grid Topology	Time (seconds)	Efficiency
32	2 × 1	2.60	83%
128	2 × 2	20.18	83%
128	1 × 4	19.00	88%
128	4 × 1	18.19	92%
128	3 × 3	9.50	78%
128	3 × 4	7.46	74%

The following results in Table 5 are the projected performance figures for the Parsytec GC system with T9000 transputers. They were obtained by assuming a peak performance of only 0.75 Mflop/s for the Parsytec T800 system (allowing for the projected factor of 2 improvement with the next compiler release) and 25 Mflop/s for the T9000. No allowance has been made for the expected reduction in communication time due to the hardware implementation of the virtual channel capability.

Table 5
Projected Results For T9000 On Parsytec MultiCluster

NMAX	Grid Topology	Time (seconds)	Efficiency
128	2 × 2	0.49	72%
128	1 × 4	0.49	84%
128	4 × 1	0.49	86%
128	3 × 3	0.21	66%
128	3 × 4	0.16	64%

The projected peak performance figures are extremely promising. It is anticipated that the T9000 β—Silicon will be available in the last quarter of 1992 and commercial shipments should take place in early 1993. We await α—Silicon benchmark figures but even if the T9000 could perform at only 50% of its theoretical limit the computational times would still be very competitive.

5.3 RESULTS ON THE KSR-1

The Kendall Square Research KSR–1 machine installed in the Centre for Novel Computing (CNC) at the University of Manchester is a 32-node, distributed-memory, MIMD parallel computer with virtual shared memory implemented in hardware. The architecture is a hierarchy of rings with up to 32 processors at level 0 and up to 34 rings at level 1. It has a 50 nanosecond clock cycle and each ring has a theoretical peak speed of 1.28 Gflop/s (i.e. 40 Mflop/s from each processor). This is the first machine to implement message passing via cache-memory references using proprietary Very Large Scale Integration (VLSI) design, and as such may indicate the direction of future scaleable super multicomputers. All processors share the same address space. Pages of memory containing data reside somewhere in the system in the 32 Mbyte node caches. Pages may be exclusively owned by a processor in whose cache they reside, or they may be read-only copies of another processor's page. Cache coherency and movement of cache pages around the system is done directly in hardware driven by address references in the software, compiler directives, or explicit subroutine calls via the presto pthread library. The application programmer therefore has a very rich resource at his disposal - the machine may be programmed using FORTRAN-77 and compiler directives to do fine-grained parallelism at the DO loop level, or it may be explicitly programmed using the pthread library to create and manipulate separate processes and to manage their shared memory regions and partially shared memory (i.e. common blocks local to one thread but not shared between threads).

5.3.1 Results for Shared-Memory Implementation

Tiling over DO-loops was done with the c*ksr*tile compiler directive. To do this one must indicate which loop index is to be tiled, and which variables are local to the loop iteration, all others being shared between iterations and processors. Some rearrangement of the loop ordering may be required for good performance but this is normally trivial. Such fine-grained programming techniques are capable of giving good performance as shown in the tables below. The programming style is relatively easy to use, and is very close in concept to proposed standards such as the High Performance Concurrent Fortran.

5.3.2 Coding to initialise a team of threads:

```
      subroutine ksrinit

c Written by S.Breit,Kendall Square Research,Jan.'92
      include 'parms'
      include 'comms'
      character*80 commandline
      call getarg(1,commandline)
      read(commandline,'(i2)') nthreads
      call pthread_procsetinfo(ncells,idummy,1,istatus)
      if (istatus.ne.0)
     & stop 'Bad status after
     & call to pthread_procsetinfo'
      if (nthreads.le.ncells) then
         print *,' Executing with',nthreads,' threads'
      else
         stop 'Number of threads exceeds
     & number of cells'
```

```
      endif
Create team of pthreads
      call ipr_create_team(nthreads,idteam)
      end
```

When the code executes loops are tiled and each tile asks for a separate thread of execution to run. Up to nthreads of these can execute concurrently in the allocated team. In this case, the operating system determines which processor executes each thread.

Table 6
Results On KSR-1 Using Compiler Directives

NMAX	Nproc	Time (s)	Speedup	Efficiency
128	2	2.54	1.96	98%
128	4	1.43	3.49	87%
128	8	0.87	5.72	72%
128	16	0.61	8.22	51%

Table 7
Results On KSR-1 Using Compiler Directives

NMAX	Nproc	Time (s)	Speedup	Efficiency
256	2	10.01	2.30	115%
256	4	5.18	4.46	112%
256	8	2.79	8.28	102%
256	16	1.75	13.20	83%

Tests were done using a beta-release compiler and operating system, Manchester 13/5/92. For the tests with nmax=256, shown in Table 7, a super-parallel speedup was measured. It is possible that this effect is caused by paging of the memory in the one-processor run. Lower memory requirements on two processors imply better cache utilisation. The effect would presumably be different on different problem sizes, and indeed on the nmax=128 case more normal behaviour was observed. Development of an efficient implementation of Fortnet on this system is underway and we aim to achieve an efficiency comparable to that of the fine-grained version discussed here.

6. CONCLUDING REMARKS

A machine-independent CFD code has been employed to investigate a range of novel architecture systems. The results from a quasi-heterogeneous system of Silicon Graphics workstations indicate the difficulties of trying to balance the computational load. The results indicate that a dynamic load balancing strategy should be employed to account for the uneven CPU balance. From the preliminary investigations it would appear that no optimal solution is available for heterogeneous systems in a multi-user environment. Results from a T800–based Parsytec Multicluster have demonstrated the scaleability available from such a system. Projected results from a T9000 system indicate a considerable improvement in performance. The results from the Kendall Square KSR-1, using fine-grained parallelism at the DO loop level, have demonstrated a super-parallel speedup for certain problem sizes. Current research is focussed on implementing Fortnet efficiently to investigate the scaleability of grid partitioning strategies within a virtual shared memory architecture.

7. ACKNOWLEDGEMENTS

Development of FORTNET was done in collaboration with Dr. Lydia Heck of Durham University and Dr. Richard Cooper of Queen's University, Belfast. We thank Dr. Robert Harrison of Argonne National Laboratory for permitting use of his tcgmsg UNIX toolset. We thank Graham Riley of the Centre for Novel Computing, Manchester, Steve Pass and Steve Breit of KSR, and Francis Wray and Hajo Meier of Parsytec for valuable discussions and providing access to their systems.

8. REFERENCES

1. Geist, G. A., Heath, M. T., Peyton, B. W., and Worley, P. H., "PICL A Portable Instrumented Communication Library." Technical Memorandum ORNL/TM - 11130, 1990. Oak Ridge National Laboratory.
2. Kolawa, A., Express Users' Guide and Reference Manual. Parasoft Inc., 1990.
3. Bomans, L., Roose, D., and Hempel, R., "The Argonne/GMD Macros in Fortran for Portable Parallel Programming and Their Implementation on the Intel iPSC/2", Parallel Computing, Vol. 15, pp. 119–132, 1990.
4. Geist, G. A. and Sunderam, V. S., "Network-Based Concurrent Computing on the PVM System", Concurrency:Practice and Experience, Vol. 4(4), pp. 293–311, 1992.
5. Harrison, R. J., "Documentation of the tcgmsg Toolset." Argonne National Laboratory, Private Communication, 1990 and 1991.
6. Allan, R. J., Heck, L., and Zurek, S., "Parallel Fortran in Scientific Computing: A New Occam Harness Called Fortnet", Computer Physics Communications, Vol. 59, pp. 325–344, 1990.
7. Emerson, D. R., Blake, R. J., and Allan, R. J., "Parallel Implementation of a Compressible Navier-Stokes Solver Using FORTNET", in Parallel Computational Fluid Dynamics '91 (Reinsch, K. G., Schmidt, W., Ecer, A., Hauser, J., and Periaux, J., eds.), pp. 141–147, Elsevier Science Publishers B.V., 1992.

8. MacCormack, R. W. and Paullay, A. J., "Computational Efficiency Achieved by Time-Splitting of Finite Difference Operators", 1972. A.I.A.A. Paper 72 - 154.
9. Emerson, D. R. and Poll, D. I. A., "High Speed Laminar Flows Over Cavities", in Applications of Supercomputers in Engineering II (Brebbia, C. A., Howard, D., and Peters, A., eds.), vol. 1, pp. 331–345, Computational Mechanics Publications, Elsevier Applied Science, 1991.

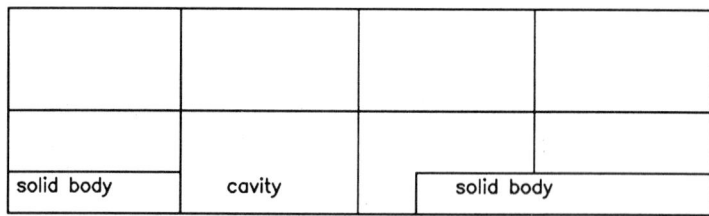

Figure 1 : Vertical and Horizontal Decomposition Strategy

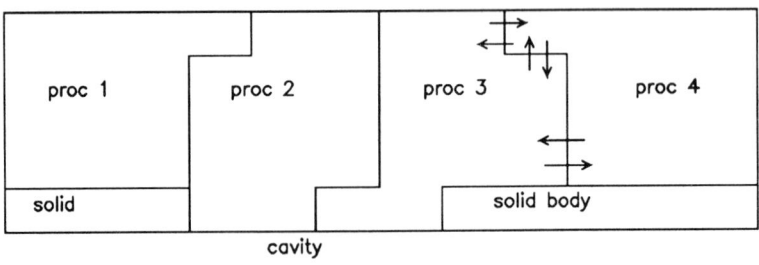

Figure 2 : Schematic of Communications for Equal Cell Allocation

Programming Paradigms for Spectral Element Methods in DM-MIMD architectures

N. Floros[a], J. Reeve[a] and O. Tutty[b]

[a]Department of Electronics and Computer Science, University of Southampton, Southampton, S09 5NH, United Kingdom

[b]Department of Aeronautics and Astronautics, University of Southampton, Southampton, S09 5NH, United Kingdom

Abstract

The Spectral Element Method is one of the few numerical methods whose parallelization is naturally occurring due to the spatial discretisation methodology. In this paper we present an implementation of an incompressible Navier Stokes solver based on this method for parallel hardware such as the Intel iPSC and Transputer arrays. The solver uses a fractional time stepping method for time discretisation, and the Mortar Element Method to allow the clustering of subdomains in regions of interest. We also describe a range of tools developed to give the user an interface independent of the underlining hardware, and present two test problems and performance analysis of these problems.

1 Introduction

The Intel iPSC and Inmos Transputer (T800 based SuperNode and Meiko architectures) are two of the most widely used parallel platforms on the US and Europe respectively. Although both machines fall within the Distributed Memory Multiple Instructions Multiple Data (DM-MIMD) architectural class, they differ significantly in their approaches to communication. The iPSC uses asynchronous communications which occur between processors, while the Transputer uses the more restrictive synchronous blocking scheme but between processes on different processors across the network [1]. On the Transputer, hardware assisted multitasking allows the application to create processes which communicate while another processes perform local work.

The aim of this paper is to present an implementation of an incompressible

Navier-Stokes solver for both architectures and a number of tools which provide a uniform user friendly interface to the solver. The spatial discretisation was done using the Spectral Element Method (SEM) with the addition of Mortar Element Method (MEM) to allow the clustering of grid points in regions of interest, while time discretisation was done using a time splitting scheme. A number of test cases are also presented for performance analysis, and for validating both the solver and programming model.

2 Incompressible Navier-Stokes

The solver described here solves the 2-D Incompressible Navier-Stokes equations

$$\frac{\partial \vec{u}}{\partial t} + (\vec{u} \cdot \nabla)\vec{u} = \nu \nabla^2 \vec{u} - \nabla p \tag{1}$$

$$\nabla \cdot \vec{u} = 0 \tag{2}$$

with (1) and (2) representing the momentum and mass conservation respectively, $\vec{u}$ the velocity vector, p the pressure, ν kinematic viscosity (set as $\frac{u_{max} H}{Re}$ and H channel height), while the density ρ has been set to 1. These were discretised in time using a time splitting scheme [2] and in space using the SEM [3].

SEM was chosen as it offers exponentially increasing accuracy with increasing number of points and parallelization is intuitive to the method. For a 2-D domain Ω, the method requires the decomposition of the geometry into a conformal grid of K subdomains within which approximate the velocity and pressure field using a tensor product of Lagrangian interpolants based on Nth order Legendre polynomials. Numerical integration is performed using a Gauss-Lobbato quadrature which generates a set of $(N+1) \times (N+1)$ linear equations which are solved using a Conjugate Gradient Solver (CGS). From the domain decomposition adjacent subdomains have grid points which occupy the same physical space, resulting in direct stiffness summations within the CGS. In addition we have implemented the MEM [4] which allows the generation of regular non-conforming decompositions.

The time splitting scheme results in decomposing equations 1 and 2 into the three computational steps. We start with the explicit treatment of the advection and forcing terms to compute an intermediate velocity $\tilde{\vec{u}}$

$$\tilde{\vec{u}}^{n+1} = \tilde{\vec{u}}^n - \Delta t \frac{\partial \tilde{\vec{u}}^n}{\partial t} + \frac{(\Delta t)^2}{2!} \frac{\partial^2 \tilde{\vec{u}}^n}{\partial t^2} \tag{3}$$

followed by the pressure step

$$\nabla^2 p^{(n+1)} = \frac{1}{\Delta t} \nabla \cdot \tilde{\vec{u}}^{(n+1)} \text{ on } \Omega \text{ , } \nabla p \cdot \vec{n} = \frac{\tilde{\vec{u}} \cdot \vec{n}}{\Delta t} \text{ in } \partial\Omega \tag{4}$$

and finally the velocity $\vec{u}$.

$$\frac{\vec{u}^{(n+1)} - \tilde{\vec{u}}^{(n+1)}}{\Delta t} = \nabla^2 \vec{u}^{(n+1)} - \nabla p^{(n+1)} \tag{5}$$

The time step Δt is determined via the usual Courant condition. To reduce errors introduced by the 2nd order Taylor expansion in the evaluation of $\tilde{\vec{u}}^{(n+1)}$ in (3) a time marching scheme has been devised which allows the evaluation of $\tilde{\vec{u}}^{(n+1)}$ through a number of intermediate steps. So (3) is written as :

$$\tilde{\vec{u}}_{m+1} = \tilde{\vec{u}}_m - \delta t \{\tilde{\vec{u}}_m \cdot \nabla \tilde{\vec{u}}_m\} + \frac{\delta t^2}{2!} \{\frac{d\vec{u}}{dt} \cdot \nabla \tilde{\vec{u}}_m + \vec{u}_m \cdot \nabla \frac{d\tilde{\vec{u}}_m}{dt}\} \text{ ; } \forall \, m \in \{0, \ldots, z\} \tag{6}$$

$$\frac{d\vec{u}}{dt} = \frac{\vec{u}^n - \vec{u}^{(n-1)}}{\Delta t} \text{ ; } \frac{d\tilde{\vec{u}}_m}{dt} = \frac{\tilde{\vec{u}}_m - \tilde{\vec{u}}_{m-1}}{\delta t}$$

$$\vec{u}_m = \vec{u}^n + (m\delta t)\frac{d\vec{u}}{dt} \text{ ; } \tilde{\vec{u}}_0 = \tilde{\vec{u}}^n = \vec{u}^n, \tilde{\vec{u}}_z = \tilde{\vec{u}}^{(n+1)} \text{ ; } \delta t = \frac{\Delta t}{z+1}$$

where a typical value of z is 19. In forming $\nabla \tilde{\vec{u}}_m$ adjacent elements have to agree on the value of the derivative for the grid points that occupy same physical space. This is done by taking the average value of their respective values.

3 Implementation

The first implementation of the solver was written in OCCAM and was targeted for the emerging T9000+C104 systems [5, 6]. It allocated each subdomain into a virtual processor and featured a modular design, Figure (1), which provided us with a number of advantages. First it maintained a two level parallelism, one at subdomain level as multiple subdomains could be allocated to the same processor, and the other at the processing and communication of the boundaries. By embedding the boundary conditions in processes we were able to provide a simple mechanism of allowing the user to develop his own boundary conditions, hence extending the use of the solver. Finally the main solver could also be replaced

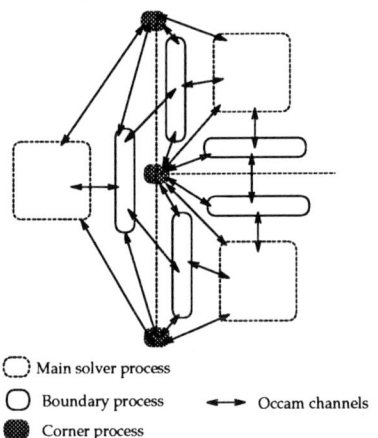

- ◌ Main solver process
- ○ Boundary process ⟷ Occam channels
- ● Corner process

Figure 1: Occam programming model

by different solver hence allowing the re-use of the tools developed to support the solver.

To ensure portability of the solver we developed a serial version in C while at the same time maintaining the modular design shown in Figure (1). In other words we converted the boundary processes to boundary data structures linked to appropriate boundary procedures; corner processes to corners data structures; data related for each subdomain were encapsulated in yet another data structure, and finally all OCCAM channels were replaced by pointers between data structures. The different solvers within the OCCAM code were re-written in C and instead of operating on one data set they operate on multiple sets connected via link lists.

Using a Single Program Multiple Data approach we ported the serial solver to both iPSC and Transputer machines, having only to add extra entries in boundary and corner data structures to cater for the fact that data may be resident on a different processor. To avoid sending multiple messages to a processor, we created another data structure which creates a vector with either all edges or corners to be send to a processor. Due to corner dependencies arising from the MEM the direct stiffness sum data exchange takes place in two parts; first the edges from which corner contribution are determined and then the corners. The Transputer implementation assumes a network of totally connected processor which is achieved via the use of the Virtual Channel Router [7].

In addition to data structure modification we had to modify three routines for the parallel implementation. The first was the inner product which in the iPSC uses the **gdsum** for determining the global sum, while for the Transputer uses a binary tree. The other two routines were the direct stiffness sum and mortar to edge projection whose communications can be overlapped with internal work. On the iPSC we used asynchronous sends and receives while on the Transputer we start processes which send and receive data while another process performs local work. A more detailed description of the implementation can be found in [8].

4 Tools

To enable the use of the solver for a wide range of problems we developed a set of tools to provide the user with an interface independent of the hardware platform on which the solver is executed [5, 6, 8]. These are written in C and use the X-Window system to provide a portable and network transparent graphical user interface.

The first tool allows the user to define the geometry of the problem and is based on a highly modified public domain CAD tool. The extra code allows the user to define boundary conditions, decompose the domain into a conformal grid of subdomains and refine regions of interest. The output of this tool is then either fed to the serial solver or to a subdomain to processor grouping tool. The output of either tools allow the respective serial and parallel implementations to create the required data structures and links for the simulation to start. For the parallel version the subdomain grouping tool creates one file for each node so that data structure creation and initialization can take place in parallel.

The graphical front end of the solver allows the user to define simulation parameters and also display at real time a 2-D plot of the magnitude of the velocity. During the simulation the solver outputs in a file copies of the velocity and pressure at a user predefined intervals, this file is then fed to a visualisation tool which allows the users to examine the computed flows in more detail. The tool can present in several display formats velocity magnitude and components, pressure, vorticity and streamlines that are calculated on the fly.

5 Test Cases

Two test cases where used to validate the implementation of the solver, the environment and also to do performance analysis of the implemented code. The first test case was to calculate the flow though a channel step expansion, determine the distance of flow re-attachment from the back of the step and compared it with results obtained from a finite difference solver [9]. The applied boundary conditions for the velocity were parabolic with $\vec{u}_{max} = (1,0)$ in the inflow, no slip on the walls and the Neumman condition $\frac{\partial \vec{u}}{\partial n} = 0$ at the outflow, while for the pressure $\frac{\partial p}{\partial n} = 0$ everywhere except in the outflow where $p = 0$. In addition we had $R_e = 100.0$ based on the upstream channel width, the Courant number set to 3.0, $K = 46$, $N = 6$ and $T = 120$. The inflow and step height was set to 1 as was the distance of the inflow to the step, while the channel length from the back of the step to the outflow was

Displayed variable : ω
N = 6 , Re = 100.00 , Cr = 3.0 , T = 120.00(sec) , Δt = 8.810e-02   Fri Apr 24 10:29:45 1992

User : nf - Dir : /home/brewery/users/nf/cfd/experiments/short_step
Session id : step   Δαιδαλos post : University of Southampton

Figure 2: Vorticity Plot of channel step

Topology name	Refinement level	K
C_0	No	40
C_1	1	50
C_2	2	98
C_3	3	242
C_4	4	458

Table 1: List of Channel Domain Decompositions

10.0. A vorticity plot of the computed flow is shown in Figure (2). The distance of the re-attachment point was found to be 3.581 in our solver, and 3.589 by the finite difference solver, a difference of only 0.2%.

The second test case is that of entry flow in a uniform channel with a uniform unit inlet velocity at $R_e = 1000.0$. This problem provides a sever test for a numerical method because of the singularities in the solution arising from the discontinuity in the velocity on the walls at the inlet. In particular the wall shear stress is infinite at the inlet. Also, boundary layers form on the walls downstream of the inlet. Thus a computational efficient solution requires clustering of grid points near the walls at the inlet, and this problem can be used to test the effects of subdomain decomposition.

Displayed variable : u
N = 5 , Re = 1000.00 , Cr = 3.0 , T = 10.00   Tue Apr 14 15:55:38 1992

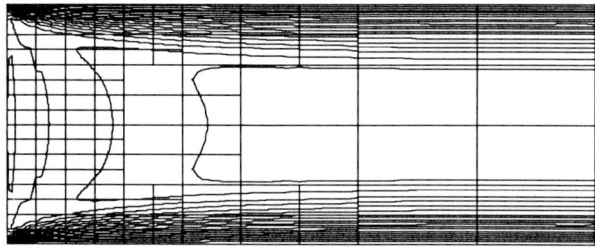

User : nf - Dir : /home/brewery/users/nf/cfd/channel_sim/C_2
Session id : channel   Δαιδαλος post : University of Southampton

Figure 3: C_2: u_x contour plot $K = 98$

Displayed variable : u
N = 5 , Re = 1000.00 , Cr = 3.0 , T = 15.00   Mon July 27 12:44:07 1992

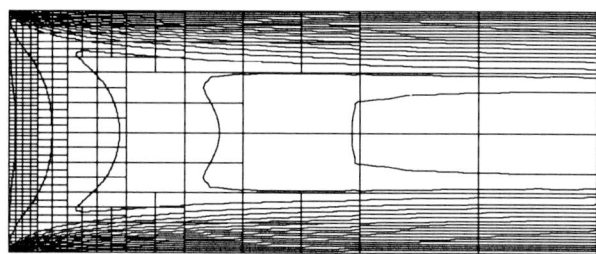

User : nf - Dir : /home/brewery/users/nf/cfd/channel_sim/C_4
Session id : channel   Δαιδαλος post : University of Southampton

Figure 4: C_4: u_x contour plot $K = 458$

We therefore tried a range of domain decompositions. Table 1 gives a list of all subdomain decompositions and Figures 3 and 4 contour plots of the x component of the velocity for the C_2 and C_4 decompositions. There is a clear benefit from the use of MEM.

The wall shear stress was calculated from the C_3 SEM solution and was compared with the shear stress from analytical Blasius solution for the boundary layer on a flat plate,

$$\tau_w = 0.322(xR_e)^{-\frac{1}{2}} \qquad (7)$$

as shown in Figure 5. The jumps in the SEM solution appear at subdomain bound-

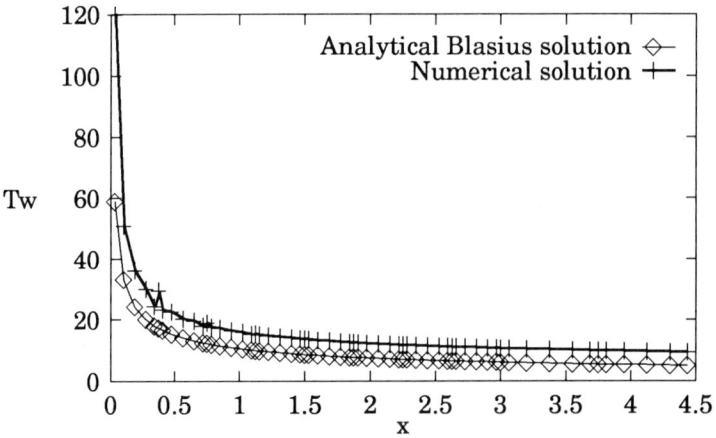

Figure 5: Wall shear stress - vs - down stream distance

aries, and not surprisingly, are most evident near the inlet where the shear stress gradient is largest. They occur because the SEM does not enforce continuity in derivatives of primitive variables. The two curves, in Figure 5, have the same shape but do not coincide. There are two (physical) reasons for this; first the Blasius solution is a similarity solution of the boundary layer equations and has an (implicit) origin shift along the x axis, and second, downstream the wall shear stress from the Blasius solution tends to zero as $x^{-\frac{1}{2}}$, while τ_w for the Navier-Stokes solution must tend to 6, the value for fully developed Poiseuille flow.

The simulation for the channel step expansion and for the C_0, C_1, C_2 and some tests run for the C_3 where performed and on a 32 node T800 box with 4 Mbytes each node. The Transputer box features a hardware assisted through routing mechanism based on a network of T222 (16bit Transputer) [10] and offers the same facilities to VCR as a network of C104s. An 8 node iPSC/860 was used for the C_3 and C_4 simulations.

6 Performance Analysis

Two of the most time consuming sections of the developed code are the inner product and the $blk(\hat{A})u$ matrix operations in the CGS. We timed these operations on all there types of microprocessors and we obtained the results shown in Table 2.

Processor	$blk(\hat{A})u$ (MFLOP)	(r,r) (MFLOP)	Clock (MHz)
i860 (-O4 icc v.3)	11.09	4.39	40
T800	0.39	0.42	20
SparcStation 1	1.523	1.112	20 (12MB)
SparcStation IPX	3.185	2.779	40 (32MB)

Table 2: MFLOP ratings for compute intensive program sections

Despite the fact that the workstation code was compiled using a public domain C compiler GNU C, it gave performances close to vendor peak performance and despite the overheads imposed by OPENWINDOWS 3. On the other hand the i860 and T800 performances are rather disappointing and attributed mainly to poor compiler optimizations and architectural peculiarities. In particular the T800 with similar peak performance as the SparcStation 1 is hit the most as all data are located at the slow main memory and so the code can not take advantage of the fast on chip memory.

Having determined the code performance for compute intensive sections of the solver we carried out a detailed performance evaluation of the T800 and iPSC parallel implementations using the C_2 and C_4 domain decompositions with 9800 ($N = 10$) and 11450 ($N = 5$) grid points respectively. We choose these two domain decompositions as we wanted to investigate what is the effect of subdomain number K and polynomial order N on the execution time. Timings were obtained from both the T800 hardware assisted through routing box and the iPSC/860 to produce an efficiency graph Figure (6). Memory restriction prevented us from obtaining a complete set of results for the T800 system. From the graph we can deduce that the T800 is more efficient than the iPSC not surprising as the T800 presents a slower but more balanced system than the iPSC. The C_4 is consistently more efficient than the C_2 for the iPSC while the same is not true for the T800 results.

The main computational steps in the developed Navier-Stoke solver are the p, $\vec{u}$ and $\tilde{\vec{u}}$ routines. Figures (7) and (8) show the times taken by each of these routines for both decomposition and parallel hardware.

The solution of problems in which Neumman boundary conditions are predominant, are notoriously time consuming processes. Figures (3) and (4) clearly show that most grid points are located in regions of the pressure solver, which Neumman conditions are applied. This causes the pressure solver to consume disproportion

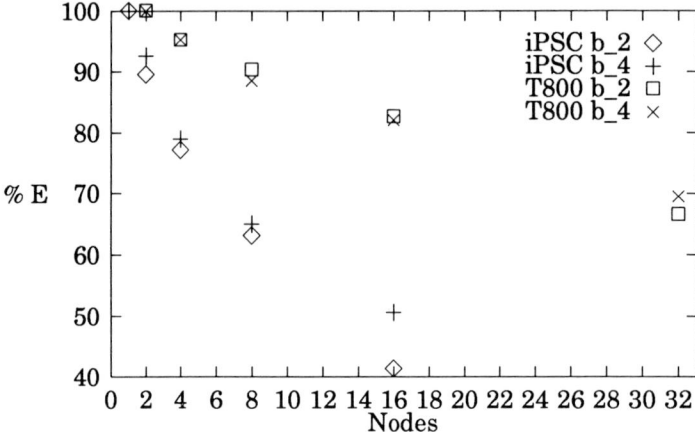

Figure 6: % Efficiency - vs - Processing nodes

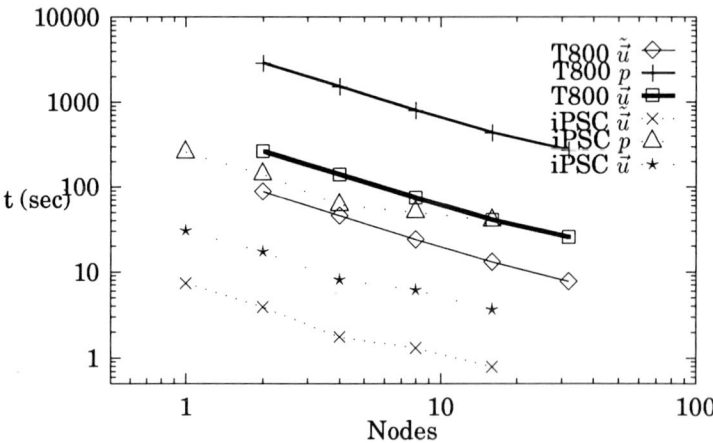

Figure 7: C_2: Log of time taken per solve - vs - Log Processing nodes

amount of time when compared to the time required for the two helmholtz solvers for determining $\vec{u}$, and two explicit solvers for $\tilde{\vec{u}}$. The $blk(\hat{A})u$ operation in the CGS have of the order $KO(N^3)$ operations which means that the C_2 requires twice the number of operations than the C_4. Because of the vector architecture of the i860 the C_2 decomposition is executed faster than the C_4, as its vector length, N=10, is twice that of the C_4, N=5. The opposite is observed in the T800 results where the number of required operations is that which determines the execution time and not the vector lengths ; similar results with the T800 were also observed for the SparcStation 1 and IPX results.

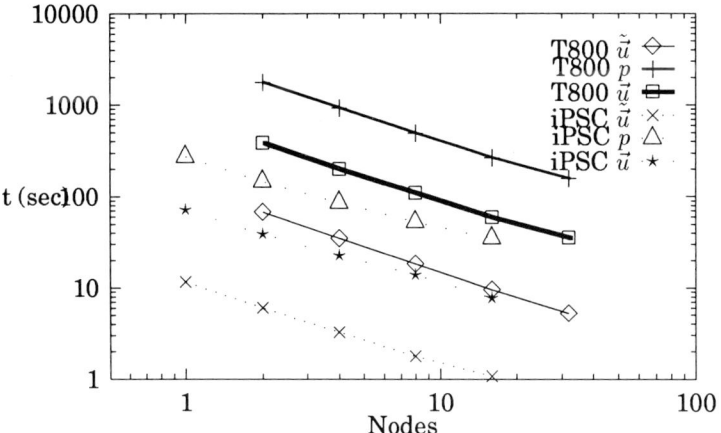

Figure 8: C_4: Log of time taken per solve - vs - Log Processing nodes

Under optimal conditions we would have expected the execution time to be inversely proportional to the number of processing nodes, instead from the plots it becomes evident that the relationship is of the form (8).

$$t = b \times Nodes^a \tag{8}$$

where b is the time taken for the solution in one processor and a the slope, which is also a measure of the efficiency of the implementation.

Close examination of the plots in Figure (7) shows a substantial change in slope between Nodes 4 - 8 and 16-32 for both iPSC and T800 results, the same is observed for the results in Figure (8) but at a very much reduced magnitude. Measurement of the slopes in Figure (7) for the iPSC shows a wide range of efficiencies from -1.134 for the $\tilde{u}$ between Nodes 2 and 4 to -0.39 for $\vec{u}$ and p between nodes 4 and 8. On the other hand the iPSC results for the C_4 decomposition, Figure 8, show a more subdue deviation ranging from -0.93 to -0.83. Similarly the T800 slopes for the C_2 range from -0.92 to -0.68 while for the C_4 -0.93 to -0.87.

Within those measured solvers there are three routines which involve communications and should be responsible for the slopes of the curves observed in the Figures (7) and (8). These routines are the inner product (r.r), direct stiffness sum (ds) and the mortar to edge projection (me), Figures (9) and (10) show the obtained results.

Examining these plots it becomes evident that the reason of the wide range of efficiencies obtained are the direct stiffness sum routine and the inner product. In particular we observe that the for the T800 C_2 results the inner product is re-

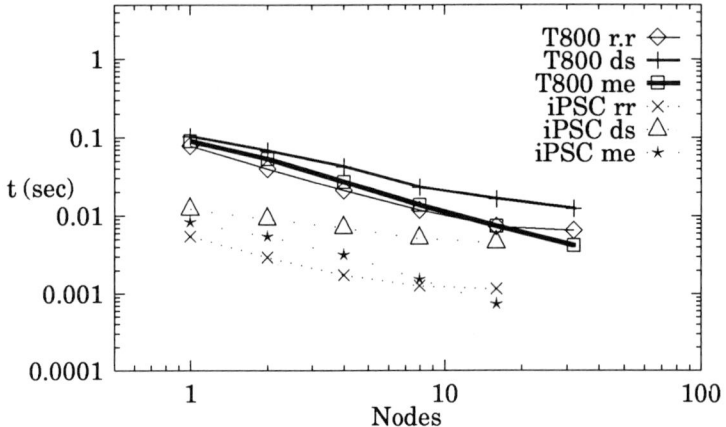

Figure 9: C_2: Log of time - vs - Log Processing nodes

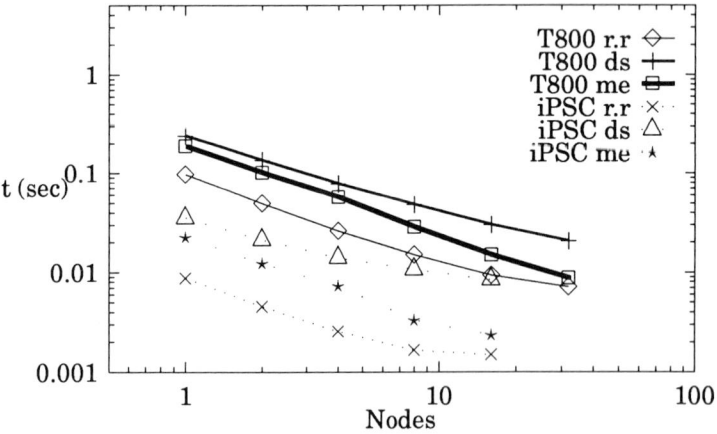

Figure 10: C_4: Log of time - vs - log Processing nodes

sponsible for the reduction in efficiency while for the iPSC it is the direct stiffness sum. The first is because the time required to perform the global sum has increased in comparison with the time to calculate the partial sum. For the iPSC results however the network size is small for it to affect the results, however the slow communications hit particular hard the ds operation (slope -0.37) as it involves the transmission of many of small messages.

The increase in subdomains per processor for the C_4 decomposition ,Figure (10), benefit all three operations for both platforms as there is increased amount of internal work in comparison with the communications. Interestingly the mortar to edge projection exhibits an almost constant efficiency as it is the only routine whose work and communications are both reduced with increasing number of nodes.

7 Conclusions

In this paper we have described the implementation of an Incompressible Navier Stokes solver for iPSC/860 and Transputer based hardware. A SEM was used for spatial discretisation with the addition of an MEM to allow the clustering of subdomains in regions of interest, while time discretisation was done using a time splitting scheme resulting in a simple, accurate and efficient solver. The solver was developed as part of an environment called $\Delta\alpha\acute{\iota}\delta\alpha\lambda o\varsigma$ (Dædalus) which hides many of the complications related with the use of parallel hardware from the end user.

The test problems shown that first our solver produces comparable results with other solvers and that has been efficiently implemented on both platforms. In particular the entry flow problem showed that MEM can be used effectively and can improve both accuracy and execution time. The performance analysis shows first that the i860 compilers have long way to go before even reaching 50% of peak performance and that the T800 has fallen behind in the war of MIPS and MFLOPS, which is not surprising considering the age of the design. Secondly the reduction in efficiency occurs because of communications in both the inner product and the direct stiffness sum routines. The first is difficult to improve on as both in the iPSC and Transputer we used what we believe is the most efficient implementation. Probably there is room for improvement in the second procedure via the use of a more efficient partitioning algorithm as described in [11].

Work in near future includes further studies for the validation of the solver, consideration of mechanisms which could enable the automatic generation of subdomains from the problem geometry and finally stabilization of boundary layers using small outflow regions near the inflow.

8 Acknowledgments

This work was funded initially by the Esprit 2701 $\mathcal{PUMA}$ project and then by the Southampton Novel Architecture Center (SNARC). Access to an 8 node iPSC/860 was provided by the Southampton Parallel Applications Center and to a 16 node by the Daresbury Laboratory and are gratefully acknowledged.

References

[1] The T9000 Transputer Product Overview Manual, First Edition 1991, INMOS.

[2] Y. Maday, A.T. Patera and E.M. Rnquist, An Operator-Integration-Factor Splitting Method for Time-Dependent Problems: Application to Incompressible Flow, Journal of Scientific Computing, Vol 5, No 4, December 1990.

[3] P. F. Fischer Spectral Element Solution of the Navier Stokes Equations on High Performance Distributed-Memory Parallel Processors, M.I.T Thesis June 1989.

[4] Y. Maday, C. Mavriplis and A. T. Patera, Nonconforming Mortar Element Methods: Application to Spectral Discretisations, ICASE Report No 88-59, October 1988.

[5] N. Floros, J. Reeve and O. Tutty, $\Delta\alpha\acute{\iota}\delta\alpha\lambda o\varsigma$ (Dædalus): An Environment for Research into Incompressible Flows, Parallel Computing and Transputer Applications '92, Barcelona September 1992 to appear.

[6] N. Floros, J. Reeve and O. Tutty, Computational Fluid Dynamics Applications Report, PUMA Deliverable 5.1.2, University of Southampton October 1991.

[7] M. Debbage, M. Hill, D.A. Nicole. Virtual Channel Router Version 2.0 User Guide. Puma Working Paper 25, University of Southampton, June 1991.

[8] N. Floros, J. Reeve and O. Tutty. $\Delta\alpha\acute{\iota}\delta\alpha\lambda o\varsigma$: Incompressible Navier Stokes Solver for Parallel and Distributed Architectures, Southampton Novel Architecture Research Center tech. rep. SNARC 92-x (to appear).

[9] O.R. Tutty and T.J. Pedly. Oscillatory Flow in a Stepped Channel. J. Fluid Mechanics (in press).

[10] M. Debbage, M. Hill and D.A. Nicole, Hardware Packet Router for VCR Version 2.0, PUMA Working Paper 42, University of Southampton October 1991.

[11] H. D. Simon Partitioning of Unstructured Problems for Parallel Processing Computing Systems in Engineering Vol 2, No 2/3, pp. 135-148 1991

Application of a Parallel CFD code to Large-Scale Practical Problems

E. R. Galea, A. Chan, M. Cross, N. Hoffmann, C. Ierotheou, S. Johnson and K. Pericleous.

Centre for Numerical Modelling and Process Analysis, University of Greenwich, Wellington Street, London SE18 6PF. United Kingdom.

Abstract
In this paper we discuss the performance characteristics of a general purpose commercial fluid flow code HARWELL-FLOW3D which has been modified to make efficient use of distributed memory parallel architectures. The parallel FLOW3D code has been applied to such diverse applications as fire simulation and mold filling. The code has successfully been implemented on both homogeneous and heterogeneous platforms.

1. INTRODUCTION

The attraction of parallel computer architectures lies in their potential to provide low cost supercomputer performance in the solution of numerically intensive problems. The CFD Navier-Stokes code FLOW3D by AEA HARWELL [1] has been mapped onto various local-memory parallel systems at the University of Greenwich (formerly Thames Polytechnic) [2] and applied to a number of industrial and environmental applications. Two of these, covering a wide range of practical interest, involve the modelling of fire/smoke movement in aircraft cabins [3,4] and the modelling of the filling process and associated free surface behaviour in containers [5].

Both cases are computationally intensive; the first because of the need to perform a large number of interactions of the flow solution procedure on a fine grid and a complex geometry in order to converge a notoriously unstable buoyancy driven flow field. The second, due to the need to use explicit time marching schemes in order to capture free surface behaviour [6].

The use of a single code to handle two problems so manifestly different demonstrates the generality of the parallel mapping strategy adopted.

2. PARALLEL IMPLEMENTATION

As the parallel implementation has been described elsewhere [2,7] it will only be briefly described here. The strategy employed is based on the systematic partitioning of the computational domain. Nearest-neighbour communications are possible if the processing elements (PEs) are set up in a one- or two-dimensional grid topology. The computational domain is partitioned by assigning to each PE an equal number of x-y slabs, in the implementation described here a one-dimensional decomposition is employed. Moderate restructuring of the original code is required to map it onto a pipeline processor topology. In the pipeline topology each PE has its own copy of the FLOW3D executable. Each PE is responsible for the computation of cell information in its assigned area of the domain. When necessary, data is transferred between neighbouring PEs. This is done in order to preserve as far as possible the data dependencies present in the original serial code.

3. HARDWARE

The parallel version of FLOW3D has been ported to two distinct types of distributed memory parallel platforms. The first utilises a homogenous collection of PEs, where each PE is a T800 Transputer from INMOS [8] while the second consists of a heterogenous collection of PEs.

In the latter case, each PE consists of a T800 processor for communications and an Intel i860 microprocessor to perform the computations. The T800 with a 20 MHz clock is capable of a theoretical performance of 1.5 Mflops and a sustained performance of 0.7 Mflops while the i860 has a peak performance of 80 Mflops and a sustained performance of 9 Mflops.

In both cases the PEs are connected in a pipeline array. In the case of the homogeneous PEs the master processor has 8 Mb off-chip memory while each of the slaves has 2 Mb whereas each of the heterogeneous processors possesses 16 Mb off-chip memory.

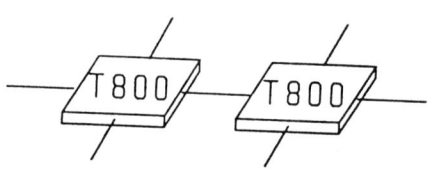
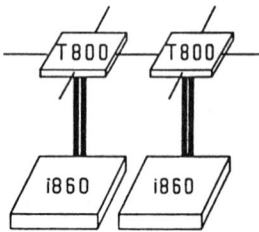

Figure 1a Homogenous Architecture.

Figure 1b Heterogenous Architecture.

Both systems are housed in a single parallel computing facility supplied by TRANSTEC. The facility is hosted by a SUN4 workstation and contains a rack of 5 vme boards. Two host 40 transputers and associated memory, the remaining three being occupied by 12 i860s. The programming environment is CTOOL SET and the i860s use the Portland FORTRAN compiler while the transputers use 3L FORTRAN [9].

4. PARALLEL APPLICATIONS

4.1. Fire Simulation

The recent success of Fire Field Modelling in uncovering details of the fire mechanism responsible for the Kings Cross tragedy [10] highlights its value as a fire analysis tool. The versatility of the technique - which it derives from its fundamental approach and minimal use of empiricism - make it an ideal design tool, useful in assessing the design of ANY inhabited enclosure for safety and in the development of fire fighting strategies. However, for all its potential, the procedure remains within the realms of a 'research' tool. The greatest inhibitor of its progress is the cost (in terms of time and money) associated with performing the simulations. Hundreds of hours may be involved in even the simplest of calculations.

The reason for this prodigious consumption of computer power is found at the very heart of fire field modelling. In its simplest form it involves the simulation of transient, three-dimensional, recirculating, turbulent, buoyant flows. While simulations of this type have the ability to predict the spread of fire hazards such as heat and smoke within an enclosure, they are amongst the most difficult problems in CFD.

Two recent examples illustrate this point. The transient calculations which were at the heart of the Kings Cross simulation required of the order of 48 hours of computation per run on a dedicated processor of a CRAY 2 supercomputer [10]. These computations were performed on a 15,000 cell mesh.

The example, discussed in this paper (see figures 2 to 4) involves a high performance workstation, the SUN SPARC-2. Three-dimensional calculations comprising 30,360 cells required in-excess of 386 hours (2.3 weeks!) of computer time to perform a 60 second simulation of fire aboard an aircraft. And using data from a CFD based benchmark [11], it is estimated that similar calculations performed on the HP-720, one of the most powerful workstations currently available, would require 180 hours.

While large, these times do not represent upper limits on run times, as the computational grids involved in each simulation are relatively coarse. In addition, these simulations were of the simplest type which can be performed. If physical phenomena such as combustion, radiation, generation of toxic species and the action of water sprays are to be included, then the run times may be expected to increase many times over.

The model described here simulates a fire aboard a single aisle commuter aircraft containing 29 seats and two exits. The aircraft cabin is 9.6m long with maximum width and height of 1.9m and 1.8m respectively. This geometry was meshed using a Body Fitted Co-ordinate (BFC) grid comprising, 22 * 23 * 60 cells over the entire domain and 16 * 20 * 54 cells for the internal domain. The time discretisation was 0.25 seconds and the simulation ran for 60 seconds.

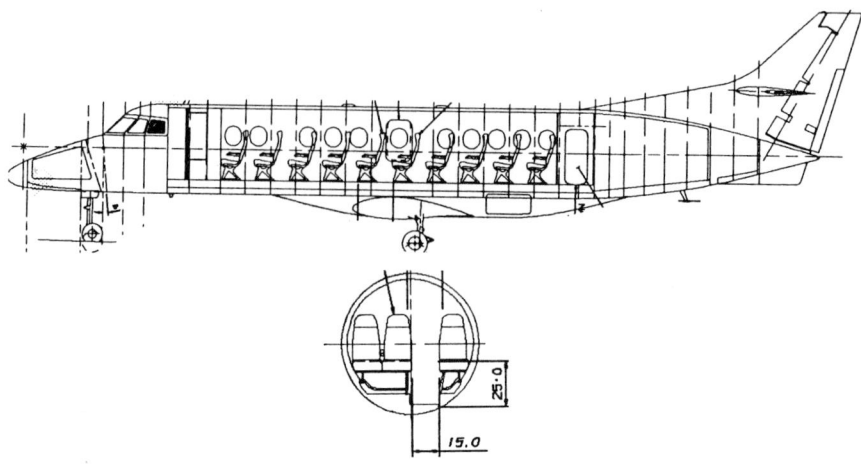

Figure 2 Aircraft geometry used in simulation.

The equations solved were the compressible Navier-Stokes equations with Pressure, Temperature, 3 Velocity components and Turbulence variables k and eps (buoyancy modified). All solid obstacles were treated as isothermal boundaries, while the external boundaries were treated as zero pressure boundaries, the initial temperature was 24° C. The fire, located on a seat top opposite the open doors, was modelled as a simple volumetric heat source which was ramped to 50 kW.

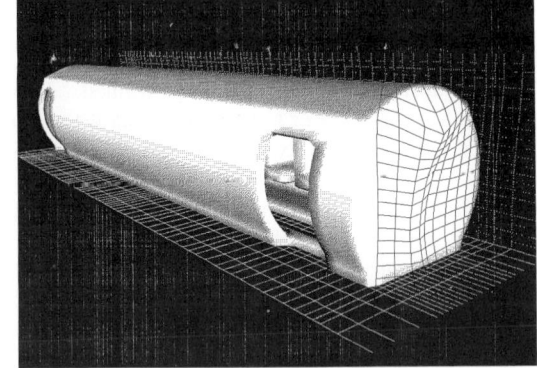

Figure 3 Computational mesh used in aircraft fire simulation.

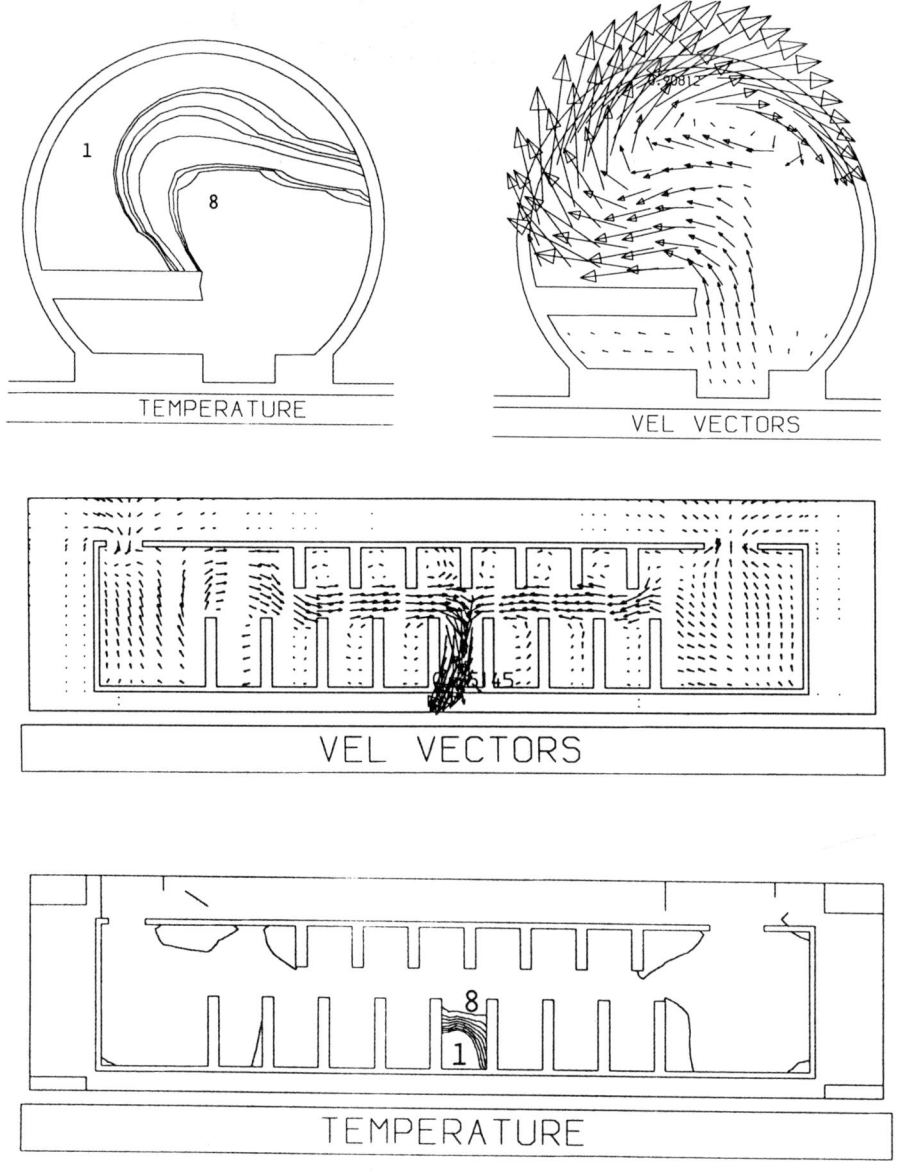

Figure 4 Temperature contours °K and velocity vectors through the heat source and a plane parallel to the floor passing through the seat tops. Figures depict the cabin environment 50 seconds after fire ignition. 1: 321 K, 8: 298 K.

Figure 4 depicts temperature contours and velocity vectors through the heat source. The thermal plume rising above the fire is clearly visible. Figure 4 depicts velocity vectors in a plane 0.92m above and parallel to the floor. Entrainment of air into the fire source is clearly depicted.

Tables 1 and 2 display results for the parallel fire simulations running on up to 8 processors. The simulation was performed using two meshes, the 30,360 cell grid described above and a finer mesh comprising almost 55,000 cells. This latter case could not run on the SPARC2 and single i860 configurations due to a shortage of memory. The runtimes indicated in the tables thus represent a minimum estimate.

Table 1
Parallel Fire Results.
30,360 Cells, 0 - 5 second simulation.

System	Time (hrs)	Speed-up	Efficiency
Sparc2	29.2	-	-
1 i860	15.6	-	-
4 i860s	5.9	2.6	66%
8 i860s	4.0	3.9	49%

Speed-up over Sparc2 using 8 i860s = 7.3

Table 2
Parallel Fire Results.
54,648 Cells, 0 - 5 second simulation.

System	Time (hrs)	Speed-up	Efficiency
Sparc2	53*	-	-
1 i860	28.1*	-	-
4 i860s	9.1	3.2*	81*%
8 i860s	6.1	4.6*	58*%

Speed-up over Sparc2 using 8 i860s = 8.7*

* indicates lower bound estimate.

4.2. Mold Filling

One of the most critical stages in casting production is the filling of the mold. The filling process has to proceed in such a way that excessive liquid metal surface waves are avoided. If wave action becomes excessive, then surface impurities, air bubbles etc, may be entrained into the metal. As the metal component solidifies, flawed castings will be formed which will have to be discarded and thus increase the costs of manufacturing of such metal components. In order to optimize the filling process and hence reduce the number of flawed components produced, mathematical modelling is vital to casting industries.

The current model [5] is based upon the solution of the transport equation of a conserved scalar marker Φ, with the van Leer TVD scheme [6] employed to suppress numerical diffusion and hence preserve the definition of the air-liquid interface. Each cell of the computational domain has a marker Φ, which has a value between 0 and 1. When Φ = 1, a cell is full of liquid, and 0 if the cell is full of air. The liquid surface is represented by Φ = 0.5. The model is three dimensional and capable of describing any multi-surface flow problem. It was first implemented in the PHOENICS code in serial. However as the van Leer scheme is extremely costly in terms of CPU time, the model has also been implemented as an add-on module to the Harwell-FLOW3D code both in serial and parallel.

The application presented in this paper concerns the filling of a thick-walled container. The geometry of the container is shown in figure 5. The dimensions of the container are 35.5 x 14.2 x 3 cm (as only half of the container is modelled).

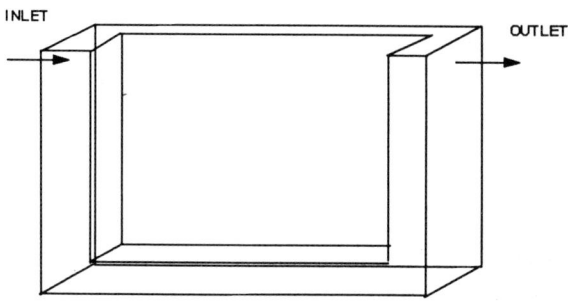

Figure 5 Geometry of the container with symmetry at the front.

The properties and boundary conditions used in the simulation are shown below:

Liquid(water):   density 1000 Kg/m^3 viscosity 10^{-3} Ns/m^2

Gas (air):   density   1 Kg/m^3 viscosity 10^{-5} Ns/m^2

Boundary conditions: Inlet velocity = 1 m/s   Outlet pressure = 0

External force: Gravity g = 9.81 m/s^2 (acting downwards)

The liquid with a horizontal velocity of 1 m/s enters via the inlet at the top left hand corner of the mold and exits via the outlet at the opposite end, after filling up the mold as shown in sequence form, figure 6 to 8. These figures clearly depict the formation of air cavities and wave actions. This run simulated a filling process of 0.8 second in real time.

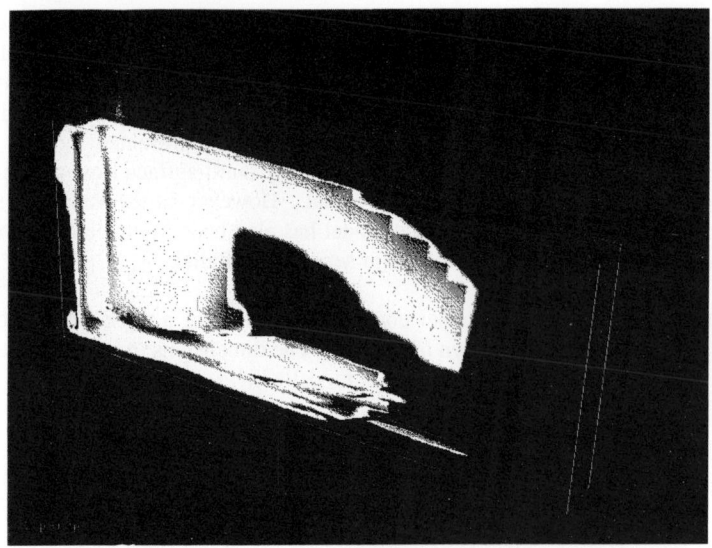

Figure 6 Liquid surface front after 0.2 second.

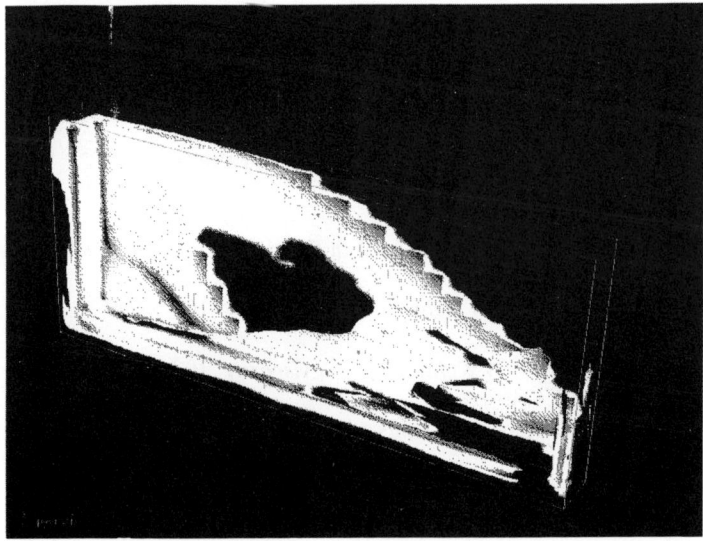

Figure 7 Liquid surface front after 0.3 second.

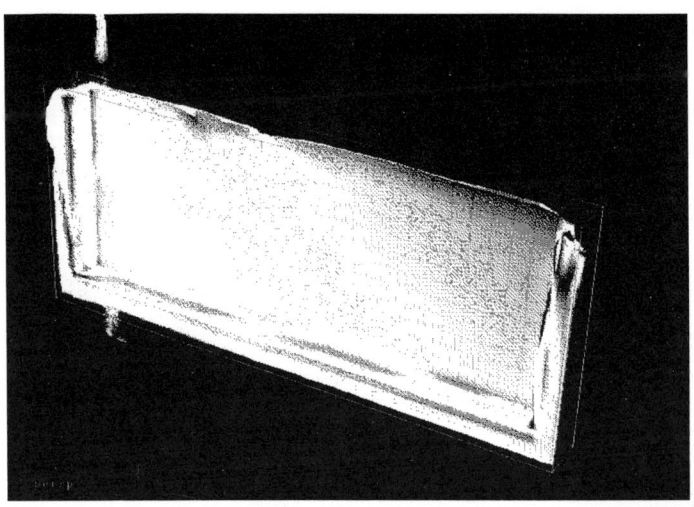

Figure 8 Liquid surface front after 0.8 second.

The mesh used in this simulation is 22 x 13 x 40 = 11440 cells and the time step 10^{-3}. The results presented here were calculated on 4 INTEL-i860 processor based parallel nodes. The overall CPU time for the simulation was about 11 hours, see Table 3 below.

Table 3
Parallel Mold Filling Results
11440 cells, 0 - 0.8 second

System	Time (hrs)	Speed-up	Efficiency
1 i860	23.5	-	-
2 i860s	14.9	1.58	79%
3 i860s	12.8	1.83	61%
4 i860s	11.3	2.1	52%

Table 4
Parallel Mold Filling Results
24882 cells, 0 - 0.8 second

System	Time (hrs)	Speed-up	Efficiency
1 i860	51.4	-	-
2 i860s	28.5	1.8	90%
3 i860s	21.5	2.39	79.8%
4 i860s	17.8	2.88	72%
6 i860s	13	3.95	66%
8 i860s	10.96	4.69	58.6%
10 i860s	10.1	5.1	51%

A finer mesh of 22 x 13 x 87 = 24882 cells with the same time step of 10^{-3} second was also run. The efficiency and speed-up measures for both the 11440 and 24882 cell meshes on 1 to 10 i860 nodes are shown in Tables 3 & 4. The model has also been tested on an INMOS-T800 transputer based parallel system and run up to 5 processors. When running on a single i860 processor with 2000 cells, the i860 is about 10 times faster than a single T800.

The speed-ups obtained for the two models are summarised in the figure 9.

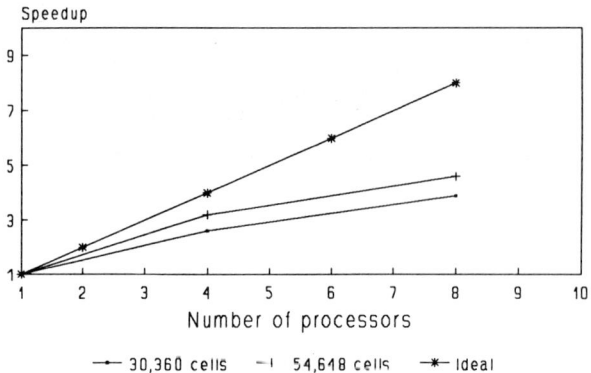

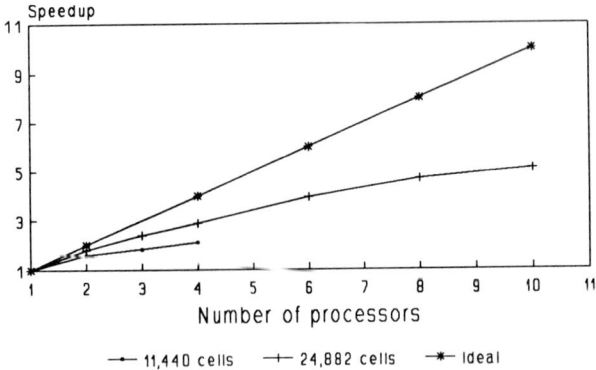

Figure 9 Speed-up curves for the mold filling and fire simulations.

5. DISCUSSION

Both the MOLD filling and FIRE simulation applications achieve similar efficiencies when implemented within the parallel FLOW3D code. As the number of PEs increases while the number of cells remains constant the efficiency decreases.

In general, efficiency increases with number of cells. To achieve maximum efficiency it is essential to solve the largest problem which can be accommodated within the available memory.

In solving fire problems, the 8 processor i860 heterogeneous system is at least 8.7 times faster than the SPARC 2 workstation. A speed-up of 3.2 out of 4 can be achieved on this system when solving a 55,000 cell problem. This reduction in run times is impressive. However, the efficiency is superficially disappointing as speed-ups of almost 4 out of 4 have been achieved on the homogeneous system with similar sized meshes [2]. The differences can be explained by examining latencies inherent in the heterogeneous PE.

Speed-up can be defined as

$$S = \frac{t_{scalar}}{t_{parallel} + t_{comm}} \qquad (1)$$

Where t_{scalar} = time to complete calculations in scalar mode. $t_{parallel}$ = time to complete calculations in parallel mode.

t_{comm} = time to perform data interchange, and

$$t_{comm} = t_{data\ transfer} + t_{latency} \qquad (2)$$

The average data transfer time (based on a 24,882 cell problem) is 150.0E-06 seconds. This is identical for the heterogeneous and homogeneous PEs as in both cases it is the T800 performing the data transfer.

For the T800 homogeneous PE the latency is simply the interrupt time ~ 10.0E-06 seconds. For the T800-i860 heterogeneous PE, the latency is made up of the T800 interrupt time and the time required to flush the i860 cache ~ 310.0E-06 seconds.

This suggests that,

$$t_{comms}\ \text{for T800-i860 PE} = 3 * t_{comms}\ \text{T800 PE} \qquad (3)$$

In CFD, using data decomposition techniques it is possible to achieve,

$$t_{parallel} = t_{scalar} / n, \tag{4}$$

where n = number of processors.

These results suggest,

$$S = \cfrac{1}{\cfrac{1}{n} + \cfrac{t_{comms}}{t_{scalar}}} \tag{5}$$

For the T800, $t_{comms} / t_{scalar} \sim 0.0025$, while for the T800-i860, $t_{comms} / t_{scalar} \sim 0.075$.

$$S_4^T = \cfrac{1}{\cfrac{1}{4} + 0.0025} = 3.96 \; : \quad S_4^i = \cfrac{1}{\cfrac{1}{4} + 0.0075} = 3 \tag{6}$$

This result is roughly inline with the observed performance of the parallel code.

6. CONCLUSIONS

An efficient parallel implementation of a **GENERAL PURPOSE NAVIER-STOKES FLOW CODE** has been achieved. The code has been demonstrated on both **heterogeneous** and **homogeneous** architectures.

High efficiencies have been achieved on two manifestly different problems of industrial relevance demonstrating the generality of the parallel mapping. The maximum benefit of fast PEs can only be realised if processor speed is matched by low system latency. Ultimately, parallel processing systems are bound by system latency, not processing power.

7. ACKNOWLEDGEMENTS

The authors are indebted to AEA Technology Harwell for making available the source code of FLOW3D, and British Aerospace, BNF, SERC for funding.

8. REFERENCES

1 A. D. Burns and N. S. Wilkes, A Finite-Difference method for the computation of fluid flows in complex three dimensional geometries, U.K. Atomic Energy Authority Harwell Report. AERE-R 12342.
2 S. Johnson and M. Cross, Mapping structured grid 3-dimensional CFD codes onto parallel architectures, App Math Mod, 15, 394, 1991.

3 E. R. Galea, On the field modelling approach to the simulation of enclosure fires, J Fire Prot Eng 1, 11-22, 1989.

4 E. R. Galea and N. Markatos, The modelling and computer simulation of fire development in aircraft, Int J Heat Mass Transfer, 34, 181-197 1991.

5 K. S. Chan, K. Pericleous and M. Cross, Numerical Simulation of Flows Encountered During Mould-filling, App Math Mod, 15 624-631, 1991.

6 B. van Leer, Towards the ultimate conservative difference scheme IV: A new approach to numerical convection, J. Computational Physics, vol 23, 276, 1977.

7 C. S. Ierotheou and E. R. Galea, A fire field model implemented in a parallel computing environment, Int J Num Meth Fluids, Vol 14, 175-187 1992.

8 INMOS T800 Architecture. Technical Note 6. Jan 1988 Inmos, Bristol, UK.

9 Parallel FORTRAN User Guide, 3L Ltd, Livingstone Scotland 1990.

10 I. Jones, S. Simcox, N. Wilkes, Modelling of the King's Cross Station fire, CRAY CHANNELS, FALL, 22-23, 1989.

11 E. R. Galea, C. S. Ierotheou, P. Leggett and M. K. Patel, CFDMARK: A Computational Fluid Dynamics based benchmark, App Math Modelling, In Press 1992.

Sparse grid multilevel methods, their parallelization, and their application to CFD

Michael Griebel

Institut für Informatik, Technische Universität München, Arcisstr. 21, D-8000 München 2, Germany

Abstract:

For the solution of PDE's, the recently developed sparse grid discretization technique and certain resulting multilevel methods involve substantially less unknowns than a standard full grid discretization. In this paper, we report on the properties and features of sparse grid multilevel methods. Furthermore, we consider their implementation on parallel computers and discuss their application to the solution of CFD problems.

1. INTRODUCTION

The realistic numerical simulation of fluid flow problems needs an immense amount of main memory and computation power which is presently beyond the possibilities of most existing computers. Especially in the higher dimensional case, the limitation of available main memory necessary to store the respective grids and solutions is a bottleneck. Furthermore, even on a supercomputer with vector facilities, the resulting computing times are extremely long.

In the last years, so-called sparse grid methods have been developed and intensively studied at the Institut für Informatik of the TU München. These discretization techniques might help to overcome main memory storage limitations to some extent. Additionally, they can be applied together with multilevel methods. Thus, the complexity of the resulting algorithms is only proportional to the number of grid points involved in the discretization. The parallelization of these methods is easy, which helps to speed up the computation even further.

So far, sparse grid methods have been shown to work successfully for simple model problems. Now, these methods are sufficiently developed and understood to be applied in the field of computational fluid dynamics. In this paper, we want to give an overview on how sparse grid multilevel methods work, how they can be implemented efficiently on parallel computers, and how they can be used for the solution of CFD problems.

2. HIERARCHICAL BASES AND SPARSE GRIDS

The sparse grid technique can be motivated by the following observation. Suppose that a one-dimensional function $u : [0,1] \to \mathbb{R}$ has to be represented on an equidistant grid Ω_k with mesh size $h = 1/2^k$. Usually, this is achieved by storing the values of the function in grid points only. To gain approximate values between two grid points, some sort of interpolation has to be applied. The simplest case of interpolation corresponds to the standard nodal basis, i.e. $u_k = \sum u(x_i) \cdot \phi_i^{(k)}(x)$, where $\phi_i^{(k)}(x)$ denote the well known hat-type basis functions (see Figure 1, left).

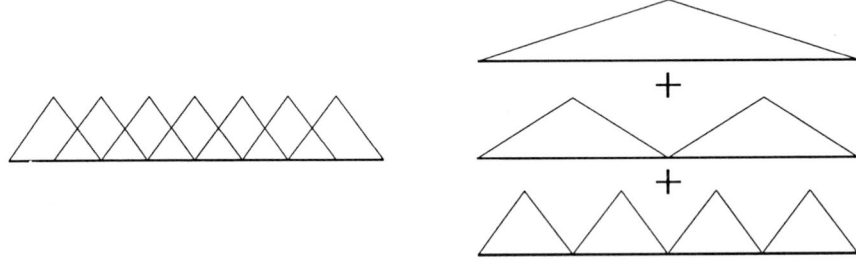

Figure 1: Standard basis functions $\phi_i^{(k)}$ and hierarchical basis functions $\tilde{\phi}_i^{(k)}$ in 1D, $k=3$.

If we use a hierarchical basis (see Figure 1, right) for the representation of our function u, then $u_k = \sum v(x_i) \cdot \tilde{\phi}_i^{(k)}(x)$. Instead of the values $u(x_i)$, now only the hierarchical values $v(x_i)$ have to be stored (see Figure 2), that contain the so-called hierarchical surplus. Note that the hierarchical basis $\{\tilde{\phi}_i^{(k)}\}$ contains just certain functions belonging to the standard bases of the coarser levels of discretization. For details, see [15].

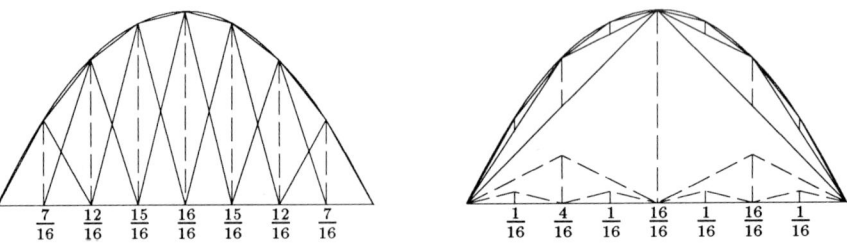

Figure 2: Standard basis coefficients $u(x_i)$ and hierarchical basis coefficients $v(x_i)$, $k = 3$.

For the example of a parabolic curve, we see in Figure 2 (right) that the hierarchical values decrease from level to level by a factor $1/4$, whereas the standard basis coefficients possess all about the same order of magnitude. In this sense, the hierarchical basis results in the reduction of redundant information for the storage of discrete functions.

For the two-dimensional case, we extend this approach in a product-type fashion. Compare Figure 4 (right). Then, we obtain the hierarchical basis as introduced in [16].

Figure 3 shows the supports of the basis functions involved. Note that extremely stretched elements are contained, and large overlaps occur.

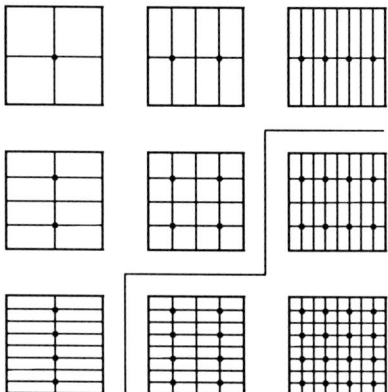

Figure 3: Supports of hierarchical basis functions in 2D, k=3.

If we represent a paraboloid, for example, by means of this hierarchical basis, we obtain the hierarchical coefficients as indicated in Figure 4. Now, the coefficients decrease in the

$\frac{49}{256}$	$\frac{21}{64}$	$\frac{105}{256}$	$\frac{7}{16}$	$\frac{105}{256}$	$\frac{21}{64}$	$\frac{49}{256}$
$\frac{21}{64}$	$\frac{9}{16}$	$\frac{45}{64}$	$\frac{3}{4}$	$\frac{45}{64}$	$\frac{9}{16}$	$\frac{21}{64}$
$\frac{105}{256}$	$\frac{45}{64}$	$\frac{225}{256}$	$\frac{15}{16}$	$\frac{225}{256}$	$\frac{45}{64}$	$\frac{105}{256}$
$\frac{7}{16}$	$\frac{3}{4}$	$\frac{15}{16}$	1	$\frac{15}{16}$	$\frac{3}{4}$	$\frac{7}{16}$
$\frac{105}{256}$	$\frac{45}{64}$	$\frac{225}{256}$	$\frac{15}{16}$	$\frac{225}{256}$	$\frac{45}{64}$	$\frac{105}{256}$
$\frac{21}{64}$	$\frac{9}{16}$	$\frac{45}{64}$	$\frac{3}{4}$	$\frac{45}{64}$	$\frac{9}{16}$	$\frac{21}{64}$
$\frac{49}{256}$	$\frac{21}{64}$	$\frac{105}{256}$	$\frac{7}{16}$	$\frac{105}{256}$	$\frac{21}{64}$	$\frac{49}{256}$

$\frac{1}{256}$	$\frac{1}{64}$	$\frac{1}{256}$	$\frac{1}{16}$	$\frac{1}{256}$	$\frac{1}{64}$	$\frac{1}{256}$
$\frac{1}{64}$	$\frac{1}{16}$	$\frac{1}{64}$	$\frac{1}{4}$	$\frac{1}{64}$	$\frac{1}{16}$	$\frac{1}{64}$
$\frac{1}{256}$	$\frac{1}{64}$	$\frac{1}{256}$	$\frac{1}{16}$	$\frac{1}{256}$	$\frac{1}{64}$	$\frac{1}{256}$
$\frac{1}{16}$	$\frac{1}{4}$	$\frac{1}{16}$	1	$\frac{1}{16}$	$\frac{1}{4}$	$\frac{1}{16}$
$\frac{1}{256}$	$\frac{1}{64}$	$\frac{1}{256}$	$\frac{1}{16}$	$\frac{1}{256}$	$\frac{1}{64}$	$\frac{1}{256}$
$\frac{1}{64}$	$\frac{1}{16}$	$\frac{1}{64}$	$\frac{1}{4}$	$\frac{1}{64}$	$\frac{1}{16}$	$\frac{1}{64}$
$\frac{1}{256}$	$\frac{1}{64}$	$\frac{1}{256}$	$\frac{1}{16}$	$\frac{1}{256}$	$\frac{1}{64}$	$\frac{1}{256}$

Figure 4: Standard basis coefficients and hierarchical coefficients, 2D case, k=3.

x- and y-direction from level to level by a factor 1/4 as in the 1D example. However, in the diagonal directions, we observe a decrease by a factor 1/16. Since, in our example, we only store hierarchical values bigger than some given threshold $\varepsilon \geq 1/4^k$, it makes no sense to take into account the grid points that possess hierarchical values smaller than this threshold, anyway.

Therefore, we omit all the grid points with hierarchical values smaller than ε. Then, we obtain a so-called sparse grid. Now, only the subset of hierarchical basis functions in Figure 3 situated above the diagonal is involved. The corresponding interpolation procedure is described in [7]. Further details can be found in [2] or [16]. The generalization of this approach to the higher dimensional case is straightforward. For examples of sparse grids, see Figure 5.

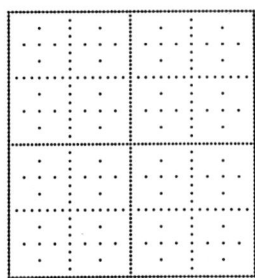 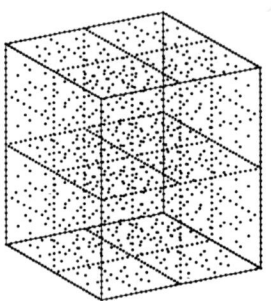

Figure 5: 2D sparse grid $\Omega^S_{6,6}$ and 3D sparse grid $\Omega^S_{5,5,5}$.

We see directly that sparse grids contain substantially less grid points in comparison to standard full grids. However, it turned out that for sufficiently smooth functions nearly the same accuracy of approximation is gained as on full grids. To be more precise, let d denote the dimension and $h = 1/N$, $N = 2^k$, the grid size. For the sparse grid, just $O(N(ld(N))^{d-1})$ grid points instead of $O(N^d)$ grid points for the full grid case have to be stored. The accuracy of the interpolation error and, as shown in [2], also of the approximation error is of the same order $O(h)$ with respect to the energy norm and deteriorates only slightly from $O(h^2)$ to $O(h^2(ld(h^{-1}))^{d-1})$ with respect to the L_2-norm, provided that the solution is sufficiently smooth to fulfill

$$\frac{|\partial^{2d}u|}{|\partial x_1^2...\partial x_d^2|} < C. \tag{1}$$

Now, our aim was to develop a solver for the discrete system that raises from the sparse grid discretization, where the resulting convergence rate is independent of h like for classical multigrid methods and where the number of operations involved is proportional to the number of grid points. We developed two different types of sparse grid solvers, first, a *multilevel method*, and, second, the so-called *combination method*, both with its distinctive advantages and drawbacks.

3. SPARSE MULTILEVEL METHODS

For the realization of multilevel methods for sparse grid problems, the basic ideas of standard multilevel approaches were adopted. First, the discretization of the problem under consideration takes place by means of finite element, finite difference, or finite volume methods using the hierarchical basis as depicted in Figure 3, where just the basis functions belonging to the sparse grid are involved. The arising sparse grid system matrix is more densely populated than a conventional full grid system associated to the standard basis functions. However, this matrix must never be assembled explicitly. Only multigrid transfer operators between the regular full grids with e.g. different mesh sizes in x- and y-direction, that are contained in the sparse grid, and the respective matrices are involved in the process.

First, based on these ideas, a hierarchical basis solver (HBMG, [1]) has been developed in [2]. It shows a convergence rate of $\rho = O(1 - 1/2^{5 \cdot k/4})$. Second, different multigrid methods and multilevel preconditioners have been designed and tested numerically. They show convergence rates independent of h. Additionally, they are robust and well suited for parallelization. For details, see [4], [5] and [9].

The irregular structure of a sparse grid needs a sophisticated data structure to represent the grid points, the values of the solution, and right hand sides. We use a binary tree approach. Each grid line is represented by a binary tree and its two adjacent boundary points. Now, the sparse grid can be implemented as a binary tree of grid lines, i.e. a binary tree of binary trees of grid points. See Figure 6 and [2] for details.

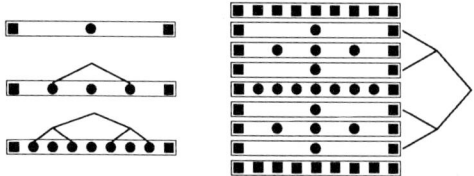

Figure 6: Binary tree data structures for the implementation of a sparse grid.

It is easy to see that adaptive refinement of the grid can be accomplished in a straightforward manner. The binary trees just have to grow locally in the region to be refined. Compare also [2]. However, the programming of general PDE's and PDE-system solvers is quite complicated. The multigrid code must be written completely new, and no existing codes that work on regular full grids can be reused. Furthermore, the parallelization of the code is not very simple, but it can be achieved by using the domain decomposition idea in a nested dissection type fashion, see [10]. These disadvantages are overcome by the following combination method.

4. THE COMBINATION METHOD

For this method, the problem under consideration is discretized and solved on different full grids with different mesh sizes in the x- and y-direction. Then, these solutions are combined linearly to produce a sparse grid solution. In some sense, this approach can be seen as a special case of multivariate extrapolation, where leading error terms are cancelled by the combination of different solutions [7].

In the 2D case, the so-called combination solution is defined as

$$u_{k,k}^C = \sum_{i+j=k+1} u_{i,j} - \sum_{i+j=k} u_{i,j}, \qquad (2)$$

where i ranges from 1 to k. See also [7].

To obtain $u_{k,k}^C$, we have to solve k different problems $L_{i,j} u_{i,j} = f_{i,j}, i = 1, .., k, j = k + 1 - i$, each with about 2^k unknowns, and $k - 1$ different problems $L_{i,j} u_{i,j} = f_{i,j}, i = 1, .., k - 1, j = k - i$, each with about 2^{k-1} unknowns, and combine their bilinearly interpolated solutions. Each problem stems from the discretization of the continuous problem on the standard grid $\Omega_{i,j}$, with grid sizes $h_x = 2^i$ and $h_y = 2^j$ in the x- and y-direction. The method is illustrated for the case $\Omega = (0,1)^2$ in Figure 7.

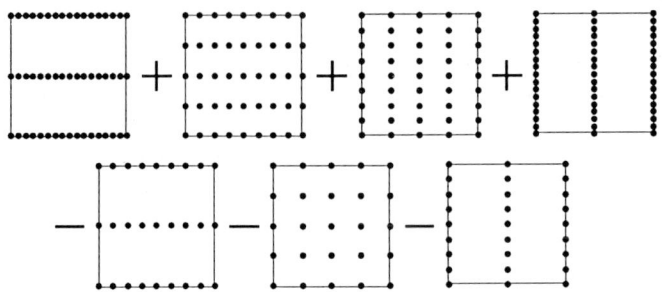

Figure 7: The linear combination of grids $\Omega_{i,j}$ and solutions $u_{i,j}$ with $i = 1,..k, j = k+1-i$, and $i = 1,..,k-1, j = k-i$, where $k = 4$.

The generalization to higher dimensions is straightforward. In the 3D case, we obtain

$$u^C_{k,k,k} = \sum_{i+j+l=k+2} u_{i,j,l} - 2 \sum_{i+j+l=k+1} u_{i,j,l} + \sum_{i+j+l=k} u_{i,j,l}. \qquad (3)$$

In this way, the sparse grid is reduced to different regular standard grids with different mesh spacing with respect to the x-, y-, and z-coordinate directions. The solution of each problem is computed on a standard grid. Therefore, the complicated data structures and algorithms necessary for the sparse grid approach are avoided, and we can use standard codes. For reasons of efficiency, a multigrid method with semi-refinement to cope with the grid distorsion is recommended. But in principle, any solver can be used.

Note that the combination method does not produce the solution of the sparse grid problem. However, the accuracy of the combination solution is of the same order as the sparse grid solution. Thus, the error is within the range of the discretization error of the sparse grid. Additionally, the combination method can be used as a preconditioner in a sparse grid multilevel method, see [6]. In numerical experiments, we obtained good results also for distorted quadrilateral and triangular domains, nonlinear problems, and PDE-systems, that are not covered by the theory presented in [7].

5. PARALLELIZATION AND SIMPLE EXPERIMENTS

Note that the solution of the different problems arising in the combination method can be computed independently. Thus, the parallelization of the method is straightforward. For the combination technique in two dimensions, k different problems each with about 2^k unknowns and $k-1$ problems each with about 2^{k-1} unknowns can be solved in parallel. In the three-dimensional case, $k \cdot (k+1)/2$ problems with about 2^k unknowns, $(k-1) \cdot k/2$ problems with about 2^{k-1} unknowns, and $(k-2) \cdot (k-1)/2$ problems with about 2^{k-2} unknowns can be solved in parallel. After the parallel solution of the different problems, the combination of the solutions can take place more or less sequentially.

In this sense, we have an algorithm with good parallelization properties on a coarse grain level and relatively moderate communication requirements. For the implementation, we used a farm-like master and slave approach. This is illustrated in Figure 8 by a p-ary tree with depth one. The master distributes the subproblems to be solved to its slaves and collects the results when they all have terminated. The combination of the solutions takes place in the master process.

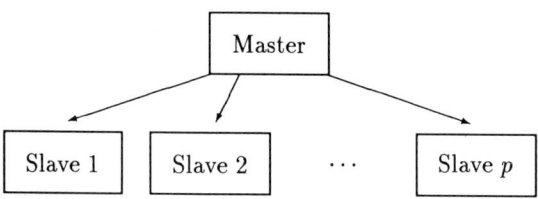

Figure 8: Master-slave concept.

Regarding the higher dimensional case, the parallelization tree remains at depth one, and only its breadth grows as the number of subproblems to be solved increases. The subproblem size remains basically the same. This simple structure is suitable, because usually the communication is very small compared to the work assigned to each processor. In extreme cases with many parallel tasks, we can avoid the bottlenecks in a $1 \leftrightarrow p$ communication by implementing the logical structure of Figure 8 with additional *submasters* each taking responsibility for a certain number of slaves. This approach can be applied recursively. Then, the process of combining the solutions is also parallelized.

The different sizes of the different problem classes suggest the following simple load balancing strategy. Assume that P processors are available. First, the k problems of size 2^k are distributed among the processors. We denote the work to solve one of these problems by W. In the case of a multigrid solver, we may assume $W = O(2^k)$. Now, the load of two processors differs at most by W. For two dimensions we are left with $k-1$ problems of size $O(2^{k-1})$ each corresponding to a load of $W/2$. As shown in Figure 9, two of these problems are used to fill the processors having less load than the others. If, after this step, not all tasks have been assigned to processors, the remaining problems are again evenly distributed among the processors. For the three-dimensional case this algorithm is continued in the canonical way with the problems of size $O(2^{k-2})$.

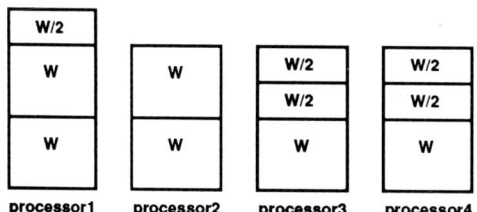

Figure 9: A simple load balancing strategy, 2D case, $P=4$.

Following this simple strategy, we achieve that the load of two different processors differs at most by $W/2$ for the 2D case and $W/4$ for the 3D case, provided that the number of problems is sufficiently large with respect to the number of processors. In the three-dimensional case, the largest efficiency with the highest speed up for a given k is obtained by using

$$P = \text{int}\left[\frac{k \cdot (k+1)}{2} + \frac{(k-1) \cdot k}{4} + \frac{(k-2) \cdot (k-1)}{8}\right] \qquad (4)$$

processors.

Now, we give the results of simple numerical experiments. We consider the three-dimensional model problem $\Delta u = 0$ in $\Omega = (0,1)^3$ with the exact solution $u = \sin(\pi x) \cdot \sin(\pi y) \cdot \sinh(\sqrt{2}\pi z)/\sinh(\sqrt{2}\pi)$ and Dirichlet boundary conditions on $\partial\bar{\Omega}$. For the solution of each subproblem arising in the combination method, we use 10 V-cycles of a multigrid method with one step of Gauss-Seidel pre-smoothing and one step of Gauss-Seidel post-smoothing. Note that our code uses full 27-point stencils that are derived from the assembly of tri-linear quadrilateral finite elements. Furthermore, each stencil is assembled explicitly at each grid point, simulating the solution of a more general, second order elliptic operator with variable coefficients.

The following results have been obtained on the Transputer SuperCluster from Parsytec with up to 256 T800 processors running under Helios. Each processor has 4 MByte memory and a peak performance of 2.2 MFlop per second. In Table 1, we show the measured times. Here, the entries of each column have to to multiplied by the associated factor in the second row. The row $P = 1$ shows the times for the sequential version. The MFlop rates are given in Table 2. We see that our code runs with 0.25 MFlop per second for one processor, which is 12% of the theoretical performance only.

Table 1: *Times (in sec.) for combination algorithm on $\Omega^S_{k,k,k}$ on the SuperCluster with P processors.*

$P\backslash k$	4	5	6	7	8	9	10	11	12	13	14
×	10^0	10^0	10^1	10^1	10^1	10^1	10^2	10^2	10^2	10^3	10^3
1	11.09	42.86	14.63	45.73	134.38	372.58	99.72	257.77	648.95	159.92	367.84
2	5.64	21.89	7.38	22.89	67.48	186.70	50.23	129.79	325.60	80.35	194.85
4	3.03	11.67	3.98	12.71	38.25	107.41	28.45	74.24	186.27	42.73	110.94
8	1.76	6.06	2.07	6.83	20.84	55.39	14.74	37.79	94.86	23.20	56.71
16	0.95	3.57	1.26	3.64	10.54	29.07	8.08	19.21	49.02	12.39	30.24
32	-	2.69	0.68	2.07	5.29	16.10	4.10	10.59	25.70	6.17	15.44
64	-	-	-	-	3.30	8.99	2.40	5.62	13.79	3.39	8.28
128	-	-	-	-	-	-	-	3.46	7.29	1.89	4.75
256	-	-	-	-	-	-	-	-	-	-	3.21

Of course, the obtained times and MFlop-rates are not very impressing. Partly, this is due to the not yet optimized version of our code and the high level approach using Helios. However, the main reason is shurely the weak computing power of the T800 processor. Much better results can be expected from the T9000 processor, which is however not yet available. Nevertheless, we see in Figure 10 that our code shows good speed up rates

Table 2: *MFlop per second for combination algorithm on* $\Omega^S_{k,k,k}$ *on the SuperCluster with P processors.*

$P\backslash k$	4	5	6	7	8	9	10	11	12	13	14
1	0.25	0.25	0.25	0.26	0.26	0.26	0.26	0.26	0.26	0.26	0.26
2	0.48	0.49	0.50	0.51	0.51	0.52	0.51	0.51	0.51	0.51	0.51
4	0.90	0.92	0.94	0.92	0.90	0.90	0.91	0.90	0.90	0.97	0.90
8	1.55	1.78	1.80	1.72	1.66	1.73	1.75	1.76	1.77	1.78	1.76
16	2.87	3.01	2.95	3.22	3.28	3.31	3.19	3.47	3.42	3.33	3.30
32	-	4.00	5.50	5.66	6.53	5.97	6.28	6.29	6.52	6.69	6.46
64	-	-	-	-	10.49	10.69	10.74	11.85	12.15	12.18	12.06
128	-	-	-	-	-	-	-	19.28	23.00	21.83	21.03
256	-	-	-	-	-	-	-	-	-	-	30.41

and an efficiency of more than 60%, which is quite satisfactory. In [11], we report in detail on the performance of our code for other MIMD computers like the iPSC/860 or the nCUBE2, for the vector computer CRAY Y-MP, and, most interesting, for a network of 110 HP720 workstations. There, it is shown that basically the same speed up and efficiency, but much higher MFlop rates are obtained for processors faster than the T800.

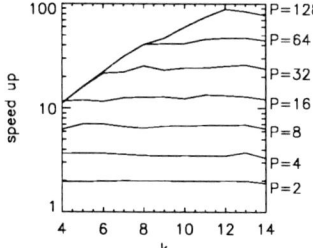
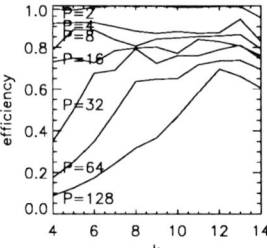

Figure 10: Speed up and efficiency on the SuperCluster SC-256 for different P.

6. APPLICATIONS OF SPARSE GRID METHODS TO CFD

Sparse grid methods require the solution of the problem under consideration to satisfy the smoothness requirement (1). In CFD-applications, however, this is usually not the case. Here, structural singularities due to reentrant corners, shocks, boundary layers, interior layers, or locally strong vortices occur that lead to a non-smooth solution behavior. Therefore, the straightforward application of sparse grid methods is severely handicapped.

This is demonstrated by the following example. We consider the Stokes equations for the lid driven cavity model problem. It is well known that at the corners where the wall turns to the driven lid, severe singularities appear in the vorticity function. Figure 11 shows the contour lines of the stream function computed on a 65×65 full grid and the corresponding stream function computed on the associated sparse grid. We clearly see that the accuracy of the sparse grid solution deteriorates in the presence of singularities, and no meaningful solution is obtained.

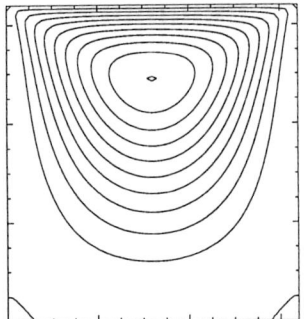 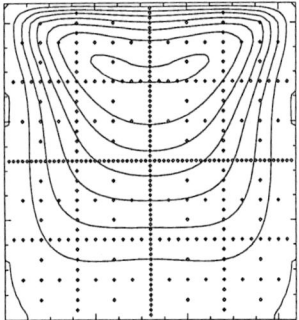

Figure 11: Stream lines of the lid driven cavity problem for the Stokes equations, full and sparse grid discretizations, k=6.

In the following, we point out, how this problem can be avoided for our two different approaches of section 3 and 4.

For the *sparse grid multilevel method* that is based on a binary tree type implementation as mentioned in section 2, adaptive refinement is fairly easy to accomplish. This helps to cope with the problem of singularities destroying the accuracy obtained with the sparse grid method. As shown in [17], an adaptively refined grid can be constructed, that resolves the singularities and still maintains the advantages of a sparse grid especially with respect to the number of grid points involved. This is demonstrated in Figure 12. The contour lines of the stream function obtained on the adaptively refined grid is indicated by bold lines, the solution obtained on the full grid is indicated by dashed lines. Basically, we see not very much of the dashed lines. This shows that the difference between the full grid solution and the adaptively refined sparse grid solution is very small. It's L_2-norm is smaller than 1.5×10^{-4}, which is below the discretization error anyway.

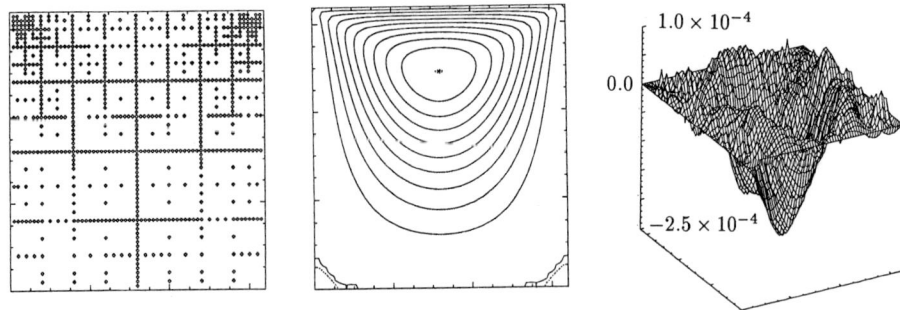

Figure 12: Adaptive sparse grid, sparse grid solution and difference to full grid solution.

For the *combination method*, the use of a graded mesh helps to cope with singularities, provided that the location of the singularity is known in advance. Figure 13 shows a graded full mesh that resolves the corner singularities of the lid driven cavity model problem sufficiently. Additionally, the associated graded sparse grid is shown.

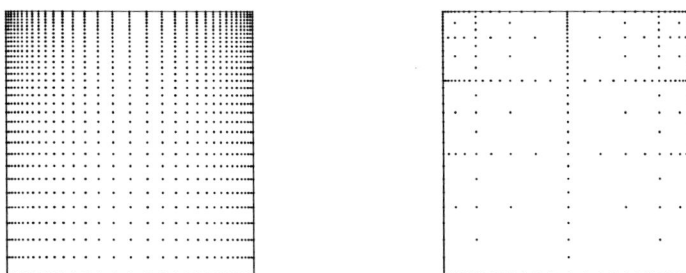

Figure 13: Graded full grid and graded sparse grid.

The combination method can be extended directly to the case of a graded mesh. Compare once more Figure 10. As demonstrated in [3], the solution obtained with the combination method for the graded grid is basically as accurate as the solution obtained on the graded full grid. For the solution of the different graded mesh problems arising in the combination method, the code COMET was used. This code is under development at the Universität Erlangen [13]. Figure 14 shows the velocity vectors of a (time-dependent) Navier-Stokes lid driven cavity problem with $Re=100$ for the graded full and sparse grid case. For details, see [3]. Regarding the difference between the two solutions, we see that it is larger than the full grid discretization error only in a very small area near the corners, where the full grid solution is polluted by the corner singularities anyway.

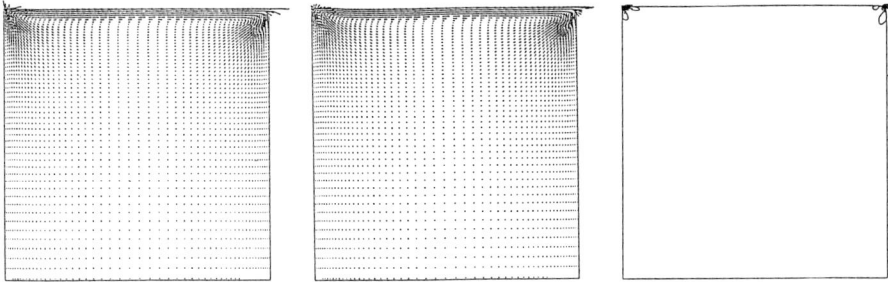

Figure 14: Velocity vectors obtained on a graded full and sparse grid and difference of the velocity in x-direction.

Another problem that appears in practical CFD-applications is the case of curved boundaries. Since the sparse grid technique is based on a product type approach that involves basically quadrilateral domains, the straightforward application of sparse grid methods to curved boundaries is not possible.

In the following, we point out how this problem can be avoided for our two different approaches of section 3 and 4..

For the *multilevel method* based on a binary tree type implementation as mentioned in section 2, adaptive refinement in an enclosing box is the remedy. As shown in [14], an adaptively refined grid can be constructed that resolves the curved boundary of the domain and still maintains the advantages of a sparse grid. See Figure 15 for two examples. Furthermore, multilevel algorithms have been developed that solve (so far simple) 2D

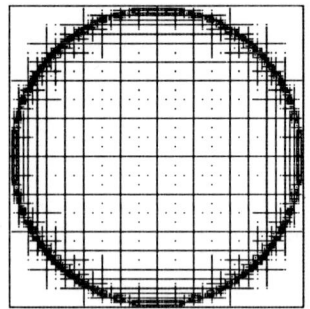 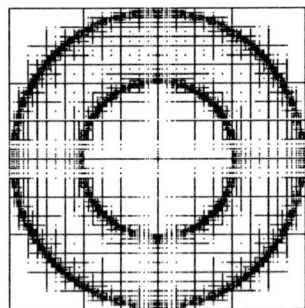

Figure 15: Resolution of curved boundary by an adaptive sparse grid, circle and torus domains.

PDE's on domains with curved boundaries with $O(N)$ operations, where N denotes the number of grid points involved in the discretization. Note that near the boundary special discretization techniques (similar to the one of Shortley-Weller) have to be applied. For further details, see [14]. Of course, on such grids, further adaptive refinement is possible to resolve singularities, boundary layers, or shocks.

For the *combination method*, isoparametric mappings can help to cope with curved boundaries. Additionally, the combination method can be extended easily to block structured grids, that are quite common in many CFD-codes. See also Figure 16 for a simple example. To show that this approach works well also for more general CDF-problems, we

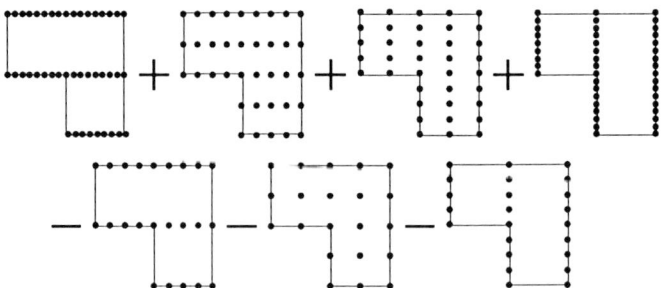

Figure 16: The linear combination of grids $\Omega_{i,j}$ and solutions $u_{i,j}$ with $i = 1,..k, j = k + 1 - i$, and $i = 1,.., k - 1, j = k - i$, for the L-shaped domain, where $k = 4$.

studied the 2D Navier-Stokes equations with complicated boundaries. For the solution of the different problems arising in the combination method, we used the L_iSS-package [12]. It involves flux-difference splitting in the discretization, block structured grids, and a multigrid method as solver. Also, a parallelized version of this package exists.

We considered the example of laminar flow over the skyline of München (2D idealization) with Reynolds number $Re = 500$. The discretization involves 33 quadrilateral subdomains with curved boundaries and graded meshes to resolve the effects caused by singularities due to reentrant corners. For each block $k = 6$ is chosen.

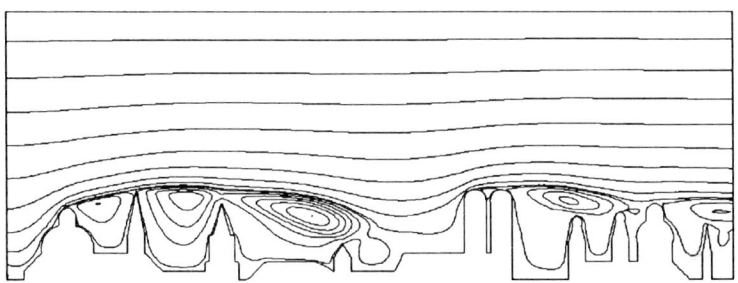

Figure 17: Laminar flow over the skyline of München, $Re = 500$.

Figure 17 shows the contour lines of the stream function computed for the sparse grid discretization by the combination method. Accuracy experiments with successive finer grids showed that the same order of accuracy was obtained with this method as for the solution obtained for the full grid approach. Further experiments and results can be found in [8].

Acknowledgement: The author is indepted to S. Zimmer for his help with the design of the figures in this paper.

7. REFERENCES

[1] R. Bank, T. Dupont, and H. Yserentant. The hierarchical basis multigrid method. *Num. Math.*, 52:427–458, 1988.

[2] H. Bungartz. *Dünne Gitter und deren Anwendung bei der adaptiven Lösung der dreidimensionalen Poisson-Gleichung*. Dissertation, TU-München, 1992.

[3] B. Durst. *Numerische Simulation einer periodischen Nischenströmung unter Verwendung dünner Gitter*. Diplomarbeit, TU München, 1991.

[4] M. Griebel. *Parallel multigrid methods on sparse grids*. Int. Series of Numer. Mathematics, Vol. 98, Birkhäuser Verlag, Basel, 1991.

[5] M. Griebel. *A parallelizable and vectorizable multi-level algorithm on sparse grids*. Parallel Algorithms for Partial Differential Equations, Proceedings of the Sixth GAMM-Seminar, Kiel, January 19-21, 1990, W. Hackbusch, ed., Vieweg-Verlag, 1991.

[6] M. Griebel. *A domain decomposition method using sparse grids*. Proceedings of the 6. International Conference on Domain Decomposition Methods in Science and Engineering, AMS, Contemporary Mathematics, 1992.

[7] M. Griebel, M. Schneider, and C. Zenger. *A combination technique for the solution of sparse grid problems.* Proc. of IMACS International Symposium on Iterative Methods in Linear Algebra, Brussels, April 2.-4. 1991, Elsevier/North Holland, Amsterdam, 1991.

[8] M. Griebel and V. Thurner. *The combination technique as a post-processor of the L_iSS-package.* TU München, Institut f. Informatik, SFB-Report, 1992.

[9] M. Griebel, C. Zenger, and S. Zimmer. *Improved multilevel algorithms for full and sparse grid problems.* TU München, Institut f. Informatik, SFB-Report 342/15/92 A, 1992.

[10] W. Hahn. *Parallelisierung eines adaptiven hierarchischen Dünn-Gitter-Verfahrens.* Diplomarbeit, TU München, 1990.

[11] W. Huber, M. Griebel, U. Rüde, and T. Störtkuhl. *The combination technique for parallel sparse-grid-preconditioning or -solution of PDE's on workstation networks.* Springer LNCS, Proc. CONPAR 92, VAPP V, Lyon, 1992.

[12] G. Lonsdale and K. Stüben. *The L_iSS-Package.* Arbeitspapiere der GMD 524, GMD, St. Augustin, 1991.

[13] M. Peric, M. Schäfer, and E. Schreck. *Numerical simulation of complex fluid flows on MIMD computers.* This volume.

[14] C. Pflaum. *Anwendung von Mehrgitterverfahren auf dünnen Gittern.* Diplomarbeit, TU München, 1991.

[15] H. Yserentant. Hierarchical bases and related preconditioners for elliptic partial differential equations. *Siam ICIAM91*, R.E.O'Malley, ed., 1992.

[16] C. Zenger. *Sparse Grids.* Parallel Algorithms for Partial Differential Equations, Proceedings of the Sixth GAMM-Seminar, Kiel, January 19-21, 1990, W. Hackbusch, ed., Vieweg-Verlag, 1991.

[17] S. Zimmer. *Lösung der Stokes-Gleichungen durch ein adaptives Verfahren mit hierarchischen Basisfunktionen.* Diplomarbeit, TU München, 1991.

Unsteady fluid flow calculations using a machine independent parallel programming environment [1]

A. Gursoy[a], L.V. Kale[a], and S.P. Vanka[b]

[a]Department of Computer Science, University of Illinois at Urbana-Champaign, Urbana IL 61801, USA

[b]Department of Mechanical and Industrial Engineering, University of Illinois at Urbana-Champaign, Urbana IL 61801, USA

Abstract

A portable parallel implementation of a fractional step method to solve 3D unsteady incompressible Navier-Stokes equations on MIMD machines is described. The context currently considered involves uniform grid spacing in one of the directions. A parallel decomposition strategy and its implementation on a portable parallel programming system (Charm) is described. Some optimizations including one that reduces the cost of parallel convergence tests are described. Performance results for calculation of flow in a driven cavity on different parallel computers are presented.

1. INTRODUCTION

The performance of MIMD multiprocessor systems has increased significantly in recent years. Many hope that massively parallel systems will be used routinely to solve difficult, computationally intensive scientific applications. However, programming and portability issues still pose a problem which could limit the expected usage of these machines.

Computational Fluid Dynamics (CFD) is one of the most computationally demanding application areas. Until recently, only the supercomputers such as a CRAY have been providing the required computing power. Powerful vectorizing compilers facilitated the task of programming on these machines. More recently, these supercomputers evolved into multiple vector processors systems delivering four to eight times more performance than their uniprocessor versions. With the availability of low-cost, high-performance nonshared memory parallel machines including intel iPSC and NCUBE hypercubes, there has been growing interest in the scientific community to use these machines [1]. Programming on these machines is obviously more difficult than sequential programming. It is necessary to simplify and support the task of writing parallel applications, and also to ensure that the investment in parallel software is protected through architectural advances and new generation of parallel machines.

---

[1]This work was supported in part by the NSF grant CCR-90-07195 and NASA-NAG 3-1208

In this paper, we will discuss a machine independent parallel implementation of 3D unsteady incompressible flows on MIMD machines. Machine independence is achieved by using Charm parallel programming environment [2]. Charm is a runtime support system which allows machine independent parallel programming across MIMD machines. It provides an explicit parallel language which uses C as its base language. It hides the details of underlying machine architecture, communication and process management from the user. It has already been implemented on various shared and nonshared memory machines including Sequent Symmetry, Alliant FX/8, intel's iPSC/2 and iPSC/860, NCUBE/2, and is currently being implemented on a network of Sun workstations.

In section 2 we will discuss briefly the relevant Charm language features. The CFD problem and its parallel implementation will be described in sections 3 and 4, followed by the performance results.

2. CHARM LANGUAGE

A Charm program consists of chare definitions, message definitions, and declarations of specifically shared objects in addition to regular C language constructs (except global variables). A chare is a medium grained process which can dynamically create other chares, send messages to other chares, and share information through specifically shared objects. The Charm system takes the responsibility of scheduling chares, dynamic load balancing, resource management, and efficient machine specific implementation of specifically shared objects on different architectures.

A chare definition consists of local variable declarations, entry-point definitions and private function definitions as illustrated in Figure 1. Local variables of a chare are shared among the chare's entry-points and private functions. Private functions are not visible to other chares, and can be called only inside the owner chare. However, C functions that are declared outside of chares are visible to any chare. Entry-point definitions start with an entry name, a message name, followed by a block of C statements and Charm system calls. Some of the important Charm system calls are:

CreateChare(chareName, entryPoint, message)
> This call is used to create an instance of a chare named as *chareName*. As all other Charm system calls, CreateChare is a non-blocking call, that is, it immediately returns. Eventually as the system creates an instance of chare *chareName*, it starts to execute the *entryPoint* with the message *message*.

SendMsg(chareID, entryPoint, message)
> This call deposits *message* to be sent to the *entryPoint* of chare instance *chareID*. *chareID* represents an instance of a chare. It is obtained by a system call *MyChareID()*, and it may be passed to other chares in messages.

The runtime system is message driven. It repeatedly selects one of the available messages from a pool of messages in accordance with a user selected queueing strategy, restores the context of the chare to which it is directed, and initiates the execution of the code at the entry point.

A BranchOffice chare (BOC) is a form of chare that is replicated on all processors. An instance of BOC has a branch on every processor. All the branches have the same ID. A BOC definition is similar to a chare definition except it contains public functions which can be called by other chares. BOC's are useful for some computations such as reduction operations (i.e., collecting some information locally on each processor, and then combining it across processors), as well as for expressing static load balancing.

```
chare chare-name {
    local variable declarations
    entry EP1 : (message MSGTYPE *msgptr) {C code block}
    ..
    entry EPn : (message MSGTYPE *msgptr) {C code-block}
    private function-1() {C code block}
    ..
    private function-m() {C code block }
}
```

Figure 1: Chare Definition

In addition to messages and BOC's, Charm provides some other ways of information sharing:

readonly A readonly variable is initialized at the beginning of a Charm program, and its value can be accessed by *ReadValue* call from any part of the program.

write-once A write-once variable is created and initialized at any point of the execution (only once). The system provides a global ID for the write-once variable, and this ID is used to access its value on any processor.

dynamic table A dynamic table is a set of entries with key and data fields. A number of asynchronous access and update calls are allowed on table entries.

The Charm system provides other information sharing mechanisms such as monotonic and accumulator variables. Details about these features can be found in [6]. It also provides a sophisticated module system that facilitates reuse, and large-scale programming for parallel software.

3. PROBLEM

3.1. Governing Equations and Numerical Method

We consider three dimensional, unsteady incompressible flows governed by the following equations:

$$\frac{\partial u}{\partial t} + \frac{\partial}{\partial x}(uu) + \frac{\partial}{\partial y}(uv) + \frac{\partial}{\partial z}(uw) = -\frac{\partial p}{\partial x} + \frac{1}{Re}\nabla^2 u \tag{1}$$

$$\frac{\partial v}{\partial t} + \frac{\partial}{\partial x}(uv) + \frac{\partial}{\partial y}(vv) + \frac{\partial}{\partial z}(vw) = -\frac{\partial p}{\partial y} + \frac{1}{Re}\nabla^2 v \tag{2}$$

$$\frac{\partial w}{\partial t} + \frac{\partial}{\partial x}(uw) + \frac{\partial}{\partial y}(vw) + \frac{\partial}{\partial z}(ww) = -\frac{\partial p}{\partial z} + \frac{1}{Re}\nabla^2 w \tag{3}$$

$$\frac{\partial u}{\partial x} + \frac{\partial v}{\partial y} + \frac{\partial w}{\partial z} = 0 \tag{4}$$

where u, v, and w are non-dimensional velocities, p is non-dimensional pressure, and Re is the Reynolds number.

These equations are discretized on a staggered grid using the central difference scheme of Harlow-Welch [3]. A fractional step method [4] is used for the time integration. The resulting discretized equations are:

$$\frac{\hat{u} - u^n}{\Delta t} = \frac{3}{2}H_u^n - \frac{1}{2}H_u^{n-1} \tag{5}$$

$$\frac{\hat{v} - v^n}{\Delta t} = \frac{3}{2}H_v^n - \frac{1}{2}H_v^{n-1} \tag{6}$$

$$\frac{\hat{w} - w^n}{\Delta t} = \frac{3}{2}H_w^n - \frac{1}{2}H_w^{n-1} \tag{7}$$

$$\nabla^2 p = -\frac{1}{\Delta t}\left(\frac{\partial \hat{u}}{\partial x} + \frac{\partial \hat{v}}{\partial y} + \frac{\partial \hat{w}}{\partial z}\right) \tag{8}$$

$$u^{n+1} = \hat{u} - \Delta t \frac{\partial p}{\partial x} \tag{9}$$

$$v^{n+1} = \hat{v} - \Delta t \frac{\partial p}{\partial y} \tag{10}$$

$$w^{n+1} = \hat{w} - \Delta t \frac{\partial p}{\partial z} \tag{11}$$

where

$$H_u^n = -\left(\frac{\partial}{\partial x}(uu) + \frac{\partial}{\partial y}(uv) + \frac{\partial}{\partial z}(uw)\right) + \frac{1}{Re}\nabla^2 u \tag{12}$$

$$H_v^n = -\left(\frac{\partial}{\partial x}(uv) + \frac{\partial}{\partial y}(vv) + \frac{\partial}{\partial z}(vw)\right) + \frac{1}{Re}\nabla^2 v \tag{13}$$

$$H_w^n = -\left(\frac{\partial}{\partial x}(uw) + \frac{\partial}{\partial y}(vw) + \frac{\partial}{\partial z}(ww)\right) + \frac{1}{Re}\nabla^2 w \tag{14}$$

There are two distinct patterns of computation in these equations: the first one deals with the evaluation of the intermediate velocity fields and the second one is the calculation of the pressure field. At time-step n, intermediate velocities $\hat{u}$, $\hat{v}$, and $\hat{w}$ are calculated by using velocity values from previous time-steps. Next, a Poisson equation is solved for the pressure field, p. For the current problem, we assumed that the grid is uniform along the z direction. Therefore, the Poisson equation is solved by applying FFT in that direction. The resulting penta-diagonal equations are solved with Stone's method [5] (a strongly implicit iterative method). Finally, the intermediate velocity fields are corrected using the pressure values. Algorithm-1 shows the basic flow of this computation.

Algorithm-1
Basic Flow of Sequential Algorithm

1. Initialization

2. Time-stepping loop

   (a) compute intermediate velocities

   (b) calculate the right-hand side of the pressure equation

   (c) FFT along the z direction

   (d) solve for pressure in all xy-planes with Stone's method

   (e) inverse FFT along z

   (f) correct velocities

4. PARALLEL IMPLEMENTATION

In this section, we will discuss parallelization of the above algorithm. A natural way to implement parallel solution of partial differential equations is to divide the computational domain into several subdomains, and distribute them to the processors. In most finite difference applications, a static and rectangular domain partitioning is implied by the inter-grid dependencies (e.g., the dependencies in a five-point stencil etc). A decomposition scheme must address some computational issues such as load balancing and minimization of communication cost. The optimum decomposition scheme depends on machine and problem characteristics such as the number of processors, inter-grid dependencies, numerical solution techniques, etc.

We chose to partition the computational domain into $n \times m$ rectangular boxes which extend along the z direction, where $n \times m$ is equal to the number of processors. In Figure 2-a, this 2D partitioning scheme is illustrated for the four processors case. The grid points along the z direction belong to one processor which makes this partitioning scheme more suitable than other partitioning schemes as explained in following sections.

Parallel implementation of Algorithm-1 requires communication at some points. The parallel version, Algorithm-2, shows the points where communication is necessary. The sequential one contains two major phases as described previously: momentum equations and the pressure equation. Boundary values for intermediate and modified velocity variables are exchanged after their calculation (Algorithm-2, steps b and h). Since the grid points along the z direction are local to a processor, no communication takes place in that direction.

The solution of the pressure equation, steps d through f in Algorithm-2, involves FFT and solution of penta-diagonal linear systems. One of the advantages of 2D partitioning over 3D appears in the FFT phase. Each FFT along the z direction is a local operation. Each processor applies FFT to its own data which results in full parallelism over the computation domain. FFT removes the inter-grid dependencies along the z direction which results in n_z independent xy-planes to be solved (each of which is a penta-diagonal linear system), where n_z is the number of grid points in the z direction, as illustrated in Figure 2-b. Due to the 2D partitioning, an xy-plane is spread across processors. These xy-

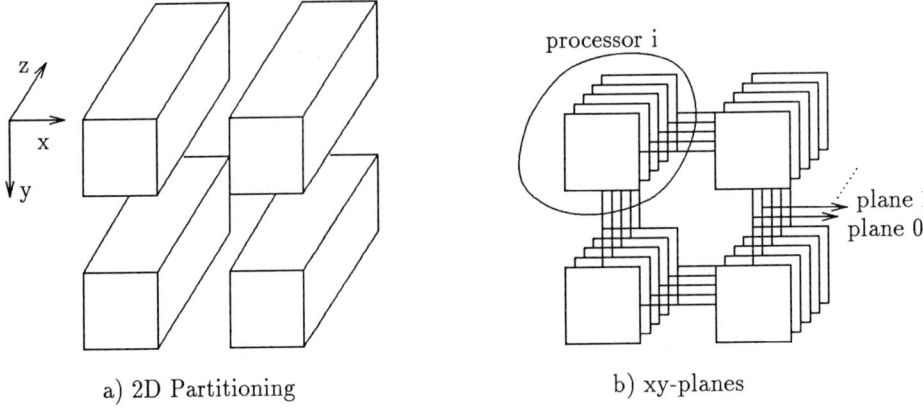

Figure 2: Decomposition of the computational domain

Algorithm 2
Basic Flow of Parallel Algorithm
---

1. Initialization

2. Time-stepping loop

    (a) compute intermediate velocities

    (b) exchange intermediate velocity boundary values

    (c) calculate the right-hand side of pressure equation

    (d) FFT along z direction

    (e) Pressure loop

        i. solve local subdomain (all xy-planes)
        ii. exchange boundary values for pressure
        iii. test global convergence
        iv. if all processors converged then exit Pressure loop

    (f) inverse FFT along z

    (g) correct velocities

    (h) exchange boundary values of corrected velocities

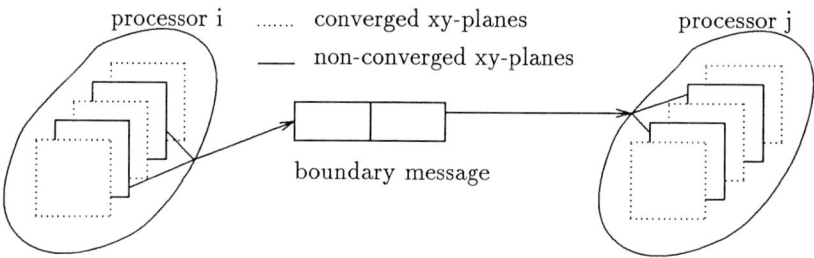

Figure 3: Dynamic messages

planes are solved iteratively with Jacobi-like relaxation across processors. The local part of an xy-plane is solved with Stone's method. Each processor first solves for local points and exchanges information at boundaries. The global convergence is then tested. This process continues until the desired global convergence is reached. The xy-plane solving phase involves two kinds of communication patterns: nearest neighbor communication for pressure values, and a spanning tree communication for global convergence check. For a particular plane, if at least one of the processors reports non-convergence, then all the processors iterate one more level for that plane. Each processor calculates its local convergence. When all of its children processors report their convergence information, the combined information is propagated to the parent processor. Finally the root processor broadcasts the result to other processors.

There are n_z planes to be solved. Iterations last until the last plane converges. In order to reduce both computation and communication cost, the converged planes are dynamically eliminated from the computation. In the first iteration, all planes are in the computation list. Therefore boundary values for all xy-planes are exchanged. After every iteration, xy-planes that are globally converged are not solved locally, and boundary messages contain only boundary values from non-converged planes as shown in Figure 3. The load is also automatically balanced among processors with a 2D partitioning scheme, because each processor owns an equal sized piece of every xy-plane. After all planes have converged, inverse FFT is applied locally to get the pressure values. Then the velocities are corrected, and new values of velocities are exchanged to be used in next time step.

4.1. Initial Results

In this part, initial performance results for the driven cavity problem will be discussed. The problem domain involves a unit cube, and the flow is driven by moving the top wall. The performance results are gathered from runs with Reynolds number set to 100 for 50 time steps. Table 1 shows the performance results on a shared memory machine, Sequent Symmetry, for fixed domain size. Shared memory results are satisfactory. The speedup is below the linear speedup because as the number of processors increase, number of iterations to solve the pressure equation increases due to the relative weakness of Jacobi-like relaxation. The Mflop rate is almost doubled as the number of processors doubled.

Table 2 and Table 3 show the performance on two nonshared memory machines: iPSC/860, and NCUBE/2. The subdomain size is kept fixed in nonshared memory runs because the number of processors is varying over a wide range. As the physical size of the cavity is still fixed as a unit cube, this leads to a finer grid. The execution time increases

Table 1
Performance of non-adaptive scheme on shared memory machines

#PE	Domain	Sequence Symmetry Subdomain	time(sec)	Mflops
1	32x32x33	32x32x33	732	0.11
4	32x32x33	16x16x33	205	0.42
8	32x32x33	8x16x33	108	0.80
16	32x32x33	8x8x33	57	1.53

Table 2
Performance of non-adaptive scheme on nonshared memory machines

#PE	Domain	Subdomain	dt	time(msec)	Mflops
		NCUBE/2			
4	16x16x33	8x8x33	0.006	36058	3.05
16	32x32x33	8x8x33	0.005	47487	9.46
64	64x64x33	8x8x33	0.002	58106	31.21
256	128x128x33	8x8x33	0.0012	73342	96.98
		iPSC/860			
4	16x16x33	8x8x33	0.006	10611	10.35
16	32x32x33	8x8x33	0.005	16709	26.89
64	64x64x33	8x8x33	0.003	21481	84.44

Table 3
Performance of sequential algorithm on nonshared memory machines

Machine	Domain	dt	time(msec)	Mflops
NCUBE	16x16x33	0.005	105380	0.96
i860	16x16x33	0.005	25498	3.96

as the number of processors increases despite the fact that the subdomain size is fixed. This is due to the slower rate of convergence of the Poisson solver. However, the Mflop rate is perhaps a better indicator for the effect of parallelization strategy. The Mflop rate is tripled as the number of processors is quadrupled as shown in the tables.

4.2. Improved algorithm

Some further improvements in the performance can be obtained by restructuring some communication strategies. Remember that there are two patterns of communication in the algorithm: nearest neighbor, and the tree communication for checking convergence. In hypercubes the cost of nearest neighbor communication is $O(1)$, whereas the cost spanning tree reduction is $O(\log n)$ where n is the number of processors. Therefore treewise reduction might be causing the performance degradation significantly. We inspected execution times of different phases of the algorithm, and we observed that there is considerable idle time in the reduction phase. Note that, the convergence test is performed after each iter-

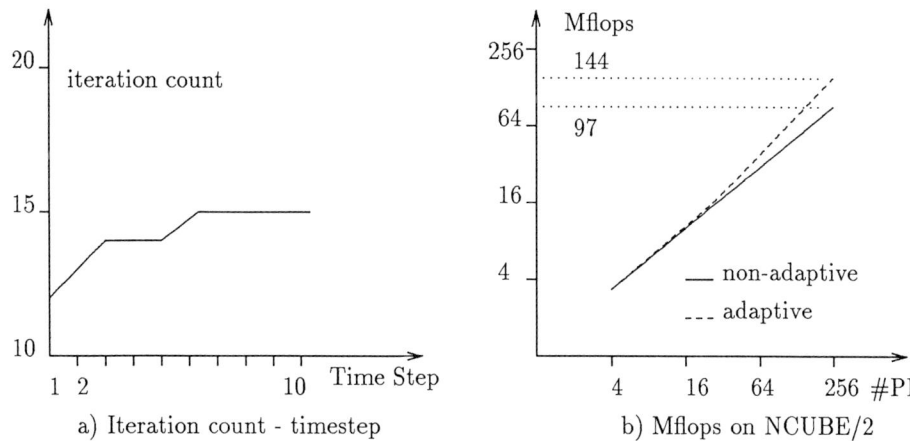

Figure 4: Iteration count, and Mflops

ation. To reduce cost of this phase, the number of convergence tests should be reduced. A close inspection of the iteration counts versus timesteps, Figure 4-a, shows that the number of iterations for a xy-plane changes only gradually from one step to the next one. Therefore an adaptive convergence test scheme is applied for consecutive time steps, as described below.

Let n_i denote the number of iterations for xy-plane i recorded from the previous time step. Based on the correlation in Figure 4-a, we can predict the number of iterations in this time step to be close to n_i. To make sure that the algorithm adapts downwards (as well as upwards), that is even when the number of iterations required decreases in succeeding time steps, we choose $\max(n_i - 1, 1)$ as the predicted number of iterations for xy-plane i. The new algorithm performs $n_{max} - 1$ iterations without any convergence check, where n_{max} is the largest among n_i. At the end of jth iteration of this phase, computation and communication is stopped for any xy-plane for which $\max(n_i - 1, 1)$ is equal to j. After $n_{max} - 1$ iterations, convergence tests for all xy-planes are performed in a single reduction operation. Iterations for xy-planes that are not converged yet are continued with the convergence test as in the base algorithm above.

In Table 4, performance results of the improved version on nonshared memory machines are listed. It is seen that this algorithm performs better than the non-adaptive case. For example, when the number of processors increases from 64 processors to 256 on NCUBE/2, the Mflop rate is tripled in the non-adaptive version whereas it is quadrupled in the adaptive version, which is optimal. The time increase from 4 to 16 to 64 processors is probably accounted for by the increase in the amount of neighbor communication. With 4 processors, each one communicates with 2 neighbors. With 16, four processors communicate with four neighbors, while others communicate with fewer. With 64, most of them have four neighbors. Comparing Table 2 and Table 4, we notice that the impact of the new scheme increases with the number of processors. This is as expected, because the time complexity of reduction increases with the number of processors. In Figure 4-b, performance difference between these two schemes is depicted.

Table 4
Performance of adaptive scheme on nonshared memory machines

#PE	Domain	Subdomain	dt	time(msec)	Mflops
		NCUBE/2			
4	16x16x33	8x8x33	0.006	35993	3.04
16	32x32x33	8x8x33	0.005	44732	9.82
64	64x64x33	8x8x33	0.002	48564	36.18
256	128x128x33	8x8x33	0.0012	48594	144.02
		iPSC/860			
4	16x16x33	8x8x33	0.006	10436	10.48
16	32x32x33	8x8x33	0.005	14890	29.49
64	64x64x33	8x8x33	0.003	16676	105.37

4.3. Pipelining

A further improvement for this algorithm is the pipelined solution of the xy-planes. The algorithm described above exchanges the boundary values of all non-converged xy-planes. After the exchange is complete, it starts to solve those xy-planes. However, the time spent in computation and communication can be overlapped by solving xy-planes in a pipelined fashion (note that xy-planes can be solved independent of each other). The pipelined execution can be performed as follows: First, xy-planes are divided into groups. As soon as a group of planes is solved, the boundary values from this group is sent while simultaneously initiating computations for the next group of xy-planes (Figure 5-a). By the time of next iteration, boundary values of the first group would have arrived, and the processor can perform computations without waiting. The performance of this approach depends on the work distribution (i.e., the amount of computation necessary to solve xy-planes), and the choice of xy-planes in the concurrent groups. In our case, however, the empirical result shows (Figure 5-b) that the work load of the first few planes is significant, while the other planes converge rapidly. Therefore, for this particular problem, pipelined execution does not provide significant performance gain. The distribution of workload for this problem also implies that a 3D partitioning will suffer from load balancing.

5. CONCLUSION

In this paper, we have described the parallelization of a fractional step method for solving unsteady Navier-Stokes equations in the context of a 3D grid that is uniformly spaced in one dimension. A 2D decomposition scheme was used. In addition to rendering the FFT a completely local operation, this decomposition also induces a uniform load distribution. Although a 3D decomposition may reduce the communication cost somewhat, we believe that the cost of load imbalance and the parallelization of the FFT far outweigh the reductions in the communication cost.

It was found that parallel convergence tests during the Poisson solver constitute a significant portion of the elapsed time. An adaptive technique for reducing the number of these expensive tests was developed and was shown to improve the performance significantly for systems with a large number of processors.

The Charm parallel programming system facilitated development of this code by providing portability, so that we could focus on the algorithm expression without getting involved in machine specific details. The program once developed ran on different parallel

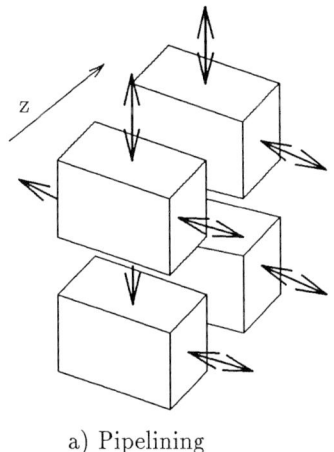
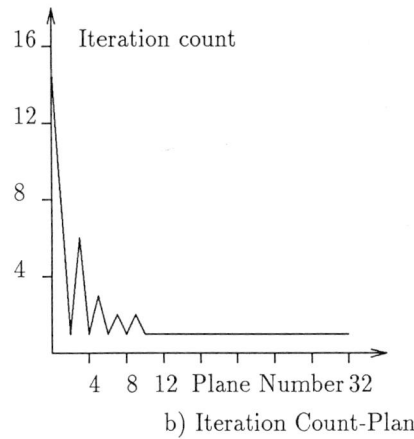

a) Pipelining b) Iteration Count-Plane

Figure 5: Pipelined Execution

machines without change. The full flexibility of this system was not used in the current implementation. Because the system is message driven, it provides many opportunities for overlapping computation and communication. This can be exploited by overlapping convergence tests and iterative computations,for example. Other algorithmic improvements in future may include replacing the block Jacobi with a red-black Gauss-seidel iterative scheme.

6. REFERENCES

[1] M.E.Braaten, "Application of parallel computing in computational fluid dynamics: A review", GE Corporate R&D report 89CRD121, July 1989.

[2] W.Fenton, B.Ramkumar, V.A.Salatore, A.B.Sinha, L.V.Kale, "Supporting machine independent programming on diverse parallel architectures", *Proceedings of the International Conference on Parallel Processing*, Vol. II, Aug 1991, pp.193-201.

[3] F.H.Harlow, J.E.Welch, "Numerical calculation of time dependent viscous incompressible flow of fluid with free surface", *Phys. of Fluids*, Vol. 8, No. 112, 1965, pp.2182-2189.

[4] J.Kim, P.Moin, "Application of a fractional step method to incompressible Navier-Stokes equations", *J.Comp.Phys*, Vol. 59, 1985, pp.308-323.

[5] H.L.Stone, "Iterative solution of implicit approximations of multidimensional partial differential equations", *SIAM J.Numer.Anal*, Vol. 5, No. 3, September 1968, pp.530-558.

[6] The CHARM(3.0) programming language manual, Department of Computer Science, University of Illinois at Urbana-Champaign, Urbana, IL, 1992.

Algorithm Modifications for Parallel Operation of a Multigrid Navier-Stokes Solver

C.S. Gwilliam and J.S. Rollett

Oxford University Computing Laboratory, Numerical Analysis Group, 11, Keble Road, Oxford, United Kingdom, OX1 3QD.

Abstract

This paper explores the modifications needed to make an existing, sequential, Fortran Navier-Stokes code efficient and effective when ported to a T800 transputer network using the 3L parallel Fortran language.

1 Introduction

Shah [4] wrote a multigrid code to solve the 2-d, laminar Navier Stokes equations for flow in a driven cavity and over a backward facing step. He used the Full Approximation Scheme (FAS) [1,5] with a smoothing algorithm called Symmetrical Coupled Gauss-Seidel/Line Solver or SCGS/LS. Our aim was to port the code to a parallel system, to investigate the efficiency of the parallel version and to make, if necessary, algorithm modifications to increase the efficiency whilst retaining effectiveness.

The Navier-Stokes equations were written in the form

$$\frac{\partial \rho u_i}{\partial x_i} = 0, \tag{1a}$$

$$\frac{\partial \rho u_i u_j}{\partial x_j} - \frac{\partial \rho u_j}{\partial x_j} u_i = -\frac{\partial p}{\partial x_i} + \frac{\partial}{\partial x_j}\left[\mu\left(\frac{\partial u_i}{\partial x_j} + \frac{\partial u_j}{\partial x_i}\right)\right], \tag{1b}$$

where p is the pressure, (u_1, u_2) is the velocity vector, ρ is the density and μ is the dynamic viscosity. The Reynolds number for reference velocity U and reference length L is defined to be

$$\frac{\rho U L}{\mu}.$$

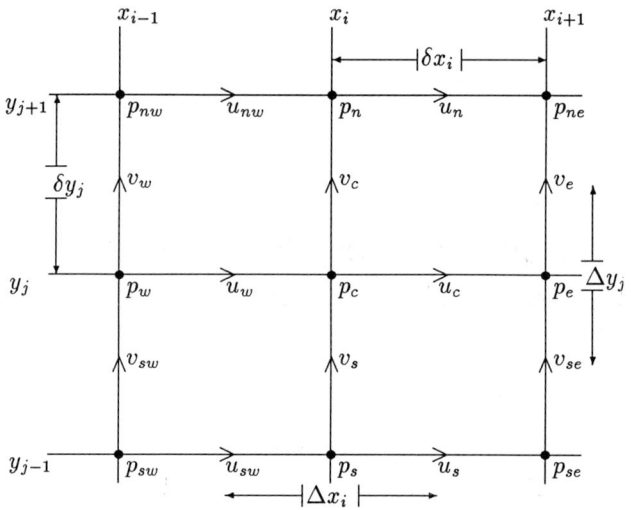

Figure 1: A typical MAC cell

Equations 1 were discretized over a Marker and Cell (MAC) mesh [3] (figures 1 and 2) to give equations 2.

$$A_c^u u_c = A_n^u u_n + A_s^u u_s + A_e^u u_e + A_w^u u_w + S_c^u - (p_e - p_c)/\delta x_i, \tag{2a}$$

$$A_c^v v_c = A_n^v v_n + A_s^v v_s + A_e^v v_e + A_w^v v_w + S_c^v - (p_n - p_c)/\delta y_j, \tag{2b}$$

$$A_e^c u_e - A_w^c u_w + A_n^c v_n - A_s^c v_s = 0, \tag{2c}$$

where

$$S_c^u = \frac{\mu}{\delta x_i \Delta y_j} (v_e - v_c - v_{se} + v_s),$$

$$S_c^v = \frac{\mu}{\Delta x_i \delta y_j} (u_n - u_c - u_{nw} + u_w),$$

$$A_n^u = \frac{1}{\Delta y_j} \max\left[\frac{\mu}{\delta y_j}, \left|\frac{\rho(v_e + v_c)}{4}\right|\right] - \frac{\rho(v_e + v_c)}{4\Delta y_j},$$

$$A_s^u = \frac{1}{\Delta y_j} \max\left[\frac{\mu}{\delta y_{j-1}}, \left|\frac{\rho(v_s + v_{se})}{4}\right|\right] + \frac{\rho(v_s + v_{se})}{4\Delta y_j},$$

$$A_e^u = \frac{1}{\delta x_i} \max\left[\frac{2\mu}{\Delta x_{i+1}}, \left|\frac{\rho(u_e + u_c)}{4}\right|\right] - \frac{\rho(u_e + u_c)}{4\delta x_i},$$

$$A_w^u = \frac{1}{\delta x_i} \max\left[\frac{2\mu}{\Delta x_i}, \left|\frac{\rho(u_w + u_c)}{4}\right|\right] + \frac{\rho(u_w + u_c)}{4\delta x_i},$$

$$A_n^v = \frac{1}{\delta y_j} \max\left[\frac{2\mu}{\Delta y_{j+1}}, \left|\frac{\rho(v_n + v_c)}{4}\right|\right] - \frac{\rho(v_n + v_c)}{4\delta y_j},$$

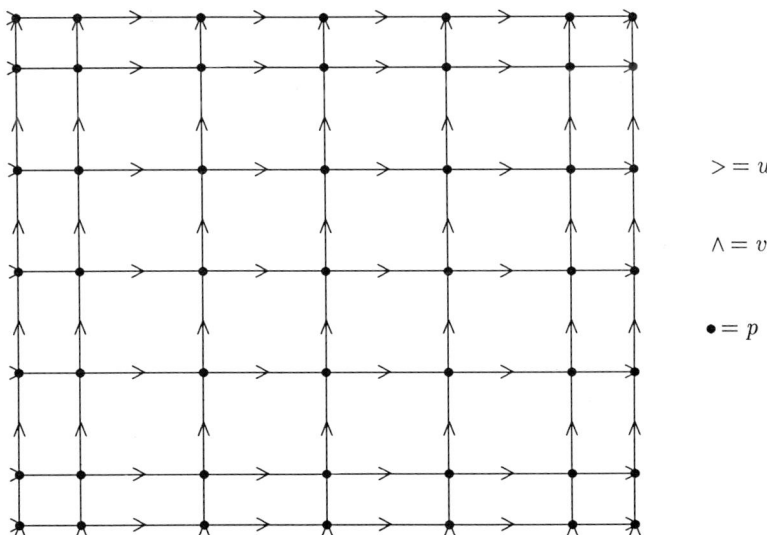

Figure 2: A typical MAC grid

$$A_s^v = \frac{1}{\delta y_j}\max\left[\frac{2\mu}{\Delta y_j}, \left|\frac{\rho(v_s+v_c)}{4}\right|\right] + \frac{\rho(v_s+v_c)}{4\,\delta y_j},$$

$$A_e^v = \frac{1}{\Delta x_i}\max\left[\frac{\mu}{\delta x_i}, \left|\frac{\rho(u_n+u_c)}{4}\right|\right] - \frac{\rho(u_n+u_c)}{4\,\Delta x_i},$$

$$A_w^v = \frac{1}{\Delta x_i}\max\left[\frac{\mu}{\delta x_{i-1}}, \left|\frac{\rho(u_{nw}+u_w)}{4}\right|\right] + \frac{\rho(u_{nw}+u_w)}{4\,\Delta x_i},$$

$$A_e^c = \frac{\rho}{\Delta x_i} = A_w^c, \qquad A_n^c = \frac{\rho}{\Delta y_j} = A_s^c.$$

2 Notation

The u_c notation is now dropped and the following notation, assuming a uniform grid, is used

$$u_{i,j} = u(x_i + h/2, y_j),$$
$$v_{i,j} = v(x_i, y_j + h/2),$$
$$p_{i,j} = p(x_i, y_j).$$

A grid point is assumed to be $\{u_{i,j}, v_{i,j}, p_{i,j}\}$ and a row of points is $\{u_{i,j}, v_{i,j}, p_{i,j}, i = 1,\ldots,N\}$.

3 The Smoothing Algorithm – SCGS/LS

The non-linear equations 2 are written

$(A^u_c)_{i-1,j} u_{i-1,j} = F^u_{i-1,j}, \quad (A^v_c)_{i,j} v_{i,j} = F^v_{i,j},$
$(A^v_c)_{i,j-1} v_{i,j-1} = F^v_{i,j-1}, \quad (A^u_c)_{i,j} u_{i,j} = F^u_{i,j},$
$(u_{i,j} - u_{i-1,j})/h + (v_{i,j} - v_{i,j-1})/h = 0.$

For a row sweep these are written in terms of corrections (δu etc.) and residuals for a fixed j to give

$$(A^u_c)_{i-1,j} \delta u_{i-1,j} + (\delta p_{i,j} - \delta p_{i-1,j})/h = R^u_{i-1,j}, \tag{3a}$$

$$(A^u_c)_{i,j} \delta u_{i,j} - (\delta p_{i,j} - \delta p_{i+1,j})/h = R^u_{i,j}, \tag{3b}$$

$$(A^v_c)_{i,j-1} \delta v_{i,j-1} + \delta p_{i,j}/h = R^v_{i,j-1}, \tag{3c}$$

$$(A^v_c)_{i,j} \delta v_{i,j} - \delta p_{i,j}/h = R^v_{i,j}, \tag{3d}$$

$$(\delta u_{i,j} - \delta u_{i-1,j})/h + (\delta v_{i,j} - \delta v_{i,j-1})/h = R^c_{i,j}. \tag{3e}$$

Manipulation leads to the following system of tri-diagonal equations

$$(A^p_c)_{i+1,j} \delta p_{i+1,j} + (A^p_c)_{i,j} \delta p_{i,j} + (A^p_c)_{i-1,j} \delta p_{i-1,j} = R^p_{i,j}. \tag{4}$$

The SCGS/LS algorithm is

1. Solve equations 4 for $i = 1, \ldots, N$ and fixed j.

2. Substitute the resulting δp into equations 3 to find the velocity corrections.

3. Update the values of $u_{i,j}$, $v_{i,j-1}$, $v_{i,j}$, and $p_{i,j}$ using relaxation parameters. For example $u^{new}_{i,j} = u^{old}_{i,j} + \omega_u \delta u_{i,j}$.

4. Repeat steps (1)–(3) for $j + 1$. The calculation of the coefficients uses the new values (i.e. a local linearization is implemented).

Having swept the grid row-by-row, the grid is swept column-by-column.

4 Geometric Parallelism

To reduce communication between processors, geometric parallelism is used. Each processor holds a complete copy of code but implements it on a portion of the domain and, therefore, only needs to hold the values for a portion of the total domain. As will be seen, only the values on the boundary of each domain need to be communicated. This is in contrast to functional parallelism where each processor is given various sub-tasks but holds all the calculation points. The data for the whole domain needs to be passed from processor to processor.

Consider solving equations 1 on a domain of $n^x \times n^y$ using a set of $p \times q$ transputers. To divide a row of n^x points between p processors let

$$n^x = p\left[\frac{n^x}{p}\right] + r_p \qquad \text{where } r_p = 0, \ldots, \left[\frac{n^x}{p}\right] - 1.$$

Each processor holds n^p points where

$$n^p = \left[\frac{n^x}{p}\right] \quad \text{or} \quad \left[\frac{n^x}{p}\right] + 1.$$

To decide where to distribute, if necessary, the extra r_p points note that

- Processors 1 and p hold the fixed boundary values.

- Processor p holds grid point $n^x - 1$ which, due to the staggered grid, has only values for v and p associated with it (figure 2).

Taking these facts into account an extra point is given to each of processors

$$\text{mod}(i, 2)p + (-1)^{\text{mod}(i,2)} \left[\frac{i}{2}\right] \quad i = 1, \ldots, r_p.$$

The n^y points are similarly divided between the q processors. Each processor will work on $n^p \times n^q$ points where

$$n^p = \left[\frac{n^x}{p}\right] \quad \text{or} \quad n^p = \left[\frac{n^x}{p}\right] + 1,$$
$$n^q = \left[\frac{n^y}{q}\right] \quad \text{or} \quad n^q = \left[\frac{n^y}{q}\right] + 1.$$

5 Halos, Storage and Communication

Each processor works on the $n^p \times n^q$ points it holds but regularly needs values of points immediately adjacent to this region which are held by neighbouring processors. Therefore, each processor holds a 'halo' of points which it uses but does not update. The division of a grid showing halo points is shown in figure 3. When the halos are included, the number of points $n_{i,j} = p_i^x \times p_j^y$ processor (i, j) holds is given by

Processor	$n_{i,j} = p_i^x \times p_j^y$
$(1,1), (1,q), (p,1), (p,q)$	$(n^p + 1)(n^q + 1)$
$(1,j), (p,j), j = 2, \ldots, q-1$	$(n^p + 1)(n^q + 2)$
$(i,1), (i,q), i = 2, \ldots, p-1$	$(n^p + 2)(n^q + 1)$
$(i,j), i = 2, \ldots, p-1, j = 2, \ldots, q-1$	$(n^p + 2)(n^q + 2)$

The pressure and velocity values are stored in 1-d arrays of length $n_{i,j}$. It is not necessary to store the whole grid on each processor. Whenever a processor updates points that lie in a neighbouring processor's halo, the halo needs to be updated. The communication pattern is shown in figure 4.

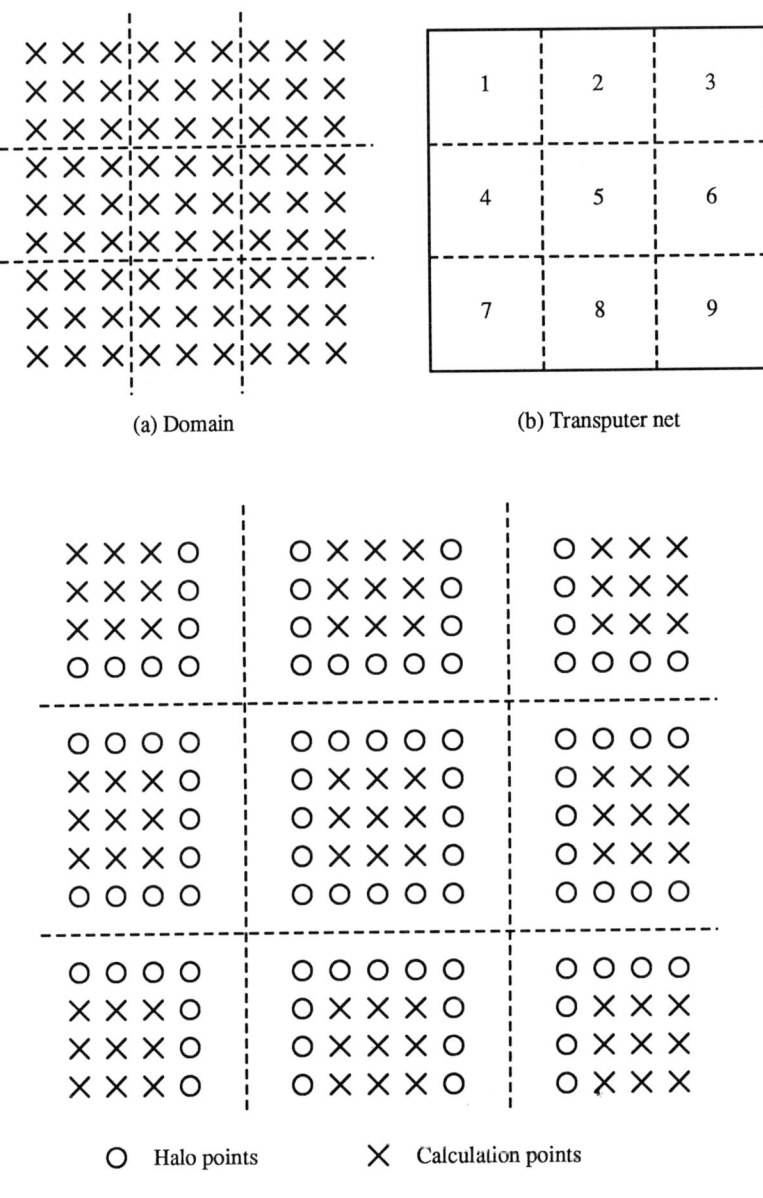

Figure 3: Domain decomposition for a 9 × 9 mesh over a 3 × 3 grid of Transputers.

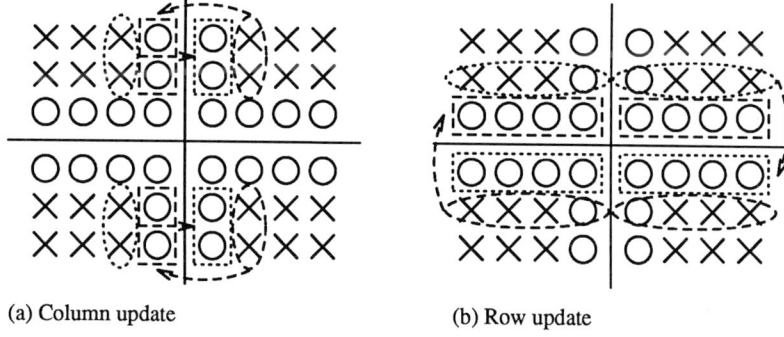

(a) Column update (b) Row update

Figure 4: Communication of updated values to the halos.

6 Inefficiencies of the Parallel Algorithm

To examine why the SCGS/LS algorithm is inefficient consider a case where 2 rows of transputers (R1 and R2) are smoothing over a grid of 4 rows of points. The main stages of the algorithm are

1. R1 solves equations 4 for row 1 of the grid and updates the points for that row. R2 is idle.

2. R1 solves the equations for row 2, using the new values from row 1, updates the points and sends these new values to R2's halo.

3. When R2 has received these values it calculates the updates for row 3 of the grid which is the first row of points it holds. These new values are sent back to R1.

4. R2 then works on the next row whilst R1 can begin to work on the first column of points ready for the column sweep.

Therefore, for most of the time, only one row of transputers is working.

Not only is the overall algorithm inefficient but the solution of the tri-diagonal equations is as well. If the forward and backward substitution method is used the following pattern of activity is seen for the solution of just one set of equations

F_i W   ...   W   W
W F_i ...   W   W
⋮   ⋮   ⋮   ⋮   ⋮
W   W   ...   W F_i
W   W   ...   W B_i
W   W   ... B_i W
⋮   ⋮   ⋮   ⋮   ⋮
B_i W   ...   W   W

Tr1	Tr2	Tr3	Tr4		Tr1	Tr2	Tr3	Tr4
F_1	W	W	W		$\vdots$	$\vdots$	$\vdots$	$\vdots$
F_2	F_1	W	W		F_i	B_{i-3}	F_{i-1}	B_{i-2}
F_3	F_2	F_1	W		B_{i-3}	F_i	B_{i-2}	F_{i-1}
F_4	F_3	F_2	F_1		F_{i+1}	B_{i-2}	F_i	B_{i-1}
W	W	W	B_1		$\vdots$	$\vdots$	$\vdots$	$\vdots$
W	W	B_1	F_2		B_{m-2}	W	B_{m-1}	F_m
W	B_1	F_3	B_2		W	B_{m-1}	W	B_m
B_1	F_4	B_2	F_3		B_{m-1}	W	B_m	W
F_5	B_2	F_4	B_3		W	B_m	W	W
$\vdots$	$\vdots$	$\vdots$	$\vdots$		B_m	W	W	W

Figure 5: Sweep pattern for the solution of the tridiagonal equations

where F_i means forward substitution on row i of the points a transputer holds, B_i backward substitution on row i and W means the processor is idle. For most of the time, only one processor in a given row will be working. The use of a more efficient, parallel solver will remove this problem but will not overcome the inefficiency of the main algorithm.

7 Overcoming the Inefficiencies

To overcome the inefficiency of SCGS/LS, a new algorithm was developed called Symmetrical Coupled Jacobi/Line Solver or SCJ/LS. The main steps in this algorithm are

1. All transputers calculate the coefficients of equations 4 for the rows of points that they hold.

2. All the resulting sets of tri-diagonal equations are solved.

3. All points are updated. All halos are updated by east-west and north-south communication.

This is then repeated for the column sweep. All processors work for most of the time as there is no need to wait for updated values from previous rows. Processors are idle only during stage 2.

The use of SCJ/LS overcomes the inherent sequentiality of SCGS/LS but the inefficiency of the tri-diagonal solver remains. By pipelining the n^q sets of equations (figure 5,[2]) held by the same processor the proportion of time spent waiting when there are p processors in a row is

$$\frac{p-1}{n^q + p - 1} \quad \text{compared with} \quad \frac{p-1}{p}$$

for the non-pipelined case. This approach keeps the transputers busy for most of a time without incurring the penalty of about 2 imposed by, for example, the Wang algorithm.

8 Results

We present results for the Driven Cavity and the Backward Facing Step. Figures 6–8 show the efficiencies for different numbers of processors for these test cases. The iteration used is a W-sweep with 2 pre- and post-smoothing iterations and 3 smoothing iterations on the coarsest grid. From these graphs it can be seen that SCJ/LS is more efficient than SCGS/LS. In figure 9 the CPU times for 1 iteration of a 3 grid W-sweep for the driven cavity test case using SCJ/LS are shown where each processor now has the same amount of data (16 processors will be doing 16 times more work than 1 processor). The times show that SCJ/LS is scalable. Tables 1-3 show how many iterations the different algorithms take to converge for the above W-sweep when different numbers of grids and different Reynolds numbers are used. From them it can be seen that the SCJ/LS algorithm is competitive with the SCGS/LS algorithm.

9 Conclusions

- Parallel SCGS/LS, though effective, is inefficient due to the local nature of the algorithm (new values from previous lines are used to calculate the updates for the line under consideration).

- The new parallel algorithm, SCJ/LS, which is found by replacing the local, line-by-line nature of the updates by a global update, is effective and efficient. SCJ/LS replaces the local linearization of SCGS/LS by a 'global' linearization. This does not significantly degrade the convergence of the non-linear problem.

- The forward and backward substitution method for the solution of the tri-diagonal equations is inefficient when used with SCGS/LS. However, the global nature of SCJ/LS, allows the equations to be pipelined and the method becomes efficient.

References

[1] A. Brandt. Multi-level Adaptive Solutions to Boundary-value Problems. *Mathematics of Computation*, 31(138):333–390, April 1977.

[2] D. J. Gavaghan. *Parallel Numerical Algorithms for the Solution of Diffusion Problems*. PhD thesis, Oxford University, 1991.

[3] F.H. Harlow and J.E. Welch. Numerical Calculation of Time-dependent Viscous Incompressible Flow of Fluid with Free Surface. *The Physics of Fluids*, 8(12):2182–2189, Dec 1965.

[4] T. M. Shah. *Analysis of the Multigrid Method*. PhD thesis, Oxford University, 1989.

[5] K. Stüben and U. Trottenberg. *Multigrid Methods: Fundamental Algorithms, Model Problem Analysis and Applications*, volume 960 of *Lecture Notes in Mathematics*. Springer-Verlag, 1981.

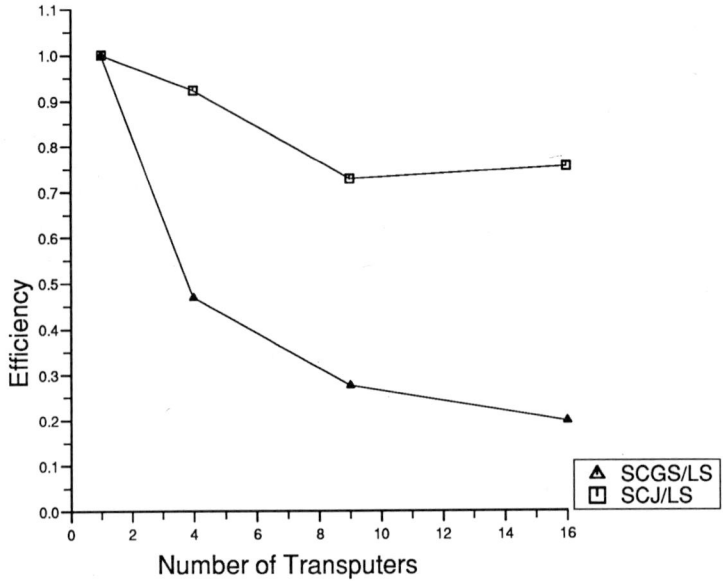

Figure 6: Efficiencies for driven cavity case (3 grids, coarsest grid = 10×10)

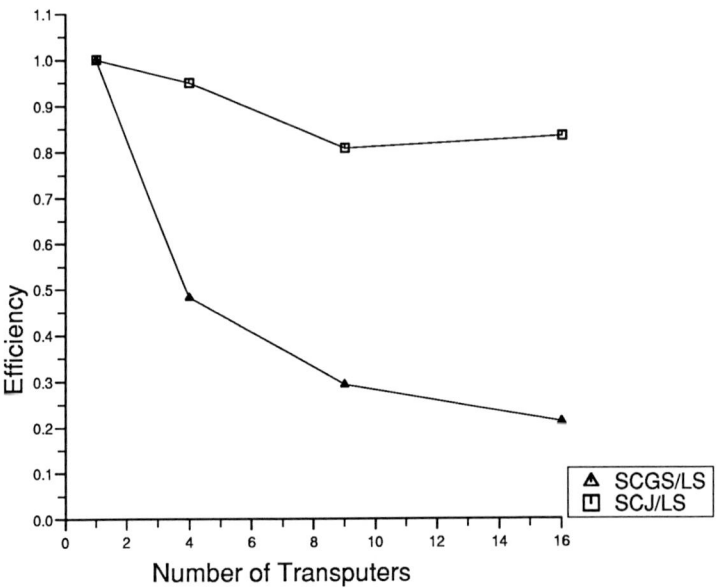

Figure 7: Efficiencies for driven cavity case (4 grids, coarsest grid = 10×10)

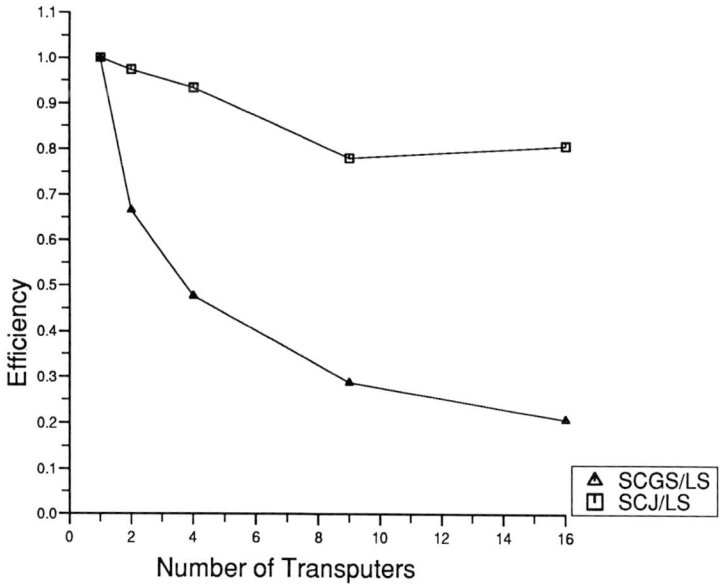

Figure 8: Efficiencies for backward facing step (3 grids, coarsest grid = 18×10)

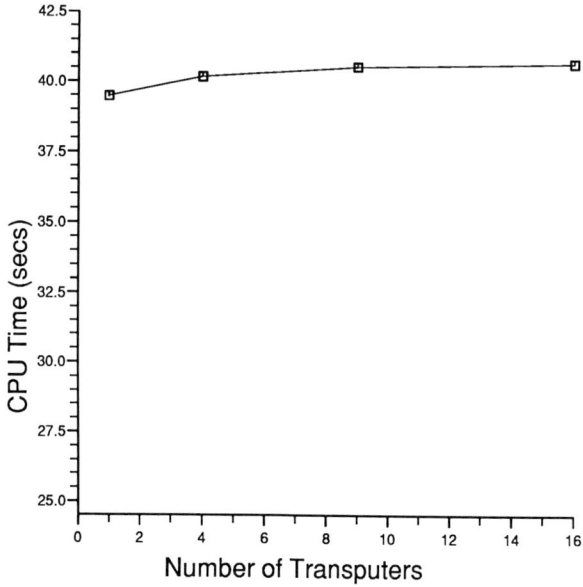

Figure 9: CPU times for driven cavity showing scalability of SCJ/LS

Algorithm	Number of Grids	Number of Iterations	Time on 16 processors
SCGS/LS	2	9	5.53
	3	6	17.98
	4	5	63.11
	5	4	203.66
	6	4	809.07
SCJ/LS	2	12	2.31
	3	8	6.61
	4	5	16.97
	5	4	54.00
	6	4	214.29

Table 1: Number of iterations and time taken for the driven cavity case to converge at Reynolds number 100, coarsest grid 10×10

Algorithm	Number of Grids	Number of Iterations	Time on 16 processors
SCGS/LS	2	12	7.37
	3	12	36.00
	4	12	151.47
SCJ/LS	2	17	3.27
	3	20	16.53
	4	19	64.49

Table 2: Number of iterations and time taken for the driven cavity case to converge at Reynolds number 1000, coarsest grid 10×10

Algorithm	Number of Grids	Number of Iterations	Time on 8 processors
SCGS/LS	2	10	4.25
	3	8	16.14
	4	10	89.45
SCJ/LS	2	10	1.61
	3	8	6.16
	4	10	32.17

Table 3: Number of iterations and time taken for the backward facing step to converge at Reynolds number 100, coarsest grid 10×6

Features of Architecture Independent Parallel CFD Software

JOCHEM HÄUSER [1], HORST D. SIMON [2], AND H.-G. PAAP [3]

Abstract

Geometrically complex, compute intensive scientific and engineering problems to be solved on massively parallel systems should possess certain features, which make them architecture independent (portable) and also allow their effective implementation on large-scale multiprocessors. Although the primary application in this paper is on CFD, the key parameters identified and the methodology developed apply to a large class of engineering and scientific problems. Many of these problems can be described by equations, which can be discretized and solved by numerical approximation. To parallelize, the discrete solution domain is partioned into subdomains, a process referred to as domain decomposition. In CFD, the solution domain is a specified subset of three-dimensional space, e.g., a certain region enclosing an aircraft. In the context of this paper, domain decomposition is achieved using an existing multi-block grid. The paper addresses the four major issues to develop and to implement CFD (engineering) applications on a variety of computer architectures, including massively parallel systems, clusters of workstations, shared memory systems, and sequential machines. The topics dealt with are the following. First, a general load balancing and communication strategy is given, followed by a description of the principles of the corresponding algorithms together with the main data structures. Second, the communication behavior of various parallel systems is investigated with regard to the load generated by a multi-block Navier-Stokes solver. Results will also be presented for the general multi-block grid generator, which serves as a testcase for a Navier-Stokes solver. Apart from the system of equations, the datastructures and the communication remain the same. All codes are written in ANSI C. Third, the convergence behavior of the implicit numerical solution scheme of the Navier-Stokes solver (ANSI C) is investigated by performing the same computations on the same grid topology, but with a widely varying number of blocks.

---

[1] European Space Research and Technology Center, European Space Agency, 2200 AG Noordwijk, The Netherlands.

[2] Applied Research Branch, Numerical Aerodynamics Simulation (NAS) Systems Division, NASA Ames Research Center, Mail Stop T-045-1, Moffett Field, CA 94035. The author is an employee of Computer Sciences Corporation. This work is supported through NASA Contract NAS 2-12961, USA.

[3] Genias GmbH, 8300 Regensburg, Germany.

1 Introduction

Many problems in science and engineering are described by physical laws, formulated by equations. Computer simulation uses a discretized form of these equations. In computational fluid dynamics, CFD, the governing equations are nonlinear-partial differential equations, describing the conservation principles for fluids in both space and time.

The concept of solution domain, SD, is used in a very general sense. Any engineering or scientific problem that can be associated with a set of objects, which can be further subdivided by some algorithm or procedure, can be parallelized by the present methodology. Parallelization is based on the decomposition of the SD into so called blocks. Each of these blocks is topologically equivalent to a simple box in the computational domain, CD. The set of all connected boxes in the CD represents the original SD. To perform the computations in the CD, the governing equations have to be transformed too.

It should be noted that the decomposition of the SD is dictated by the physics, and that in general the resulting grid topology does not lead to a loadbalanced application. The automatic blocking of complex 3D SDs is currently a major field of research. These topics are further discussed in Sec. 2.

In general, the domain of interest is very complex (irregular), e.g. it may be the outer region of an aircraft or a car. In order to resolve the various temporal and spatial scales, grids comprising several million points are needed when complex physical phenomena are encountered; for example, if turbulent flow problems are simulated. Moreover, to produce flow solutions that have design quality, a certain accuracy has to be achieved, resulting in very large computing times. The realization of a computer windtunnel, apart from the difficulties in modeling the physical processes, can only be obtained by using the most powerful massively parallel systems. Even with the fastest vector supercomputers of today, the turnaround time of the computer windtunnel will be higher than for a physical windtunnel. Two shots per day for a complex configuration cannot yet be computed.

However, parallel systems and parallel computing are not a mature field yet. Over the past 5 years the market has seen a variety of very different architectures, requiring different parallelization strategies by the application programmer. Although no standards exist both for parallel hard- and software, the multiple instruction multiple data, MIMD, architecture along with message passing seems to be the most flexible platform. It is now being used in the systems of Intel, Ncube, Thinking Machines, Parsytec etc. Therefore, communication algorithms for CFD software have been developed based on message passing. This does not exclude that the code also runs on sequential machines, shared memory systems, and on clusters of workstations.

First, in order to achieve the desired software portability, it was found necessary to extend the definition of a massively parallel system to any system having a finite number of nodes that can communicate by exchanging messages. Second, the definition of message passing had to be extented, too. This concept is explained further in Sec.2. In the above context, sequential or shared memory machines are treated as a massively parallel system, too. Message passing on shared memory systems is straightforward, since mailboxes can be directly read because of the shared memory. This allows to use exactly the same communication mechanism, regardless of the underlying hardware platforms.

2 AIPS: Architecture Independent Parallel Software

The parallel grid code presented here runs on sequential and on parallel architectures as well as on a cluster of workstations, using PVM or Express. The system on which the code is intended to be run is specified at run time. The structure of this code can serve as a template for a general parallel code, employing the multi-block concept with overlap. If another engineering problem has to be solved, the set of nonlinear Poisson equations used in the grid generation process has to be replaced. The overall data structures and the communication approach will remain intact.

In this paper we follow the terminology of Fox et al.[5]. As was mentioned in the introduction, the parallelization approach is based on the idea of partitioning or subdividing the solution **domain**, SD, into a number of **blocks (members)**. With regard to parallelization these members cannot be decomposed further. However, a block is a complex object, containing six faces. A face is also an object, which contains its own mailbox. Hence, regarded from the communication level, a face is the most basic entity or atom.

The important point is that this domain can be completely **irregular**. In addition, a domain generally is highly **inhomogeneous**, meaning that each of its blocks involves different amounts of computation. Although it has been shown in [10] that a complete 3D configuration like the Hermes Space plane or the Space Shuttle Orbiter can be decomposed into 512 blocks, each having identical amounts of computation, the underlying flow physics may dictate a decomposition leading to widely different block sizes. It should be emphasized here, that multi-block grids are constructed to best represent the flow phenomena, thus completely determining the block topology. In general, this topology will not lead to a loadbalanced application. In the next step, the computational work for each of these blocks has to be determined. It will then be necessary to collect blocks into larger **grains** or **groups**, such that the number of floating point operations is almost the same for each group. This set of groups is then mapped one-to-one onto the processor topology. The algorithm for the grouping of the blocks is based on graph theory.

For that purpose it may be necessary that blocks of the existing grid topology are further subdivided, without changing the topology itself. Blocks of the original grid topology are referred to as **superblocks**. Of course, all members of these superblocks should reside on the same processor, as far as loadbalancing allows.

This approach, dictated by the physics, has far reaching consequences for the communication algorithms. However, it will be needed for almost all real applications in science and engineering, and in particular in CFD. In general, results from **regular domains** obtained from a **homogeneous** problem cannot be used to make any predictions for applications on irregular domains. Hence, it is not useful trying to get a theoretical understanding by solving such problems.

Clearly, this approach leads to **random communication** among blocks. Therefore, parallel architectures must not rely on some kind of nearest neighbor communication, i.e. communication bandwidth should be independent of the destination node. The communication model is based on **loose synchronization**, that is after one time step or one iteration has been performed within a block, the proper receive and send messages are issued for each **matching face** also denoted as **guard ring** or **overlap face**, i.e. a face that

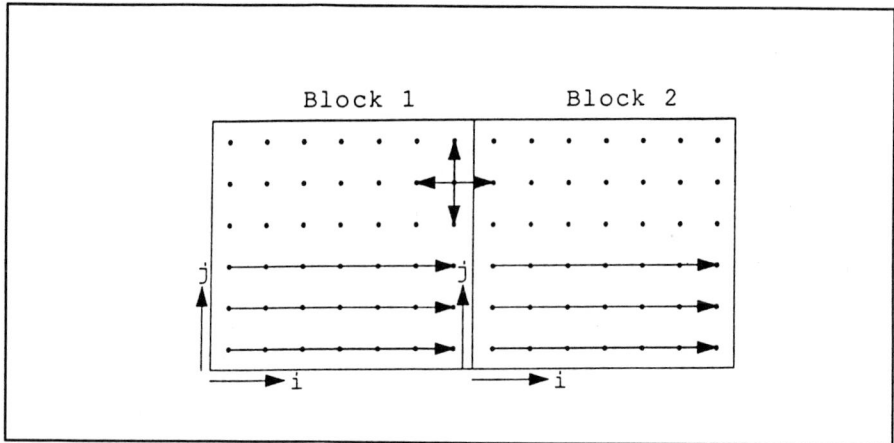

Figure 1: Update of block boundaries.

is not a physical boundary but serves as an interface to the respective neighboring block. It can be specified at run time whether blocking or nonblocking send and receive commands should be selected. The best results have been obtained for nonblocking receive, followed by blocking send. Computation is resumed upon having received all information to update the matching faces of the block. Because groups are loadbalanced, a large number of messages at the same instant of time should be expected, since each block may receive and send up to 6 messages. In contrast to [5], measurements have shown that the bandwidth strongly depends on the message traffic and may vary by an order of magnitude. This has to be considered when the efficiency of a N-S solver is evaluated.

This methodology is applicable to a large class of problems in science and engineering; for example, it comprises all problems that can be described by equations in both space and time.

The following figure explains the update procedure among neighboring blocks. The communication overhead is, of course, larger for those blocks that have to communicate outside their group, since messages have to be sent via the communication channels. If blocks reside on the same node, the send buffer of one face is used as the receive buffer of the corresponding neighboring face, i.e. the faces share the same mailbox. The actual packing and sending of the message does not take place in that case. Hence, it is obvious how the communication on a shared memory system is handled. Each node of a massively parallel system is already used as as a single shared memory system.

3   AIPS: Algorithms and Data Structures

In this section some of the major algorithms and data structures for the parallel grid generator are presented. So far the engineering community has not given much attention to software design techniques in general and to data abstraction and algorithms in particular.

These techniques are crucial in the development of general purpose parallel codes. So far, numerical discussions have been prevailing. Although important, it is believed that questions of softwareengineering are the key issues in succeeding in the implementation of real applications on modern parallel systems. Choosing a comparison from aerospace, efficient engines are necessary but by far not sufficient for the success of a new aircraft.

It should be clear that new hardware concepts have to be complemented by new software concepts. New codes should be written in object-oriented languages. C++ will provide the functionality needed for applications in science and engineering [9]. The currently used practice to port large Fortran packages, written in the sixties and seventies to the hardware platforms of the nineties will most likely turn out to be a costly undertaking. Modification and maintenance of these codes will demand substantial manpower. In addition, these codes will also require more resources than their C++ counterpart. Moreover, they will not be faster but less safe. One of the programming rules given in [11] reads: "Don't patch bad code. Rewrite it". Naturally, a certain part of the old code could be used, but the overall design of the package has to be done from scratch. This also would allow to produce reusable code. In particular, the physical submodels which are programmed in a linear manner can be retained. Thus the major part of the investment can be saved.

3.1 Program Structure of Parallel Version

The parallel version of both the grid code and the N-S solver uses the same communication algorithm to update the boundary. It works as follows. Each block sends out its update data to its neighbors as soon as they become available. Only if the neighboring blocks are not on the same node, message passing calls are needed. If blocks are on the same node, send and receive is reduced to a copy between blocks. The connectivity information for all blocks, i.e. the so called control file is needed in each processor and thus has to be read by each node. In addition, a processor-map-file has to be provided, which is read by each node. This file contains the mapping of blocks to processors (nodes), i.e. relates block numbers to node numbers. The geometry data for each block are stored in separate files. Since several blocks may reside on a single node, the processor-map-file has to be screened and the corresponding data for these blocks have to be read. It is no longer feasible that all geometry data is stored in a single file.

Within each block the following tasks are performed. If the code is used to generate grids and not to test parallel systems, the elliptic solver is used for smoothing purposes only, i.e. only a few iterations are needed.

- read parameter for SOR

- check commandline arguments, if (parallel) check own node, read processor-map-file, termed (**block_node file**)

- read control file (**cntrl2d**) and geometry file (**line2d or plane2d**)

- initialize solution (own blocks)

- if (parallel) place **irecv** for all external boundaries

- send all boundaries (intern + extern: **isend, csend**)
- start 'time-loop'
    start 'block-loop'
        get boundary-data from neighboring blocks
        do computation
        send boundaries
    end 'block-loop'
    print 'block-local' error
    if necessary (Parallel && !FixedIteration) **gdhigh**
- end 'time-loop', if
    - maximum iteration reached or - 'residual' small enough
- output (**plane2d**)

The variable gdhigh is a global variable, denoting the magnitude of the residual. The meaning of **send** is:

- copy 'face' into sendbuffer: internal send is finished
- csend or msgwait + isend: only extern, except first time

The meaning of **get** is:

- crecv or msgwait: only extern
- do some operations with data in receive buffer
- insert receive buffer into corresponding face
- irecv: only extern

For internal block connections, i.e. blocks that reside on the same node, the update procedure is as follows. Send buffer and neighboring receive buffer are linked, using the same memory.

3.2 Data Structures of the Parallel Grid Generator

The major data structures for the parallel grid generation code are explained. The programming philosophy is oriented toward object- oriented programming. In an upcoming version blocks, faces etc. will be modeled as classes in C++ instead as structures. It is then straightforward to add new features to objects (an object is an instance of a class), e.g. to convert an object block2d into block3d. It was straightforward to parallelize the 3D grid generator based on the datastructures of the 2D version. In the following the pseudo structures for a 2D block and a 1D face (edge) are presented, followed by the C implementation.

    block2d {

        /*a pool for everything a block needs to know about*/
        Dimensions;
        Addresses: where to find data-fields;
        local error;
        Face2d[4];

}

Object Face2d is used as a mailbox, sending and receiving faces to and from neighboring blocks.

    Face2d { /* the mailer */

        Dimensions;
        buffer(send + receive);
        /* only parallel version */
        where to find the neighbors;
        operations; /* what to do with the message */

}

The following structures are implemented in parallel **Grid⋆**. Similar structures are used in 3D.

```
typedef struct {
    int I, J; /* block dimensions without overlap */
    pnt2d **sp; /* pointer to block coordinates */
    pnt2d **cp; /* pointer to control functions */
    BlockName[32]; /* block name can be number or character */
    int Istart, Jstart; /* 0 or 1 depending on overlap */
    int Ir, Jr; /* block dimensions with overlap */
    int NodeId; /* 0 for sequential system */
    double LocalError; /* maximum residuum within block */
    int IsInit; /* initialization flag for algebraic grid */
    Face * fa[4]; /* pointers to face structure */
    int ExtrCorn[4]; /* 1 if corner point needs interpolation */
    int IsRestart, IsConverged, CF; /* CF control functions */
} Block2D;
typedef struct {
    int I; /* face dimension */
    pnt2d *rb; /* mailbox: pointer to receive buffer */
```

```
    pnt2d *sb; /* mailbox: pointer to send buffer */
    int Istart; /* 0 or 1 depending on overlap */
    double LocalError;
    int CF;
    double hsdist; /* distance of first grid point from wall */
    int Myface, MyBlock, typ,
        nnode, nblock, nface, nrot,
        cnode, cblock, cface, cop;
    char cblockname[32];
    int op;
    int ActivConnection; /* 1 for block on different node */
    char XYZchar; /* x, y, or z normalvector of face */
    int sendcut, reccut; /* planenumber to be sent or received */
    long rmsgid, smsgid; /* message ids */
} Face2D;
```

3.3 Internal Data Structure

Since communication among neighboring blocks, which may reside on the same or on a different processor, is needed, a doubly linked list of all blocks and all faces belonging to a single processor is built, see Fig. 2. To facilitate bookkeeping, a list of selfreferential datastructures, termed inputcontrol, is constructed. The pointer *laststructure refers to the previous inputcontrol structure. These structures are dynamically allocated. Using this structure, one can loop over all blocks and faces, needed for sending and receiving messages. An example is given in Fig.2, which assumes that three blocks are on this specific node. Two blocks have matching faces only, while one block has three **fixed faces** (physical boundaries), resulting in a total of five inputcontrol structures.

```
inputcontrol {
    int BLOCK, FACE;
    struct inputcontrol *nextblock, *nextface, *laststructure;
}
```

4 Communication Results

First, communication timing results are presented for the Intel Gamma Touchstone (NASA Ames Research Center), the Intel Delta Touchstone (Caltech, [10]) and for nCUBE (Foster City, Ca.). All codes were run on the Intel simulator on a Unix PC, prior to their implementation on the parallel system. This allows to simulate cube management as well as message passing and global operations. The number of processes is limited to 16. No host program is used in the parallel grid generator.

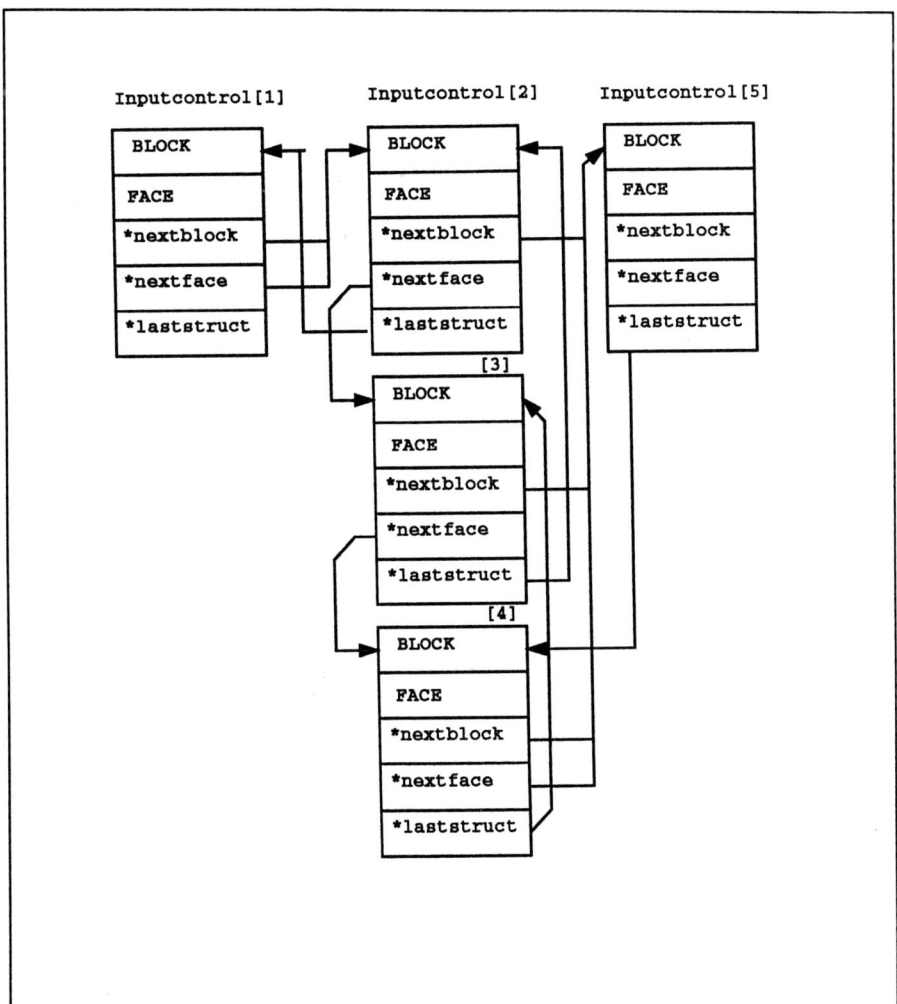

Figure 2: Example data structure that allows several blocks to reside on a single processor.

64 nodes in 4 * 4 * 4 configuration (min)				64 nodes in 4 * 4 * 4 configuration (max)			
Length	iPSC/860	Delta	nCube	Length	iPSC/860	Delta	nCube
10	0.000689	0.000725	0.000633	10	0.000718	0.000739	0.003335
100	0.000978	0.000815	0.000752	100	0.001041	0.000836	0.003524
1000	0.007160	0.001722	0.002067	1000	0.007706	0.001837	0.005034
10000	0.057722	0.011199	0.016900	10000	0.060596	0.015812	0.038941
100000	0.493383	0.082819	0.181864	100000	0.615864	0.135239	0.372721
500000	1.615960	0.530237	0.779777	500000	2.154621	0.760869	1.637275

Table 1: Communication timing for a N-S code on a 64 node configuration for Gamma, Delta, and nCUBE systems.

512 nodes in 8 * 8 * 8 configuration (min)			512 nodes in 8 * 8 * 8 configuration (max)		
Length	Delta	nCube	Length	Delta	nCube
10	0.000734	0.018141	10	0.000738	0.018945
100	0.000815	0.018110	100	0.000826	0.019894
1000	0.003974	0.002508	1000	0.004687	0.021627
10000	0.038880	0.025152	10000	0.042775	0.052975
100000	0.441635	0.245217	100000	0.488687	0.516369
500000	2.050477	1.186949	500000	2.483765	2.695112

Table 2: Communication timing for a N-S code on a 512 node configuration for Delta and nCUBE systems.

4.1 Communication Timing Results

Communication measurements are based on the simulation of the communication behavior of a Navier-Stokes code on an irregular solution domain, using the code described in [10]. There is only one block per node. The same code and input data on the three systems in Table 1 and 2 are used.

Table 3 shows the communication results for the CM5 of Thinking Machines. The code of [10], however, has been substantially modified. Therefore, further investigations will be needed before comparisons between the CM5 and the other systems can be made.

Results for the Telmat Concerto are shown in Table 4. Only 8 nodes were used, so no comparisons can be made.

4.2 Speedup Results for Parallel Grid*

The parallel **Grid*** has been implemented on nCube, Intel as well as under PVM, and Express. The major testing has been on nCUBE and Intel. A collection of 5 workstations block 2D grid. Exactly the same grid topology can be modeled with 5 blocks. Although the **NSS★** code is used in the 2D mode, calculations are actually performed in 3D, using 4 points in the third direction. In Figs. 3 and 4 the 5 and 37 block grids are shown.

The computations have been performed using the following free stream values: $Ma_\infty = 10$, $T_\infty = 50K$, $Re_\infty = 10^6/m$, $AoA = 18deg$, and $T_{wall} = 293K$. These values have been taken from an experiment performed with a European Shuttle model at the R3 windtunnel at Chalais. After 1000 iterations the results for the 5 and 37 block grids, depicted in Figs. 5 and 6 show virtually no difference. The convergence history has been plotted in Fig. 7. Again, no difference in convergence speed could be observed. Information exchange to update block boundaries takes place after each iteration step and not after each sweep. The numerical scheme uses an overlap of two gridpoints. It should be noted that in Fig. 7 the maximum residuum multiplied by the time step size is shown. The CFL number has been increased during the computation several times, causing a jump in the residuum. Eventually, a CFL number of 50 was used. After 3000 iterations the residuum is down 12 orders of magnitude. All computations have been performed on an IBM 6000/560 workstation. For this kind of convergence investigation, the implementation on a parallel system is not needed. Any deterioration in convergence would result from the blocking itself. Of course, speed up on a massively parallel system could be further reduced, depending on the communication overhead. However, this has to be clearly distinguished from the convergence issue.

Although no formal convergence multi-block theory for the N-S computations has been established, heuristic arguments derived from the governing physics can be used. A hyperbolic problem possesses a finite propagation speed for information transport, generally determined by the wave speeds encountered. It therefore can be considered to fall into the class of hierarchical problems, that is, a point in the SD requires progressively less information from other parts the further they are away. A classical example is the gravitational force, which falls of with distance. The range of coupling between grid points in a hyperbolic system is determined by the product of maximum wave speed and time step size. No coupling beyond that distance exists. Therefore, the simultaneous solution of a large system of linear equations, comprising all grid points in the solution domain is not necessary. In the present example, the SD was subdivided by performing 11 cuts in the streamwise direction and 1 additional cut in the radial direction. Obviously, the actual physical coupling was not affected by this approach. It should be investigated whether hierarchical methods could also be used in fluid dynamics.

5 Conclusions and Future Activities

First, a strategy for the parallelization of geometrically complex irregular solution domains has been presented, based on domain decomposition employing the multi-block concept. Since the block topology is determined by the physics, groups of blocks have to be formed to achieve loadbalancing. For irregular domains, random communication among the processors is needed to update block boundaries. Loose synchronization is used for communication among processors. The issue of of code portability has been discussed in detail. Second, the results of measurements have been shown, which were performed on several

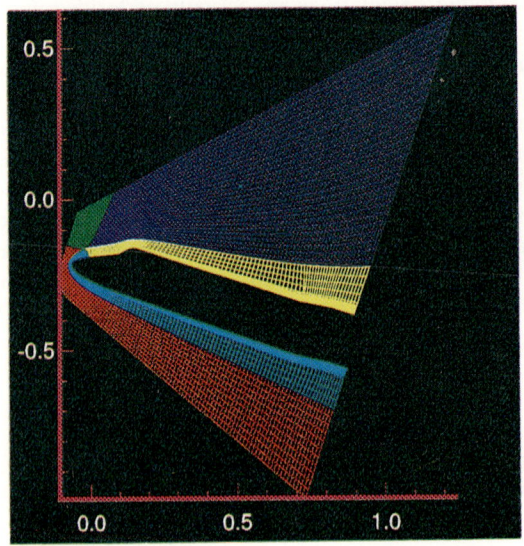

Figure 3: 5 block N-S grid for the Space Shuttle symmetry plane.

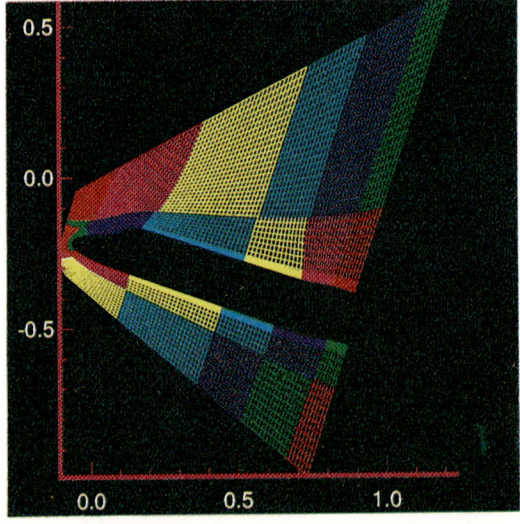

Figure 4: 37 block N-S grid for the Space Shuttle symmetry plane.

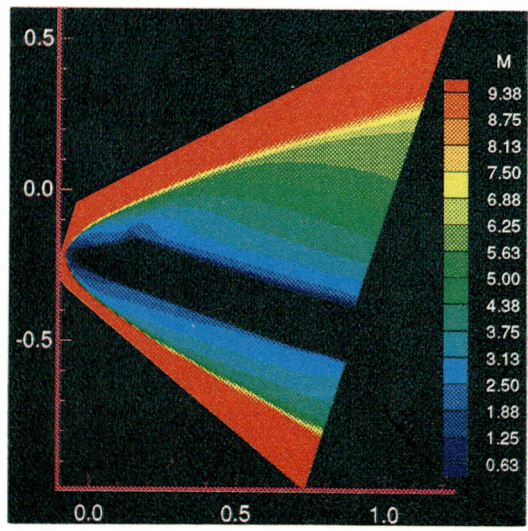

Figure 5: Ma distribution for 5 block N-S grid after 1000 iterations.

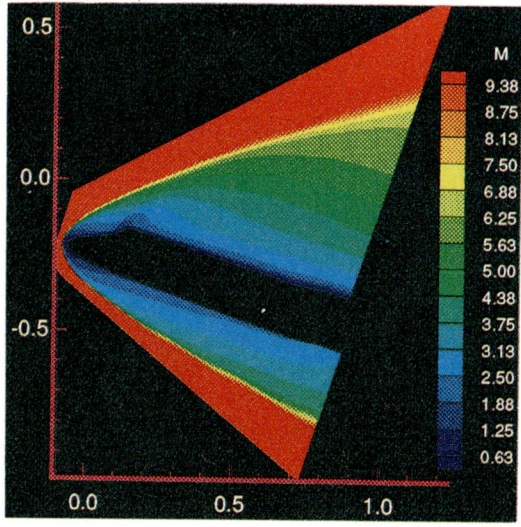

Figure 6: Ma distribution for 37 block N-S grid after 1000 iterations.

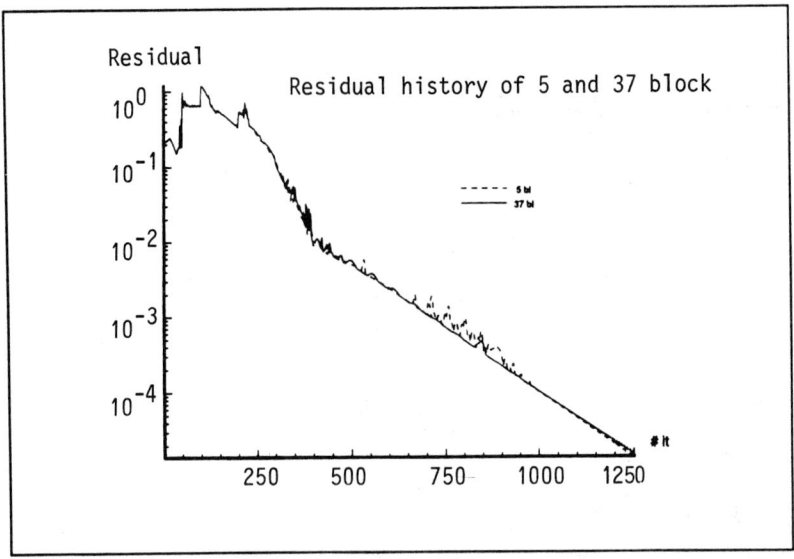

Figure 7: Convergence history for the 5 and 37 block grids. After about 1250 iterations the residual is down by 5 orders of magnitude.

systems simulating the communication behavior of a 3D multi-block N-S code. It was found that communication bandwidth varies up to an order of magnitude with message traffic. Communication bandwidth depends upon the number of processors. This has to be considered when evaluating the performance of a N-S solver. The main data structures for the communication algorithms have been presented, emphasizing the object-oriented approach. Software codes are written in ANSI C. Third, computational results of the convergence behavior of an implicit N-S solver have been presented. Convergence speed versus number of blocks has been investigated. Computations have been performed for the symmetry plane of the Space Shuttle. Euler calculations have been done comparing a 3 versus a 25 block grid as well as N-S computations using 5 versus 37 blocks. The solution domain was exactly the same in all runs. Simulation runs have been performed for Ma 2 up to Ma 20. No significant changes in convergence history have been found.

Future activities will concentrate on the implementation of an algorithm for automatic loadbalancing, based on graph theory. Second, with the information gathered, all codes will be rewritten in C++ to better reflect the object-oriented approach. Third, forthcoming parallel systems may have hundreds or thousands of processors. Therefore further investigations on the dependence of convergence history for implicit solvers versus number of blocks will be carried out. All parallel codes will be platform independent, including workstation clusters, shared memory systems as well as sequential machines.

6 Acknowledgements

The authors are grateful to W. Berry, ESA-ESTEC, for his continuous support and encouragement. We also appreciate the helpful discussions with R. Williams, Caltech and E. Krause, RWTH Aachen.

References

[1] Gentzsch, W., Häuser, J., 1988: Mesh Generation on Parallel Computers in S. Sengupta et al. (Eds.), Numerical Grid Generation in Computational Fluid Dynamics, Pineridge Press, pp. 113-124.

[2] Häuser, J. et al.: Parallel Computing in Aerospace Using Multiblock Grids, Part I: Application to Grid Generation, Concurrency: Practice and Experience, Vol.4(5), pp. 357-376.

[3] Bailey, D.H. et al., 1992: NAS Parallel Benchmark Results, RNR Technical Report RNR-92-002, NASA Ames Research Center, 13pp.

[4] Bailey, D.H., 1992: How Useful are Today's Parallel Computers?, Computers in Physics, VOL. 6, NO.2, p. 216.

[5] G. Fox et al, 1988: Solving problems on Concurrent Processors, Vol. 1, Prentice Hall, 592 pp.

[6] Gustafson, J.L. et al, 1988: Development of Parallel methods for a 1024-Processor Hypercube, SIAM, J. of Scientific and Statistical Computing, Vol. 9, No. 4, July, pp. 609-638

[7] Lin, A. : Parallel Numerical Algorithms for Fluid Dynamics Simulation, AIAA90-0333.

[8] Krause,E., et al. 1992: Memorandum zur Initiative: High Performance Scientific Computing (HPSC) 5pp.

[9] Häuser, J., Simon, H.D., 1992: Aerodynamic Simulation on Massively Parallel Systems; in Parallel Computational Fluid Dynamics '91, eds. W. Reinsch et al., North-Holland, pp. 207-227.

[10] Häuser, J., Williams, R.D., 1992: Strategies for Parallelizing a Navier-Stokes Code on the Intel Touchstone Machines, J. Num. Methods in Fluids, Vol.6, pp.

[11] Kernighan, R. 1977: The Elements of Programming Style, Prentice Hall.

Data parallel finite element techniques for computational fluid dynamics on the Connection Machine systems

Zdeněk Johan[a], Thomas J.R. Hughes[a], Kapil K. Mathur[b] and S. Lennart Johnsson[b]

[a]Division of Applied Mechanics, Stanford University, Stanford, CA 94305–4040, USA

[b]Thinking Machines Corporation, 245 First Street, Cambridge, MA 02142–1264, USA

Abstract

A finite element method for computational fluid dynamics has been implemented on the Connection Machine systems CM-2 and CM-200. An implicit iterative solution strategy, based on the preconditioned matrix-free GMRES algorithm, is employed. Parallel data structures built on both nodal and elemental sets are used to achieve maximum parallelization. Communication primitives provided through the Connection Machine Scientific Software Library substantially improved the overall performance of the program. Computations of three-dimensional compressible flows using unstructured meshes having close to one million elements, such as a complete airplane, demonstrate that the Connection Machine systems are suitable for these applications. Performance comparisons are also carried out with the vector computers Cray Y-MP and Convex C-1.

1. Introduction

The finite element method has taken the lead as an industrial numerical tool because of its ability to handle complex configurations through the use of unstructured meshes. However, there has been some scepticism in the community about how well finite element methodologies would perform on massively parallel computers. Our objective is to demonstrate that such computers are suitable for finite element techniques in large-scale computational fluid dynamics. We have chosen to work on the Connection Machine systems CM-2 and CM-200 built by Thinking Machines Corporation because they appeared as the most mature massively parallel computers both in terms of hardware and software. Finite element methods for structural analysis have been implemented on the Connection Machine system CM-2 by Johnsson and Mathur [1, 2], Belytschko et al. [3] and Farhat et al. [4], among others. Two-dimensional CFD codes using finite element (see [5]) and finite volume techniques (see [6] and references therein) have also been implemented on the Connection Machine system CM-2. These investigations demonstrated the potential of the CM-2 for finite element applications.

An outline of this paper follows: The solution strategy is presented in Section 2. Implementation and communication issues are discussed in Sections 3 and 4, respectively. Numerical examples in Section 5 illustrate the techniques we have used on the Connection Machine systems. Finally, conclusions are drawn in Section 6.

2. Implicit iterative finite element solver

The time-dependent compressible Navier-Stokes equations are discretized using a space-time Galerkin/least-squares variational formulation. This finite element method has been introduced and analyzed by Hughes and Johnson and their respective co-workers. The reader is referred to [7] and references therein for a description of the formulation implemented in our finite element program. Convergence to steady-state is achieved through a time-marching process. At each discrete time t_n, finite element discretization of the compressible Navier-Stokes equations leads to the following nonlinear problem:

Given the solution vector $\widetilde{v}_{(n-1)}$ at time t_{n-1}, and a time increment Δt, find the solution vector $\widetilde{v}$ at time t_n, which satisfies the nonlinear system of equations

$$\widetilde{G}(\widetilde{v}; \widetilde{v}_{(n-1)}, \Delta t) = 0 \tag{1}$$

$\widetilde{G}$ is a system of nonlinear functionals of $\widetilde{v}$ and of parameters $\widetilde{v}_{(n-1)}$ and Δt. This system is solved for $\widetilde{v}$ by performing a succession of linearizations through a truncated Taylor series expansion of $\widetilde{G}$. This leads to a set of linear systems of equations of the form

$$\widetilde{J}^{(i)} \widetilde{p}^{(i)} = -\widetilde{R}^{(i)} \tag{2}$$

where

$$\widetilde{J}^{(i)} = \frac{\partial \widetilde{G}}{\partial \widetilde{v}}(\widetilde{v}^{(i)}; \widetilde{v}_{(n-1)}, \Delta t) \tag{3}$$

$$\widetilde{p}^{(i)} \stackrel{\text{def}}{=} \widetilde{v}^{(i+1)} - \widetilde{v}^{(i)} \tag{4}$$

$$\widetilde{R}^{(i)} = \widetilde{G}(\widetilde{v}^{(i)}; \widetilde{v}_{(n-1)}, \Delta t) \tag{5}$$

$\widetilde{v}^{(i)}$ and $\widetilde{v}^{(i+1)}$ being the approximations of $\widetilde{v}$ at iterations i and $i+1$, respectively. $\widetilde{R}$ is the residual of the nonlinear problem and $\widetilde{J}$ is the consistent Jacobian associated with $\widetilde{R}$. The consistent Jacobian is often replaced by a Jacobian-like matrix $\widetilde{\mathcal{J}}$ leading to a more stable time-marching algorithm (see Johan et al. [8]). A residual-like vector $\widetilde{\mathcal{R}}$ associated with $\widetilde{\mathcal{J}}$ can be defined as

$$\widetilde{\mathcal{J}} \stackrel{\text{def}}{=} \frac{\partial \widetilde{\mathcal{R}}}{\partial \widetilde{v}} \tag{6}$$

For clarity, the supercript (i) will be dropped in the remainder of this paper.

A scaling (or preconditioning) transformation is first applied to the system of equations $\widetilde{\mathcal{J}} \widetilde{p} = -\widetilde{\mathcal{R}}$ to nondimensionalize it and improve its conditioning. We have used a block-diagonal preconditioner as it has been shown to be both inexpensive and efficient (see Shakib et al. [9]). The size of the blocks equals the number of degrees of freedom per node. Let $\widetilde{W}$ be the nodal diagonal blocks of the left-hand-side matrix

$\widetilde{\mathcal{J}}$. Since our finite element formulation generates symmetric positive-definite nodal diagonal blocks, $\widetilde{\boldsymbol{W}}$ accommodates a Cholesky factorization

$$\widetilde{\boldsymbol{W}} = \widetilde{\boldsymbol{U}}^T \widetilde{\boldsymbol{U}} \tag{7}$$

A two-sided preconditioning step is then applied to $\widetilde{\mathcal{J}}\,\widetilde{\boldsymbol{p}} = -\widetilde{\boldsymbol{R}}$, leading to the scaled system of equations $\mathcal{J}\,\boldsymbol{p} = -\boldsymbol{R}$ with

$$\mathcal{J} = \widetilde{\boldsymbol{U}}^{-T}\,\widetilde{\mathcal{J}}\,\widetilde{\boldsymbol{U}}^{-1} \tag{8}$$

$$\boldsymbol{p} = \widetilde{\boldsymbol{U}}\,\widetilde{\boldsymbol{p}} \tag{9}$$

$$\boldsymbol{R} = \widetilde{\boldsymbol{U}}^{-T}\widetilde{\boldsymbol{R}} \tag{10}$$

This preconditioned system of equations is solved using the Generalized Minimal RESidual (GMRES) algorithm. This algorithm was introduced by Saad and Schultz [10]. Its effectiveness for computational fluid dynamics problems has been demonstrated by several research groups (see, for example, [8, 9, 11, 12]). Since the matrix $\mathcal{J}$ is a Jacobian-like matrix, the matrix-vector products $\mathcal{J}\,\boldsymbol{u}$ of the GMRES algorithm can be replaced by the one-sided finite difference stencil

$$\mathcal{J}\,\boldsymbol{u} \approx \frac{\mathcal{R}(\boldsymbol{v} + \delta\,\boldsymbol{u}) - \mathcal{R}(\boldsymbol{v})}{\delta} \tag{11}$$

where $\boldsymbol{v}$ is the current solution and δ is a small scalar. This approximation circumvents the need for computing and storing the left-hand-side matrix $\mathcal{J}$, thus saving a substantial amount of storage. Matrix-free techniques for finite element applications have been analyzed by Johan et al. [8].

Note that this implicit iterative scheme reduces to computing a succession of block-diagonal preconditioners $\widetilde{\boldsymbol{W}}$ and residual vectors $\widetilde{\boldsymbol{R}}$, or residual-like vectors $\widetilde{\mathcal{R}}$. The classical technique for evaluating $\widetilde{\boldsymbol{W}}$ and $\widetilde{\boldsymbol{R}}$ is first to compute the element arrays $\boldsymbol{w}^e$ and $\boldsymbol{r}^e$ and then obtain the global preconditioner and residual by performing an assembly operation, i.e.,

$$\widetilde{\boldsymbol{W}} = \mathop{\mathbf{A}}_{e=1}^{n_{\text{el}}} \boldsymbol{w}^e \quad \text{and} \quad \widetilde{\boldsymbol{R}} = \mathop{\mathbf{A}}_{e=1}^{n_{\text{el}}} \boldsymbol{r}^e$$

where n_{el} is the number of elements. The basics of finite element programming can be found in [13]. A description of the parallel implementation of the above techniques are presented in the following section.

3. Implementational aspects

Our Fortran 90 finite element program for the Connection Machine systems was derived from a highly vectorized Fortran 77 program written by Shakib and Johan [14]. The conversion to the new Fortran standard was necessary to achieve parallel execution on the Connection Machine systems, since the Connection Machine Fortran (CMF) compiler does not recognize Fortran 77 constructs as parallel instructions. The parallel data structures chosen for the implementation and the work required for the conversion process are described in the following sections.

3.1. Parallel data structures

Appropriate data structures are essential to achieve good performance on a massively parallel computer. A description of possible data structures for finite element methods and a detailed analysis of their storage and arithmetic requirements can be found in [15]. Different data structures have also been analyzed by Farhat et al. [6] for finite volume and finite element applications in computational fluid dynamics. A reduced number of data structures will limit the amount of communication required between the different data sets. Some authors have proposed a single data structure. However, having only one data set seemed cumbersome in our implementation, with a possible loss of finite element generality. Therefore, we have adopted the following two data structures:

1. At the element level, i.e., during the computation of the element arrays w^e and r^e, the elements are assigned to the processing nodes of the Connection Machine systems. It is possible to have several elements per processing node, leading to the notion of virtual processing: Each processing node performs operations on a certain number of elements in a sequential fashion. The allocation of multiple elements to a node is handled by the CMF compiler, and the scheduling of the execution by the Connection Machine run-time system.

2. At the GMRES algorithm level, we assign the nodes of the mesh to the processing nodes with possible virtual processing. All the dot product and DAXPY operations [16] of the GMRES algorithm are then executed in parallel over the nodes (with an additional global sum for the dot product operation).

Note that these data structures are both the most "natural" and the simplest to use in a general finite element program. Experience has shown that simplicity and efficiency are tightly coupled in data parallel programming.

All the arrays of the two data structures are allocated in the distributed memory of the Connection Machine systems through the dynamic allocation capability of Fortran 90. The mapping of array elements to the memory of the processing nodes is controlled by the LAYOUT directives in order to minimize communication needs. No communication is necessary in referencing array elements to the same node. As an example of the use of LAYOUT directives, consider the array RES(NDOF,NUMNP) containing the global residual. NDOF and NUMNP are the number of degrees of freedom per node and the number of nodes, respectively. The directive RES(:SERIAL,:NEWS) for the layout will ensure that the finite element nodes are spread as evenly as possible across the processing nodes (the :NEWS directive), while the degrees of freedom of each node are stored on the same processing node (the :SERIAL directive).

3.2. Fortran 77 to Fortran 90 conversion

The absence of dynamic memory allocation in Fortran 77 has led programmers to simulate their own dynamic allocation. This is usually done by defining a large one-dimensional array and then storing all the data in it. This feature had to be eliminated during the conversion to Fortran 90 to achieve the proper layout of the data structures described above. Each array is dimensioned in the routine where it is needed and then passed to subsequent subroutines.

The CMF compiler parallelizes only Fortran 90 array operations. Therefore, all the DO loops enumerating the nodes and the elements and the operations thereupon had to be replaced by array operations. This change involved merely editing work on the vectorized code. The simplicity of the conversion showed us that vectorization and data parallelism are both based on the concept of nonrecurrence in the operations. An operation which can be vectorized can also be parallelized.

The overall structure of the program remained identical throughout the conversion process, implying that the initial phase of the port was a fairly trivial operation. However, rewriting some parts of the code to take advantage of the Fortran 90 constructs and to optimize memory usage and execution rate [17] was a more lengthy process. The final version of the parallel code has several advantages over the Fortran 77 version: it is shorter by about 30% and easier to read due to the array syntax, promising simplified maintenance of the program. Adding new features to the program is now an easier task. We hope all hardware vendors will provide Fortran 90 compilers, and C compilers possessing similar attributes, on their computers in the near future.

4. Communication issues

Inter-processing node communication on parallel computers is often viewed as the major difficulty for programmers. Communication can represent a substantial part of the total run-time, thus affecting the overall efficiency of the program. The current CMF compiler does not recognize special forms of communication to translate them into optimal code. Optimizing communication is indeed a nontrivial task, and the subject of leading-edge compiler research [18]. Meanwhile optimizing routing of data between processing nodes requires specialized routines written in a low-level language. But, many communication operations are generic, and communication libraries have emerged as means of providing both efficiency and portability. The following sections describe the gather and scatter operations, which are the two types of communication performed by our finite element program.

4.1. Gather operation

The computation of the element data w^e and r^e presented in Section 2 requires the knowledge of the current solution at the element nodes v^e. The vector v^e is obtained by gathering the values from the nodal solution vector $\tilde{v}$. This simply consists of an indirect addressing via the mesh connectivity array. The gather operation is sometimes referred to as localization, or accumulation. In Fortran 90, this operation can be written

```
DO I = 1, NEN
   DO N = 1, NDOF
      VL(I,N,:)  = V(N,IEN(I,:))
   END DO
END DO
```

where VL(NEN,NDOF,NUMEL) is the array containing v^e; V(NDOF,NUMNP) contains $\tilde{v}$ and IEN(NEN,NUMEL) is the mesh connectivity array (see Hughes [13]). The scalar NUMEL is the number of elements; and NEN is the number of element nodes, e.g., NEN = 4 for a linear tetrahedron.

The above code fragment is recognized for parallel execution over the elements by the CMF compiler. However, the executable code generated by the compiler will call for the router to compute the addresses of the data to be gathered each time such an operation is required. The trace of the routing activity, i.e., the paths of all elements being moved, is the same for every gather operation as long as the connectivity and its layout do not change. Hence, the addresses and the routing information only need to be computed once for a given connectivity, and the information stored and reused for subsequent gather operations. The time expended in computing the routing information amounts to a couple of gather operations. Hence, saving the routing information also yields a performance improvement for slowly changing connectivities. The saving in communication time is achieved at the expense of some additional storage required to save the trace.

The CMSSL (Connection Machine Scientific Software Library) [19] primitives sparse_util_gather_setup and sparse_util_gather are used to compute and save the trace, and to perform the actual communication. The performance of these routines for general finite element applications is given in [20] along with a description of the methodology implemented in the primitives. The gather algorithm used in this implementation is a two-step process: First, the nodal data are duplicated as many times as there are elements connected to each node. Then, these duplicated data are sent through a one-to-one mapping to the corresponding elements. The preprocessing step computes this one-to-one communication pattern. In our finite element program we use a slightly modified version of the CMSSL sparse_util_gather routine. The modification facilitates the simultaneous handling of several degrees of freedom per node.

4.2. Scatter operation

Once the element data $\boldsymbol{w}^e$ and $\boldsymbol{r}^e$ are computed using a so-called "embarrassingly parallel algorithm" (i.e., one for which *no* communication is required), their components are scattered to the nodes (also said to be "assembled at the nodes") to evaluate the global preconditioner $\widetilde{\boldsymbol{W}}$ and residual $\widetilde{\boldsymbol{R}}$. The scatter operation is a send operation with addition of the colliding data at the nodes. The assembly of the residual can be written

```
      DO I = 1, NEN
        DO N = 1, NDOF
          FORALL (NEL = 1:NUMEL) RES(N,IEN(I,NEL)) = RES(N,IEN(I,NEL))
    &                                              + RL(I,N,NEL)
        END DO
      END DO
```

where RL(NEN,NDOF,NUMEL) is the element residual array.

The scatter operation presented above is not parallelized by the current CMF compiler. However, several alternatives are available to the programmer:

1. A coloring technique often used on vector computers to implement the scatter operation can also be used here. Such techniques are described in [9] and [21]. The idea is to decompose the mesh into blocks of disjoint elements. It can be easily shown that the number of blocks equals the maximum number of elements connected to a node. Consequently, this method is not suitable on the Connection Machine systems when the number of blocks becomes large (large tetrahedral meshes often require up to 100 blocks), because the routing activity corresponding to each block is not load-balanced.

2. The CMF utility library provides a `CMF_send_add` routine. This routine uses the combining facility of the communication hardware, allowing all data to be scattered in parallel while the required additions are being performed in parallel.

3. As with the gather operation, the routing activity corresponding to the nodal connectivity can be precomputed and the information used subsequently. The CMSSL routines `sparse_util_scatter_setup` and `sparse_util_scatter` [19] perform the preprocessing and the assembly, respectively. The scatter operation is done in a fashion similar to the gather operation: First, the element data are sent to the nodes using a one-to-one mapping, i.e., two or more data values arriving in the same node at the same time are stored in different memory locations on the same processing node. Then, all the values at a node are added up. For some finite element applications, this two-step procedure has been shown to be more efficient than the combining feature of the Connection Machine systems CM-2 and CM-200 router. The CMSSL `sparse_util_scatter` routine was modified in a way similar to the gather routine to simultaneously handle multiple degrees of freedom.

These gather/scatter communication procedures are very general and can be used for any finite element application.

5. Numerical examples and benchmarks

Three-dimensional compressible flow problems were solved using the techniques presented in the previous sections to further evaluate their performance. A local time-stepping strategy associated with the matrix-free GMRES algorithm was used for all cases. The dimension of the Krylov space was set to 5 and the tolerance equaled 0.1. All examples were computed in double precision. The gather/scatter operations were performed using the CMSSL primitives. A comparison with the Fortran 77 version of the code running on vector computers was made when possible. All computations were initiated with the free stream flow. No attempt was made to project coarse mesh solutions on refined meshes. It is believed that this approach would have considerably shortened the solution times. In addition, it is believed that multigrid methods are capable of dramatically reducing solution times. These will be studied in future work.

5.1. Three-dimensional blunt body

This example consists of a Mach 3 viscous flow around a blunt body made of a half-sphere extended by a cylinder. The angle of attack is 0 degrees and the Reynolds number is 1,000 based on the radius of the sphere. The computation was only performed for half the body since the flow is symmetric. The mesh contains 3,566 nodes and 13,280 tetrahedra. A 4-point integration rule was used in the elements. A view of the symmetry plane and half-body is shown in Figure 1. This problem was solved on a 128-processing node Connection Machine system CM-2 running CMSS version 6.0 and CMF version 1.0. The Mach number contours in the symmetry plane after 250 time steps are depicted in Figure 2. One can note the bow shock and the development of the boundary layer on the body. Timings for the first 20 time steps are presented in Table 1. The same problem was also solved on a Convex C-1 using the vectorized version of the finite element program. No timings are reported for the gather/scatter on the Convex C-1, because their vectorization has made them a negligible part of the total time. Several remarks can be made:

1. The 128-processing node Connection Machine system CM-2 is about 12 times faster than the Convex C-1, bringing this Connection Machine system configuration to the performance level of a one-CPU Cray 2.
2. The computation time accounts for two-thirds of the total time. The ratio between computation and communication times is a function of the number of integration points in the elements. If 1-point element quadrature had been used, the computation time would have been substantially smaller. However, the communication time would have remained the same since it is only a function of the number of nodes and elements. This fact is evident in some of the following performance results.

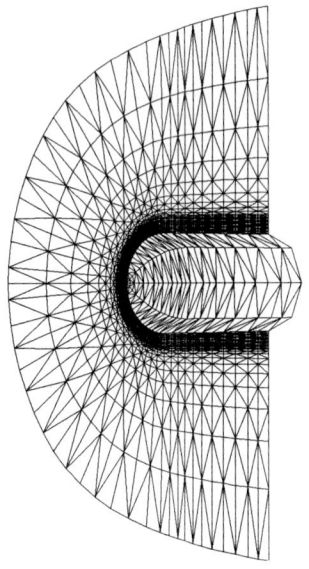 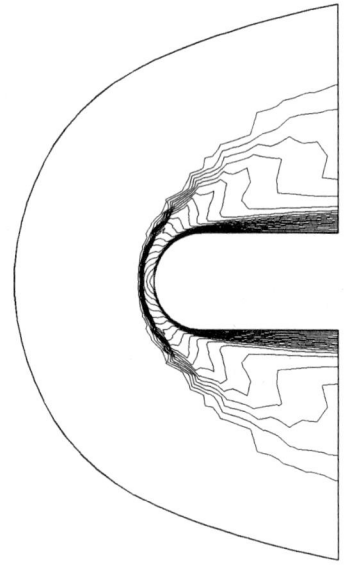

Figure 1. 3-D blunt body. Surface mesh of body and plane of symmetry.

Figure 2. 3-D blunt body. Mach number contours in the plane of symmetry.

Table 1. 3-D blunt body. Computation, communication and total elapsed times on a 128-processing node Connection Machine system CM-2 and a Convex C-1 using a 4-point integration rule.

	Gather	Computation	Scatter	Total
CM-2 (CMSS 6.0)	39 s	170 s	44 s	253 s
Convex C-1	—	—	—	3003 s

5.2. Falcon Jet at level cruise

A transonic inviscid flow at Mach 0.85 was computed around a generic Falcon Jet at a 1 degree angle of attack. The Falcon Jet airplanes are designed and built by Dassault Aviation. The calculation was only done on half the airplane since the flow is symmetric. The first mesh used had 10,202 nodes and 54,957 tetrahedra. The surface mesh for the whole jet can be seen in Figure 3. A 4-point integration rule was used on the elements. A converged solution was obtained after 50 time steps at a CFL number of 10. This problem was solved on a 512-processing node Connection Machine system CM-2 as well as on a one-CPU Cray Y-MP and a Convex C-1. The timings are presented in Table 2. First, a comparison was made between versions 6.0 and 6.1 of CMSS, showing a 15% speedup of the total execution time when upgrading the software. Most of the gain is due to faster communication. CMSS 6.1 yielded a 19-minute run time on the Connection Machine system CM-2 versus 39 minutes on the one-CPU Cray Y-MP. The execution rate on the Cray Y-MP, measured using the hardware performance monitor, is 178 MFlops/s. Hence, an effective performance of 370 MFlops/s is achieved on a 512-processing node Connection Machine system CM-2 running CMSS version 6.1. A mini-supercomputer like the Convex C-1 is not an option for these types of computations.

The same problem was solved on a 2,048-processing node Connection Machine system CM-2 running CMSS 6.0 using a finer mesh. It has about 8 times more data than the previous mesh, with 77,279 nodes and 439,272 elements (see the surface mesh for the complete airplane in Figure 4). The number of equations equals 386,395. The Mach number contours are shown in Figures 5 and 6. Note the supersonic pockets on the outward part of the wings followed by recovery shocks (adaptive mesh refinement is necessary to better resolve the latter). The quality of the calculation can also be deduced from the Mach number contours at the top of the cockpit. For well-designed airplanes cruising at transonic speeds (the Falcon Jet family of airplanes falls in that category), the Mach number on the cockpit remains just below the sonic point during cruise. Supersonic pockets would be followed by recovery shock waves, generating additional unwanted drag. The computation, done under the same conditions as the one on the coarse mesh, took 47 minutes and 50 s, indicating that scalability is achieved since the ratio (elapsed time)×(number of processing nodes)/(number of elements) is very close to that for the previous case.

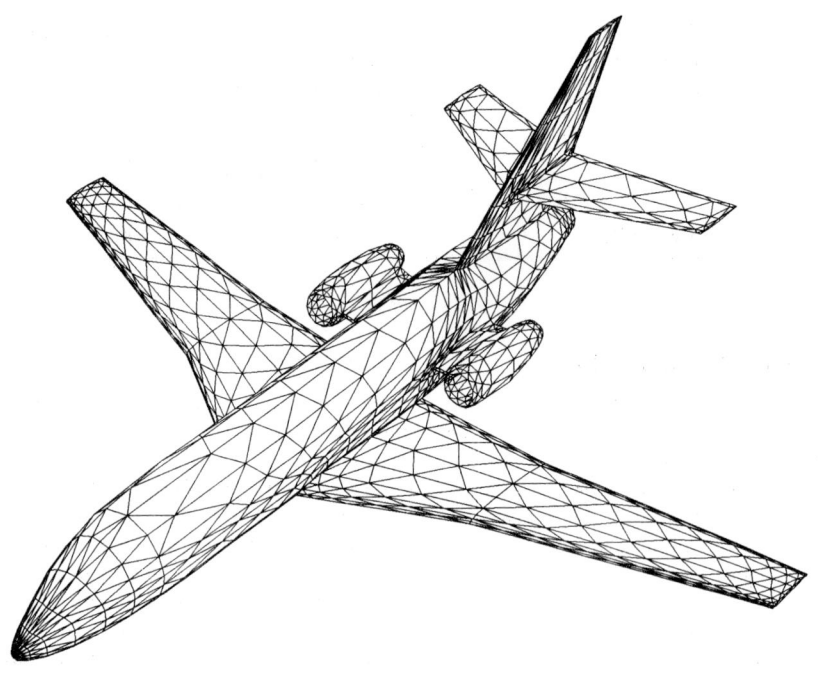

Figure 3. Falcon Jet. Coarse surface mesh.

Table 2. Falcon Jet at level cruise (coarse mesh). Computation, communication and total elapsed times on a 512-processing node Connection Machine system CM-2, a one-CPU Cray Y-MP and a Convex C-1 using a 4-point integration rule.

	Gather	Computation	Scatter	Total
CM-2 (CMSS 6.0)	236 s	897 s	234 s	22 min 47 s
CM-2 (CMSS 6.1)	148 s	833 s	150 s	18 min 51 s
Cray Y-MP	—	—	—	39 min 26 s
Convex C-1	—	—	—	20 h 42 min

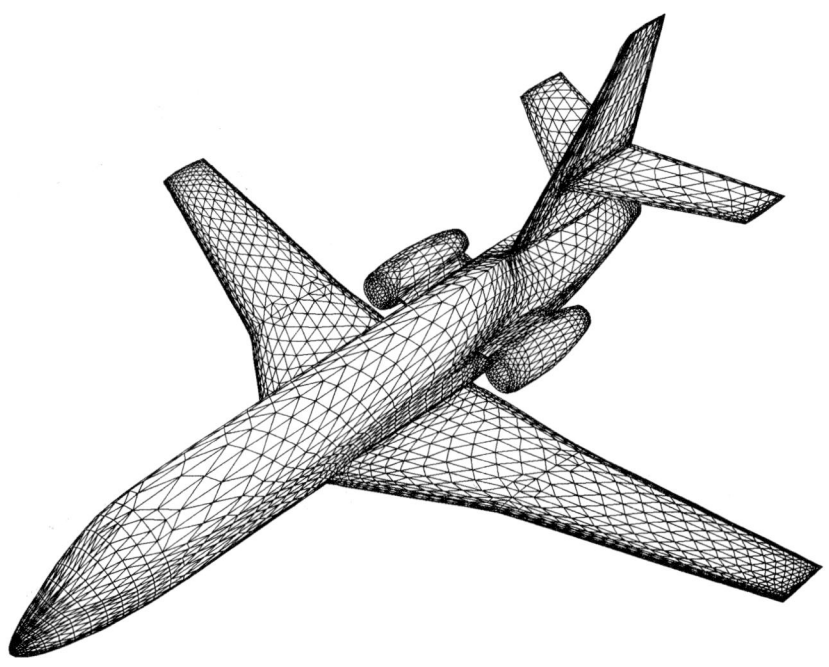

Figure 4. Falcon Jet. Fine surface mesh.

Finally, the fine mesh was used to compare the performances of a 2,048-processing node Connection Machine system CM-2 running CMSS version 6.0 and a system CM-200 running CMSS version 6.1. A 1-point element integration rule was used and 60 time steps at a CFL number of 20 were calculated. The timings are reported in Table 3. The CM-200 running CMSS version 6.1 offered a performance about twice that of the CM-2 running CMSS version 6.0. Note that the ratio between computation and communication has decreased compared to previous computations. This decrease is due to the reduction in the number of integration points per element, as described in Section 5.1.

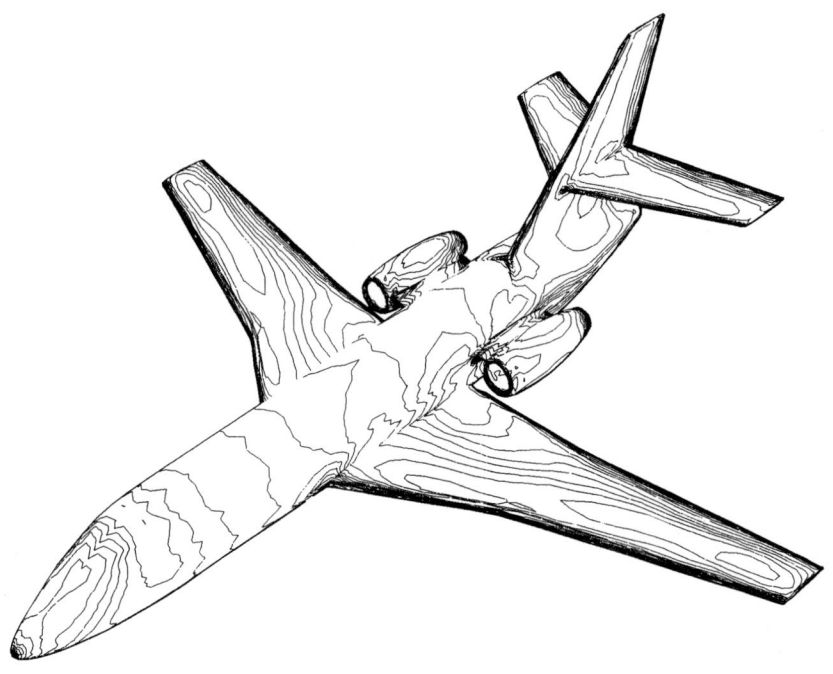

Figure 5. Falcon Jet at level cruise. Mach number contours.

Figure 6. Falcon Jet at level cruise. Mach number contours.

Table 3. Falcon Jet at level cruise (fine mesh). Computation, communication and total elapsed times on 2,048-processing node Connection Machine system CM-2 and CM-200 using a 1-point integration rule.

	Gather	Computation	Scatter	Total
CM-2 (CMSS 6.0)	1051 s	1326 s	967 s	55 min 44 s
CM-200 (CMSS 6.1)	453 s	800 s	415 s	27 min 48 s

6. Conclusions

An implicit iterative finite element solver for computational fluid dynamics has been ported to the Connection Machine systems CM-2 and CM-200 using a data parallel style of programming. The use of Fortran 90 constructs simplified the implementation process, leading to a clearer and better structured program. Communication routines provided through the Connection Machine Scientific Software Library improved substantially the overall efficiency. Performance benchmarks on industrial examples having close to one million degrees of freedom are the final proof that data parallel programming on the Connection Machine systems is suitable for solving CFD problems on unstructured meshes.

All techniques presented in this paper are open for considerable improvement. During the course of the implementation and benchmarking of our finite element code, the performance of some of the system software functions improved by 50% or more. Moreover, improved CMF compilers will appear in the near future, generating faster code at the processing node level. Improvements in mapping and communication procedures are also possible. It is the current thrust of our research. Last, but not least, is the evolution in hardware technology. A recent example is the announcement of the massively parallel Connection Machine system CM-5, which is expected to give an order of magnitude improvement in performance over the CM-2. Initial work on this system has shown great promise for our methodology. In the coming years, we anticipate developments in massively parallel supercomputing to completely revolutionize computational engineering analysis and design.

Acknowledgments

The authors would like to express their appreciation to Steve Daly, John Kennedy and Arthur Raefsky for their interest in this work and their helpful comments. This research was supported by the NASA Langley Research Center under Grant NASA-NAG-1-361 and Dassault Aviation, Saint Cloud, France. The blunt body and Falcon Jet meshes were generated by Dassault Aviation. Access to Connection Machine systems was provided by Thinking Machines Corporation, the Los Alamos Advanced Computing Laboratory and the Department of Geophysics at Stanford University. Access to a Cray Y-MP was provided by the San Diego Supercomputer Center.

References

1 K.K. Mathur and S.L. Johnsson, "The finite element method on a data parallel computing system," *International Journal of High Speed Computing*, **1** (1989) 29–44.

2 S.L. Johnsson and K.K. Mathur, "Experience with the conjugate gradient method for stress analysis on a data parallel supercomputer," *International Journal for Numerical Methods in Engineering*, **27** (1989) 523–546.

3 T. Belytschko, E.J. Plaskacz, J.M. Kennedy and D.L. Greenwell, "Finite element analysis on the Connection Machine," *Computer Methods in Applied Mechanics and Engineering*, **81** (1990) 229–254.

4 C. Farhat, N. Sobh and K.C. Park, "Transient finite element computations on 65,536 processors: The Connection Machine," *International Journal for Numerical Methods in Engineering*, **30** (1990) 27–55.

5 R.A. Shapiro, "Implementation of an Euler/Navier-Stokes finite element algorithm on the Connection Machine," *AIAA 29th Aerospace Sciences Meeting*, AIAA-91-0438, 1991.

6 C. Farhat, L. Fézoui and S. Lantéri, "Computational fluid dynamics with irregular grids on the Connection Machine," *Computer Methods in Applied Mechanics and Engineering*, to appear.

7 F. Shakib, T.J.R. Hughes and Z. Johan, "A new finite element formulation for computational fluid dynamics: X. The compressible Euler and Navier-Stokes equations," *Computer Methods in Applied Mechanics and Engineering*, **89** (1991) 141–219.

8 Z. Johan, T.J.R. Hughes and F. Shakib, "A globally convergent matrix-free algorithm for implicit time-marching schemes arising in finite element analysis in fluids," *Computer Methods in Applied Mechanics and Engineering*, **87** (1991) 281–304.

9 F. Shakib, T.J.R. Hughes and Z. Johan, "A multi-element group preconditioned GMRES algorithm for nonsymmetric systems arising in finite element analysis," *Computer Methods in Applied Mechanics and Engineering*, **75** (1989) 415–456.

10 Y. Saad and M.H. Schultz, "GMRES: A generalized minimal residual algorithm for solving nonsymmetric linear systems," *SIAM Journal of Scientific and Statistical Computing*, **7** (1986) 856–869.

11 L.B. Wigton, N.J. Yu and D.P. Young, "GMRES acceleration of computational fluid dynamics codes," *AIAA 7th Computational Fluid Dynamics Conference*, AIAA CP854, 1985.

12 M. Mallet, J. Périaux and B. Stoufflet, "Convergence acceleration of finite element methods for the solution of the Euler and Navier-Stokes equations of compressible flow," *Notes on Numerical Fluid Mechanics*, Vieweg, **20** (1988) 199–210.

13 T.J.R. Hughes, *The finite element method: Linear static and dynamic finite element analysis*, Prentice-Hall, Englewoods Cliffs, NJ, 1987.

14 F. Shakib, Z. Johan and T.J.R. Hughes, "ENSA-3C: A space-time Galerkin/least-squares finite element program to analyze the compressible Euler and Navier-Stokes equations for general divariant gases," *User's Manual*, Stanford University, Stanford, CA, 1990.

15 S.L. Johnsson and K.K. Mathur, "Data structures and algorithms for the finite element method on a data parallel supercomputer," *International Journal for Numerical Methods in Engineering*, **29** (1990) 881–908.

16 C.L. Lawson, R.J. Hanson, D.R. Kincaid and F.T. Krogh, "Basic Linear Algebra Subprograms for Fortran Usage," *ACM TOMS*, **5** (1979) 308–323.

17 G. Sabot, J. Marantz and D. Gingold, "CM Fortran optimization notes: Slicewise model," *Technical Report*, TMC-166, Thinking Machines Corporation, Cambridge, MA, 1991.

18 M. Chen, Y.-I. Choo and S.L. Johnsson, *Compiler Technology for Massively Parallel Architectures*, DARPA Contract to Yale University and Thinking Machines Corporation, 1991.

19 *CMSSL for CM Fortran, Version 2.2*, Thinking Machines Corporation, Cambridge, MA, 1991.

20 K.K. Mathur, "On the use of randomized address maps in unstructured three-dimensional finite element simulations," *Technical Report*, TMC-37/CS90-4, Thinking Machines Corporation, Cambridge, MA, 1990.

21 R.M. Ferencz, "Element-by-element preconditioning techniques for large-scale, vectorized finite element analysis in nonlinear solid and structural mechanics," *Ph.D. Thesis*, Stanford University, Stanford, CA, 1989.

MAPPING UNSTRUCTURED MESH CFD CODES ONTO LOCAL MEMORY PARALLEL ARCHITECTURES

B W Jones, M G Everett and M Cross

Centre for Numerical Modelling and Process Analysis, University of Greenwich
London, UK.

Abstract

A brief review of the techniques available for decomposing an unstructured mesh for the parallel implementation of CFD codes, leads to the conclusion that those based on recursive spectral bisection and recursive clustering have the most potential. The conventional recursive clustering procedure is extended to yield optimal mappings onto any number of processors with a specified topology. The technique appears as both effective and robust, and particularly suited to internal flow domains where the external boundary may be complex in shape.

1. INTRODUCTION

Computational fluid dynamics calculations often require prodigious amounts of computer processing power. Since parallel architectures provide significant raw processing power it is not surprising that there has been a significant effort by the CFD community to exploit such systems. It is self-evident that in mapping any scalar application onto a local memory parallel architecture it is vital to decompose it in such a way as to

- keep all the processors busy for as large a proportion of the time as possible (ie. balance the computational load),
- minimise the amount, level and frequency of communication between processors and wherever possible constrain the distance of communication to nearest neighbours, and
- distribute the data evenly over the whole processor array.

Initial work on mapping CFD codes onto parallel systems focused upon software which employed structured meshes. Effectively, the approach involved decomposing the mesh and mapping the constituent submeshes onto the processor array in such a way that communication was usually restricted to nearest neighbour. Various approaches for two and three dimensional mapping are described by a number of groups in reference (1). One such approach is described by Johnson and Cross[2] who implemented the commercial CFD code, FLOW3D, onto transputer and i860 based systems. Here, the mesh is decomposed into (i,j) slabs such that each processor has an equal number. The processors are then configured as a simple pipeline and the solution proceeds as in scalar, except that periodically information is exchanged between neighbouring processors to send or receive latest values of solution variables on adjacent (i,j) slabs. Because, each processor has (roughly) the same amount of work to do at each stage, the communications can be synchronised at very little cost to the efficiency of the parallel implementation. On transputer based systems efficiencies of 80%+ have been reported on 50 processor systems running problems with 40,000+nodes[2].

The next stage in exploiting parallel architectures for CFD involves codes based upon unstructured meshes. Although, the approach for such codes should be analogous, the key new problem to be addressed involves the strategy for decomposing the mesh. For structured meshes (even block structured meshes) the strategy is fairly obvious. However, for unstructured mesh codes the decomposition is problem dependent and so algorithms are required which will partition the mesh onto a given processor topology in an optimum fashion. In this paper, we provide a brief overview of the techniques available and summarise progress on an algorithm based recursive clustering.

2. OVERVIEW OF EXISTING APPROACHES TO UNSTRUCTURED MESH COMPOSITION

In the past few years a number of approaches to mesh decomposition have been considered, with:

- simulated annealing
- recursive graph bisection
- nearest neighbour
- recursive spectral bisection, and
- recursive clustering

being the most prominent.

Simulated annealing procedures have been investigated in some detail by Williams[3]. Although this technique has been successful in some optimisation problems, it has not proved to be an effective tool in mesh decomposition[3]. Recursive graph bisection procedures have been explored by Simon and he has established that such techniques are generally inferior to those based upon recursive spectral bisection[4]. Hence, we will focus only on the three remaining procedures:

2.1. Nearest Neighbour approaches

Though often rather crude in implementation, nearest neighbour algorithms can be useful for mapping certain unstructured meshes with a "fairly organised" structure onto a multi-processor configuration[5]. Since total communication costs need to be low one strategy would be to map nearest neighbour elements onto the same or adjacent processors. Typically nearest neighbour algorithms proceed in two phases:

(i) An initial mapping is generated by grouping tasks into clusters, by only placing connecting elements in a cluster.
(ii) The initial mapping is modified using a boundary-refinement procedure to improve the load balancing.

The initial mapping is usually done by a technique known as strip partitioning. The mesh is divided uniformly into horizontal and vertical strips. The two orthogonal strip partitions can then be overlapped in such a way as to generate the same number of regions as processors. The nature of the construction guarantees that the generated partitions satisfy the nearest-neighbour property.

Unfortunately, the strip partitioning procedure does not always work well, particularly on meshes that are genuinely unstructured. As such, its partitions can be arbitrarily poor. The key reason for this poor performance, is that essentially the nearest neighbour strategy is a geometric approach, whereas in reality the task of mesh decomposition is a topological one. This is demonstrated in the example shown in Figure 1. Figure 1(b) shows the simple split into two equal sub-meshes produced by the nearest neighbour approach of the mesh shown in Figure 1(a). Unfortunately, this split, though viable, does not minimise the communication cost between the sub-meshes. The split that achieves this is shown in Figure 1(c). Although, the two arms are located on the same processor and have no connection, the total amount of communication is less than the nearest neighbour split; of course, each processor still has the same size mesh to deal with. Although the mesh in Figure 1 is trivial, it demonstrates the one type of limitation experienced by all approaches which are essentially based upon geometrical rather than topological considerations.

2.2. Recursive Spectral Bisection

The recursive spectral bisection procedure (RSB) has been developed and explored separately by both Williams[3] and Simon et al[4,6]. The approach is based upon the computation of a specific eigenvector of the Laplacian matrix of the connected graph, G.

The Laplacian matrix $L(G) = -D + A$ where A is the adjacency matrix of the graph and D is the diagonal matrix of vertex degrees. This matrix has a number of important algebraic properties which reflect some of the basic structure of the graph. For example all the cofactors are equal and have a value whose modulus equals the number of spanning trees. Since the matrix is obviously singular then zero is an eigenvalue, (in fact, zero is the largest eigenvalue). The eigenvector associated with the second largest eigenvalue has some interesting geometric properties. This eigenvector assigns an axis to the graph and all the nodes may then be located at co-ordinates along the axis. The differences in these co-ordinates gives a distance between the nodes. This information can then be used to partition the vertices in a nearest neighbour type strategy.

The RSB algorithm works as follows:

1. The second largest eigenvalue λ_2 and corresponding eigenvector $\overline{x}_2$ (the Fielder vector) gives some directional information on the graph.
2. Sort the vertices of the graph so that they correspond monotonically to their entries in the Fielder vector.
3. Half of the vertices are assigned to each subdomain.
4. Repeat recursively.

The mesh decomposition produced by the RSB algorithm have been impressive, notably for external flow regions[6].

The RSB algorithm assumes that the decomposition will produce connected sub-meshes (or domains). However, this is not guaranteed and it is not then clear how the algorithm would proceed since the underlying theory only applied to connected graphs. Moreover, when the domains are connected then the resultant partition is obviously nearest neighbour and has all the advantages and suffers the same disadvantages as the general nearest neighbour strategy. So, for example, if the RSB is applied to the graph in Figure 1(a) then it cuts the graph into two blocks down the central block as in Figure 1(b).

2.3. Recursive Clustering

An alternative topological approach which does not suffer from the disadvantages of the RSB, has been developed by Sadayappan et al[7]. The recursive clustering algorithm (RCA) proceeds as follows:

1. Arbitrarily assign each element to one of two clusters A and B such that there is an approximately equal number of elements on each.
2. Evaluate the communication cost of this partition and find which pair of elements when swapped give the maximum reduction in cost. (Note this could be an overall increase in cost).
3. Temporarily removing the previously swapped pair, find the next best pair and continue until no more pairs remain.
4. From the set of all swaps, find the subset which minimises the communication cost. Provided this reduces the cost make the swap.

The procedure can then be repeated in a recursive manner, so that we can obtain 2^n partitions.

The cost function is calculated by counting the number of nodes that are shared between the two partitions. For example, in Figure 2(a) we have eight nodes shared between the two partitions, therefore the communication cost is equal to eight. If element A and element B are swapped, then by looking at Figure 2(b), we can see that the cost has been reduced to six, hence it would be advantageous to implement this swap.

When applied to the simple problem in Figure 1, the RCA produces the desired split in Figure 1(c). Also, when applied to the simple mesh in Figure 3(a), the division in four domains is straightforward and that into eight is also optimum. Indeed, the RCA appears to work well on a wide variety of complex unstructured meshes. However, the RCA does have a number of limitations:

- The mesh can only be partitioned into 2^n clusters.
- Each split, even if perfect (ie. optimal) does not imply an overall optimal solution.
- The optimisation procedure tends to get caught in local minima.

3. EXTENDED RECURSIVE CLUSTERING ALGORITHM

From the survey outlined above it appears that despite its limitations (some of which are shared by the other approaches anyway), the recursive clustering algorithm has the potential to be the most effective at mesh decomposition with minimal inter-processor communication.

In what follows, we describe modifications to the standard RCA to provide a more flexible and robust mesh decomposition software tool.

As published the RCA is limited to splits of 2^n clusters. However, it is a straightforward matter to eliminate this constraint. Say, for example, we have a parallel system with 5 processors. Initially, the mesh is then split arbitrarily into five clusters; then every pair of clusters is operated on to minimise the communication cost as in the conventional RCA. Unfortunately, this method is still susceptible to local minima trap. For example, looking at Figure 4(a), if we wanted to partition this mesh into five, the minimum cost solution would be 12 and the optimum result can be seen in Figure 4(b). However, when using the method above, the result of partitioning into five can be seen in Figure 4(c). The cost of this split is 15. By looking at this split, we can see that the elements that have been assigned to processor 3 are not all connected. The reason for this is that the recursive clustering method only looks at a pair at a time and does not take into account the other partitions. It finds the local minima between the pair but does not find an overall global minimum. Unfortunately, it is the optimisation of the local cost measure using the swapping strategy that leads to the undesirable disconnected domains in Figure 4(c). Moreover, the modified RCA does not account for a specific processor topology. Hence, a version is required which can avoid disconnected domains when they are undesirable (yet exploit them when they are!) and account explicitly for a given processor topology.

One way of modifying the RCA to account for the processor topology is to change the cost function in the optimisation procedure. The simplest function to minimise is the total inter processor distance travelled over the topology which enables all relevant communications to take place. For example, suppose the processor topology in the above illustration was configured as a simple chain as in Figure 5(a). Obviously we would not want an element on processor 1 to have its neighbour on processor 5 as they would have to communicate via 4 other processors. Figure 5(b) shows a simple 40 - element mesh together with an assignment of element numbers. The element numbers have been randomly generated and we assign elements 1-8 to the first processor, elements 9-16 to the second, etc. The unmodified cost function is based upon the number of shared nodes and hence takes no account of the location of neighbouring elements. In contrast the modified cost contains a distance measure of communication cost which becomes part of the optimisation procedure. For example, suppose an element on processor 1 shares a node with an element on processor 5, then the cost of this node is now 4 (not 1) since it has to be communicated via 4 other processors. A further example is given by elements 33 and 12 which have to communicate via 3 other processors and share 2 nodes; we therefore assign each node a cost of 2. The cost of each node in the mesh is calculated and summed up to give what we shall call the global cost.

The basis of this global cost is to force elements to be on the same or neighbouring processors. If two neighbouring elements are on different processors which are some distance away, then it obviously makes the value of the global cost much higher. However, if one of the elements was moved to the same processor as its neighbour, then the global cost is greatly reduced. The global cost now reflects the processor topology and we proceed with a standard minimisation procedure. If we use the mesh given in Figure 4(a), then we can see by looking at Figure 5(c) that by using this global cost method, we generate an optimum decomposition where all neighbouring elements are on the same or neighbouring processors and the minimum cost of 12 is achieved.

Figure 6(a) shows a simple example of a geometry that has a key feature of internal flows (ie. the external boundary is much more complex than in external flows). Figure 6(b) shows the decomposition of the mesh for a pipeline of four processors produced by the extended RCA. We also attempted to perform a decomposition of the domain with the RSB algorithm. Unfortunately, in the first split into two clusters, one contained sub-meshes that were disconnected. As pointed out above, the RSB algorithm can only deal with connected domains and so could not decompose the mesh in Figure 6(a) onto four processors adequately.

4. CONCLUSIONS

From a review of the existing approaches to unstructured mesh decomposition for mapping CFD codes onto local memory parallel architectures, we have concluded that whilst the Recursive Spectral Bisection algorithm may well be adequate for external flow domains, it will not be robust enough for internal flow domains. An alternative approach to RSB is based on recursive clustering. The conventional RCA, despite some limitations, appears to have the best potential for producing optimal and robust mesh decompositions for complex shaped external boundaries, such as are typical of internal flows. In this paper, we have shown how that the conventional RCA may be extended to yield a tool which gives optimal decompositions of 2D unstructured meshes for pipeline processor topologies.

The extended RCA can handle 3D meshes without modification (in a manner similar to the RSB) and future work will focus upon the extension of the software tools to map 3D meshes onto any specified processor topology.

5. REFERENCES

1 K G Reinsch et al (eds), Parallel Computational Fluid Dynamics '91, pub North-Holland (1992).
2 S P Johnson and M Cross, Mapping structured grid 3D CFD codes onto parallel architectures, Appl Math Modelling, 15, 394-405 (1992).
3 R D Williams, Performance of dynamic load balancing algorithms for unstructured mesh calculations, Concurrency: Practice and Experience, 3, 457-481 (1991).
4 H D Simon, Partitioning of unstructured problems for parallel processing, Computing Systems in Engineering (in press).
5 P Saddayappan and F Ercal, Nearest-neighbour mapping of finite element graphs onto processor meshes, IEEE Trans on Computers, C-36, 1408-1421 (1987).
6 V Venkatakrishnan, H D Simon and T J Barth, A MIMD implementation of a parallel Euler solver for unstructured grids, Jnl of Supercomputing (in press).
7 P Saddayappan et al, Cluster partitioning approaches to mapping parallel programs onto a hypercube, Parallel Computing, 13, 1-16 (1990).

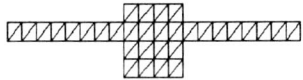

a) Original mesh

b) Split into two using nearest neighbour

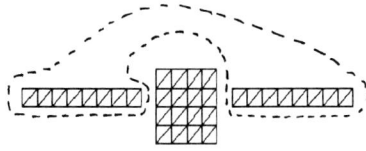

c) Ideal decomposition with minimal communicaton cost

Figure 1 A simple mesh illustrating the limitation of nearest neighbour technique

a) Cost=8   b) Cost=6

Figure 2 An example of the local cost function used in the recursive clustering method

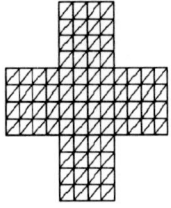

a) The total mesh

b) Decomposition onto four processors

c) Decomposition onto eight processors

Figure 3 A mesh to illustrate the potential of the recursive clustering algorithm

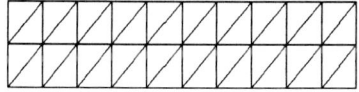

a) Simple mesh

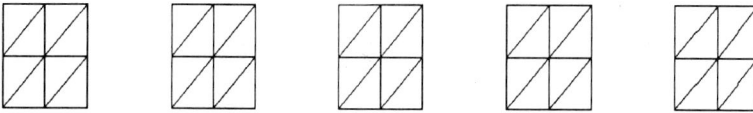

b) Optimal solution cost=12

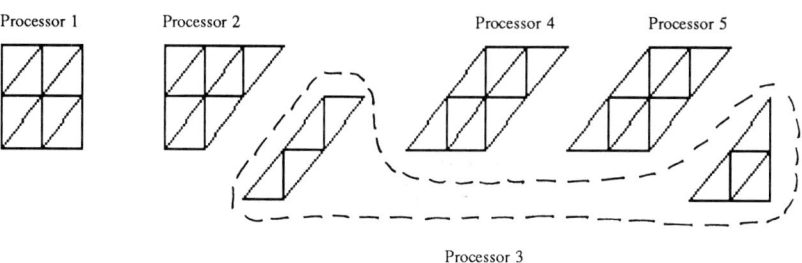

c) Solution using RCA, cost=15

<u>Figure 4</u> Limitation of the recursive clustering algorithm on an arbitrary number of processors

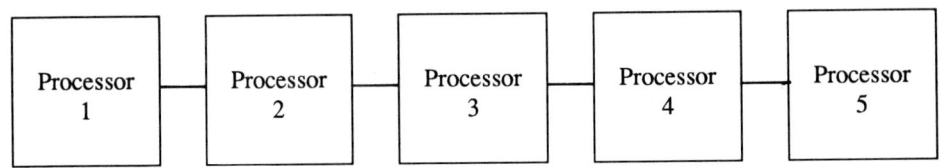

a) Processor pipeline

b) Original mesh with element numbers randomly generated

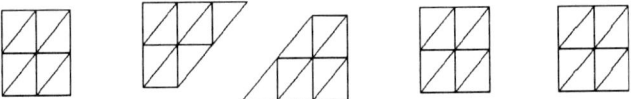

c) Split into 5 using the global cost method, cost=12

Figure 5 An improved version of the recursive clustering algorithm for an arbitrary number of processors

a) Original mesh   b) Partitioned into 4

Figure 6 Using the improved recursive clustering algorithm on a more complex problem

Parallel Adaptive Navier-Stokes Algorithm on the Cray Y-MP/8

Y. Kallinderis[*] and A. Vidwans [†]

Dept. of Aerospace Engineering and Engineering Mechanics

The University of Texas at Austin

Austin, TX 78712

Abstract

A novel parallel adaptive grid algorithm for simulation of viscous flows has been developed and implemented on the Cray Y-MP/8. The algorithm solves the full Navier-Stokes equations in two dimensions. The numerical scheme is a typical time-marching Finite-Volume central-type differencing method. The evolving flow field is monitored and the adaptive algorithm places additional points in regions of relatively large gradients, while it removes points from regions in which the field variation is relatively small. The adaptive grid algorithm was made parallel by employing a novel data structure, which uses shifting of cells within the tree data structure each time a cell is divided or undivided. The parallel algorithm was implemented on an eight-processor Cray Y-MP system. Substantial speedup compared to the corresponding sequential algorithm was realized. Speeds of approximately one Gigaflop were attained.

1 Introduction

Computational fluid dynamics (CFD) has advanced rapidly over the last two decades and it is recognized as a valuable tool for engineering design. However, numerical simulation of viscous flow fields remains very expensive even with use of current vector computers. Advances in numerical algorithms are not expected to reduce the cost of those computations to the extent that they can routinely be applied for design.

Vector computers consisting of few powerful processing units that work independently accelerated computations by one or two orders of magnitude compared to scalar machines, which is not sufficient for efficient large scale flow simulations. Another approach to

---
[*]Assistant Professor
[†]Graduate Research Assistant , Dept. of Electrical and Computer Engineering

computer architectures has been employment of a number of processors that work in parallel executing the same job. Parallel computing appears to be a promising approach for future design applications of CFD. Novel numerical algorithms that are suitable for parallel execution are required. Those are generally very different from the corresponding sequential algorithms.

Euler solvers in 3-D, and Navier-Stokes schemes in 2-D for structured grids have been implemented on the CM-2 [4, 5]. The Intel iPSC/860 has been used for 2-D unstructured grid generation [6], as well as for 3-D Euler solver with unstructured grids [7]. The Cray Y-MP8 has been employed for the 2-D Maxwell's equations of electromagnetics on structured grids. [8].

Adaptive algorithms are flexible in adjusting the grid during the solution procedure without intervention by the user. Several of those adaptive grid algorithms have been developed for sequential execution and have reached a level of maturity. However, work on parallel adaptive algorithms is just starting, and the area is basically unexplored.

The present work developes a novel parallel adaptive grid refinement / coarsening algorithm. Creation as well as updating of the data structure after each grid adaptation is done in parallel. Furthermore, a parallel Finite-Volume algorithm, which uses explicit time-marching and central differencing type of spatial discretization combined with second and fourth order artificial dissipation, has been developed. The parallel adaptive scheme is implemented on an eight-processor Cray Y-MP system. Parallel execution speeds of approximately one Gigaflop were attained.

In the following, the adaptive Navier-Stokes method is presented first. Then the corresponding parallel algorithms for both the Finite-Volume scheme, and the adaptive-grid method are described. Finally, results with parallel execution on the Cray Y-MP8 are presented and discussed.

2 Adaptive Finite-Volume Algorithm

The system of the two-dimensional Navier-Stokes equations is written in cartesian two-dimensional conservation form as :

$$\frac{\partial \mathbf{U}}{\partial t} + \frac{\partial \mathbf{F}}{\partial x} + \frac{\partial \mathbf{G}}{\partial y} = \frac{\partial \mathbf{R}}{\partial x} + \frac{\partial \mathbf{S}}{\partial y} \tag{1}$$

where $\mathbf{U}$ is the state vector, $\mathbf{F}, \mathbf{G}$ are the convective flux vectors in the x and y- directions, respectively. Also, $\mathbf{R}, \mathbf{S}$ are the corresponding viscous flux vectors.

A one-step Lax-Wendroff-type integration scheme [2, 3] has been employed. The method integrates the above relation over the cell-area, leading to the following discrete relation for the inviscid part:

$$\frac{\Delta \mathbf{U}_c}{\Delta t} S + \frac{\mathbf{F}_{se} + \mathbf{F}_{sw}}{2}(y_{se} - y_{sw}) - \frac{\mathbf{G}_{se} + \mathbf{G}_{sw}}{2}(x_{se} - x_{sw})$$
$$+ \frac{\mathbf{F}_{ne} + \mathbf{F}_{se}}{2}(y_{ne} - y_{se}) - \frac{\mathbf{G}_{ne} + \mathbf{G}_{se}}{2}(x_{ne} - x_{se})$$

$$+ \frac{\mathbf{F}_{nw} + \mathbf{F}_{ne}}{2}(y_{nw} - y_{ne}) - \frac{\mathbf{G}_{nw} + \mathbf{G}_{ne}}{2}(x_{nw} - x_{ne})$$
$$+ \frac{\mathbf{F}_{sw} + \mathbf{F}_{nw}}{2}(y_{sw} - y_{nw}) - \frac{\mathbf{G}_{sw} + \mathbf{G}_{nw}}{2}(x_{sw} - x_{nw}) = 0 \qquad (2)$$

where S is the cell area, and Δt is the corresponding time-step. The subscripts se, ne imply values of $\mathbf{F}, \mathbf{G}, x, y$ at the corners se, ne, etc.

The state-vector change in time $\Delta \mathbf{U}$ at the center of the cell is distributed to the corners ($\delta \mathbf{U}$) using the following formulas [2]:

$$(\delta \mathbf{U})_{sw} = \frac{1}{4}\{\Delta \mathbf{U} - \Delta f - \Delta g\} \qquad (3a)$$

$$(\delta \mathbf{U})_{nw} = \frac{1}{4}\{\Delta \mathbf{U} - \Delta f + \Delta g\} \qquad (3b)$$

$$(\delta \mathbf{U})_{ne} = \frac{1}{4}\{\Delta \mathbf{U} + \Delta f + \Delta g\} \qquad (3c)$$

$$(\delta \mathbf{U})_{se} = \frac{1}{4}\{\Delta \mathbf{U} + \Delta f - \Delta g\} \qquad (3d)$$

where

$$\Delta f \equiv \frac{\Delta t}{S}(\Delta \mathbf{F} \Delta y^l - \Delta \mathbf{G} \Delta x^l) \qquad (4a)$$

$$\Delta g \equiv \frac{\Delta t}{S}(\Delta \mathbf{G} \Delta x^m - \Delta \mathbf{F} \Delta y^m). \qquad (4b)$$

and $\Delta \mathbf{F} \equiv (\frac{\partial \mathbf{F}}{\partial \mathbf{U}})\Delta \mathbf{U}$, $\Delta \mathbf{G} \equiv (\frac{\partial \mathbf{G}}{\partial \mathbf{U}})\Delta \mathbf{U}$

The subscripts sw, nw, ne, se denote the four corners of a grid cell. The cell-metric terms $\Delta x^l, \Delta y^l, \Delta x^m, \Delta y^m$ denote the cell dimensions in the l and m cell-directions.

The term $\Delta \mathbf{U}_{vis}$ representing the change in time of the state-vector due to the viscous terms only is calculated by performing a line integration around cell $ABCD$ (Figure 1) using the midpoint rule:

$$\Delta \mathbf{U}_{vis} = \frac{\Delta t}{S} \oint_{ABCD} (\mathbf{R} dy - \mathbf{S} dx) = \frac{\Delta t}{S}\{ + \mathbf{R}_{da}\Delta y_{da} - \mathbf{S}_{da}\Delta x_{da}$$
$$+ \mathbf{R}_{cd}\Delta y_{cd} - \mathbf{S}_{cd}\Delta x_{cd}$$
$$+ \mathbf{R}_{bc}\Delta y_{bc} - \mathbf{S}_{bc}\Delta x_{bc}$$
$$+ \mathbf{R}_{ab}\Delta y_{ab} - \mathbf{S}_{ab}\Delta x_{ab}\}. \qquad (5)$$

In the above expression, the $\mathbf{R}$ and $\mathbf{S}$ terms are evaluated at midfaces da, cd, bc, ab of the *primary* cell $ABCD$. The metric terms are:

$$\Delta y_{da} = y_d - y_a \; , \; \Delta x_{da} = x_d - x_a \; , etc.$$

The terms represented by **R, S** include stress and heat conduction terms, and involve first order spatial derivatives. Green's theorem is applied in order to discretize the derivatives in a similar way as in equation 5.

The standard second and fourth order smoothing operators due to [1] are employed in order to capture shock waves, as well as to damp spurious oscillations.

2.1 Adaptive Grid Refinement / Coarsening

Grid adaptation consists of adjusting the grid spacing so that the numerical error is relatively small and equally distributed throughout the solution domain. This is accomplished by increasing grid resolution locally in regions in which flow features exist. Initial coarse grid-cells are divided by inserting additional points in between the initial points, thus creating a local embedded grid. Several levels of such finer grids are allowable, and they can be limited to those regions of the domain in which important features exist. Conversely, excessive resolution is removed by deleting grid-points locally over regions in which the solution does not vary appreciably. Details of the adaptive algorithm can be found in [2].

3 Parallel Algorithm for the Solver

The Finite-Volume solver consists of the following steps:

- Flow of data from a set of entities called the node-set to the set of entities called the cell-set.

- Independent processing of this data at every member of the cell-set.

- Transfer of data back from members of the cell-set to the corresponding members of the node-set.

- convergence test based upon data accumulated globally over all members of the node-set.

The computational grid is represented as a cell-set, a node-set, and a "mapping" (a set of pointers) from the cell-set to the node-set which defines the physical association between the members of the two sets. Every member of the these sets has a group of "attributes" associated with it which represent different computational quantities such as pressure, velocity etc.

Based upon the above representation, the actual algorithm can be described as follows :

- Rout the data held in attributes of the node-set to the members of the cell-set based upon the mapping defined from the cell-set to the node-set. This is termed as the "gather" operation.

- Perform the cell-based computations at the level of every member of the cell-set. This operation inherently parallel and free of any form of communication between the cells.

- Rout the newly calculated attributes for every cell back to the nodes using the mapping defined previously. This is termed as the "scatter" operation.

- Perform global operations such as average, maximum on the node-attributes and determine whether a state of convergence has been achieved.

- Repeat the above steps if convergence has not been achieved.

The overall worst-case complexity of the solver algorithm is determined by the worst of the worst-case complexities of the constituent operations. While the cell-based operations are essentially $O(1)$ and consequently not a factor, the others can be easily seen to be $O(\log n)$. This is the case because the mapping between the cells and the nodes can be thought to be a permutation of complexity $O(\log n)$ where n is the maximum of the cardinalities of the node and cell sets [9]. The convergence test is a simple binary tree operation on the node-attributes and is consequently also $O(\log n)$.

A potential mutual exclusion problem can arise in the scatter step if the mapping from cells to nodes is not one-to-one resulting in more that one cells attempting simultaneous writes on the same node. This can be eliminated by insisting that all such mappings be one-to-one.

It should be noted that depending upon the cardinality of the mapping from the cells to the nodes (4 for a 2-D quadrilateral grid), the gather and scatter steps are performed multiple times. However, since this cardinality is independent of the size of the input viz. the number of nodes/cells, it merely affects the constant factor in the complexity leaving the overall order unchanged.

3.1 The Model of Parallel Computation

The above algorithm defines an abstract "model of parallel computation" which is a high-level representation of the above steps and is independent of the underlying architectural platform. This can be pictorially represented as a graph where the nodes are the "units of computation". While the edges represent flow of data among units of computation.

The model consists of the following four types of units of computation:

- Nodes : These represent the operations performed at every member of the node-set. These typically consist of calculating some of the attributes such as viscosity and pressure.

- Gather : This unit of computation represents the "gather" operation or the routing of data from the nodes to the cells. Data flows in on the incoming arcs and is permuted according to the mapping before being sent out on the outgoing arcs to the cells.

- Cells : These perform the cell-based operations after the gather operation has been completed. Incoming arcs carry data from the nodes permuted by the gather unit and outgoing arcs carry data to the scatter unit to be routed back.

- Scatter : This is similar to the gather except that it applies the inverse permutation of the mapping used earlier to rout newly calculated attributes from the cells to the nodes.

- Convergence Test : This is an operation performed globally on one of the node-attributes. Incoming arcs carry this data in, while the outgoing arc carries a boolean value used to decide whether the algorithm is to be terminated.

A schematic for the model appears in Figure 2.

4 Parallel Algorithm for the Adapter

Adaptation consists of locally refining or coarsening the computational grid based upon the properties of the current flow field. Following steps can be identified in this context:

- Feature detection based upon the current flow field.

- Adaptation constraints on introduction/deletion of new nodes and cells.

- Grid restructuring based upon the feature detection.

The feature detector merely examines the current solution represented by attributes of the cell-set members and calculates a new attribute for every cell which indicates whether the cell is to be divided, deleted or left unchanged during the adaptation.

This is followed by a "constraint imposition" step where heuristic constraints are imposed on the adaptation attribute so as to avoid problems such as double interfaces and also to determine, for every cell, which of the associated nodes are to be deleted in case of a cell marked for deletion and added in case of a cell marked for refinement.

The next step is to actually rearrange the grid by introducing additional cells locally wherever necessary and deleting the cells which are not required. The algorithm itself can be described as follows:

- Attributes indicating the status of every cell during adaptation are calculated. This step is essentially similar to the cell-based operations step of the solver algorithm and involves no communication between cells.

- The adaptation constraints are imposed by having every cell communicate to its neighbours, its own status vis-a-vis the adaptation. This is achieved by using a modified form of the gather operation with the mapping used being from the cell-set to itself.

- Grid restructuring is done by performing a parallel prefix operation over the cell-adaptation attribute [10]. This is followed by a monotone rout of the cells based upon the results of the prefix operation. The output of the prefix operation indicates the relative displacement of every member of the cell-set due to adaptation and the monotone rout actually displaces the cells accordingly creating "holes" corresponding to the cells to be added. These are then filled in by calculating the mappings and other cell attributes afresh.

The overall complexity of the operation is once again determined by the worst of the complexities of the individual components. The gather and the monotone rout by virtue of being permutations are O(logn) and so is the parallel prefix algorithm. Since the cell-based operations are essentially O(1), it is clear that the overall complexity of adaptation is also O(logn).

4.1 The Model of Parallel Computation

The above algorithm defines a model of parallel computation which consists of the following units of computation :

- The Feature Detector : Attributes that determine the status of a cell during adaptation flow in over the input arcs, and the adaptation attribute flows out over the output arcs.

- The Modified gather : This unit receives the adaptation attribute as input and modifies it based upon the information obtained at every cell from its neighbours.

- The Parallel Prefix : This unit receives the constrained adaptation attribute and calculates the prefix values.

- The Monotone Rout : This routes the cells and their attributes based upon values calculated received from the parallel prefix.

- The Cell-based operations corresponding to filling up of holes in the restructured grid.

A schematic of the Data flow graph is shown in Figure 3. The models of parallel computations are independent of any particular architecture employed for their implementation. The following sections describe the various machine-specific issues which are associated with the implementation of these algorithms on two principal parallel architectures - the shared memory MIMD and the SIMD machines.

5 Parallel Implementation on the Cray Y-MP/8

The parallel processing capabilities on the CRAY-YMP can be exploited by the use of the "Autotasking" feature wherein the user indicates points of potential parallelism in the implementation by the use of directives which instruct a preprocessor to reconfigure the source program in such a way so as to enable maximum speedup to be obtained.

The usual method employed by the parallelising compiler on the CRAY is to partition a parallel work unit such as a DO-LOOP in Fortran on the processors available with the segment within each processor being vectorised. This enables the user to exploit the processing capabilities of a single high-performance CPU as well as the parallelsim offered by multiple such CPUs.

The parallelising compiler has the capability to convert multiply assigned names to single assignment. This is of particular significance in case of highly parallel sections of code utilising such names as temporaries since user memory is saved to a large extent by not having these names declared as arrays.

5.1 Parallel Solver

The standalone solver (without the adapter) was found to give a maximum speed of approximately 160 Mflops on a single CPU on the eight processor CRAY and a speed of about 1.1 Gigaflops on all the eight processors. While the speed in the former case is entirely due to the vectorisation capabilities of a single CPU the speedup obtained over that in the latter case is because of the parallelism offered by multiple such CPUs.

It is interesting to note that the increase in speed is not in exact proportion to the number of processors employed because of the additional overhead of partitioning work among several processors and some communication overhead involved.

Figure 4 illustrates the attained speed as a function of the number of processors that were utilized in parallel. The obtained speed depends on the number of grid-points. A speed of approximately 1.065 Gigaflops was obtained in the case of 60 K points.

5.2 Parallel Adapter

The case of parallel simulation of supersonic viscous flow ($M_\infty = 1.4$) through a channel with a bump was among the cases that were completed. Figure 5 illustrates the complex flow field in terms of Mach number contours. The computed field includes shock wanes, expansion fans, as well as a shear layer formed at the region of the lower boundary. Two levels of embedding were employed by the parallel adaptive algorithm. Figure 6 shows the two-level adapted grid. It is observed that additional resolution is automatically placed by the adaptive algorithm within the local regions of the flow features. The result was obtained in about 150 seconds when executing the code in parallel on the eight processors of the Cray Y-MP. Parallel execution speeds on the Cray-Y-MP8 are shown in Fig. 7 as a function of number of processors from one to eight. It is observed that the increase in speed is almost linear.

The solver coupled with the adapter was found to give a speed of 140 Mflops for the case when adaptations were done every 3 time steps and the speed with all eight processors approached 0.95 Gigaflops.

It should be noted that the parallel algorithm described above has inherent in it an implicit form of load balancing where the load on every unit of computation is adjusted as soon as the grid is modified - i.e. the load balancing strategy employed here is a "greedy" strategy where the workload is adjusted as soon as the load is perturbed by addition or deletion of a grid point.

6 Summary

A novel parallel adaptive grid algorithm for 2-D Navier-Stokes simulations was developed and implemented on a shared memory MIMD architecture.

The numerical scheme is a typical time-marching Finite-Volume central-type differencing method. Development of the parallel solver was facilitated by splitting the operations of the scheme so that they are local and restricted to within each cell, thus minimizing the communications cost.

The grid is adaptively refined and/or coarsened during the solution process in order to resolve local features efficiently. The adaptive grid algorithm was made parallel by employing a novel data structure, which uses shifting of cells within the tree data structure each time a cell is divided/undivided.

The parallel algorithm was implemented on an eight-processor Cray-Y-MP system. Substantial speedup compared to the corresponding sequential algorithm was realized. Speeds of approximately one Gigaflop were attained.

7 Acknowledgements

This work was supported by NSF Grant ASC-9111540, and monitored by Dr. M. Patrick. This support is gratefully acknowledged. Computing time on the Cray Y-MP8 was provided by the CHPC center of the University of Texas at Austin.

References

[1] A. Jameson, W. Schmidt and E. Turkel. *Numerical Solutions of the Euler Equations by Finite-Volume Methods Using Runge-Kutta Time-Stepping Schemes.* AIAA Paper 81-1259, 1981.

[2] Y. Kallinderis, and J. R. Baron "Adaptation Methods for a New Navier-Stokes Algorithm," *Journal of the American Institute of Aeronautics and Astronautics*, vol. 27 , no. 1 , pp 37-43 ,1989.

[3] Y. Kallinderis, "Adaptation Methods for Viscous Flows," Ph.D. Thesis, MIT, Dept. of Aeronautics and Astronautics, CFDL-TR- 89-5, May 1989.

[4] R. K. Agarwal, "Development of a Navier-Stokes Code on a Connection Machine", AIAA Paper 89-1938-CP, 1989.

[5] L. N. Long, M. M. Khan and H. T. Sharp, "A Massively Parallel Three-Dimensional Euler Method", AIAA Paper 89-1937-CP, 1989.

[6] R. Lohner, J. Camberos, and M. Merriam, "Parallel Unstructured Grid Generation", AIAA Paper 91-1582-CP, 1991.

[7] R. Das, D. J. Mavriplis, J. Saltz, S. Gupta, and R. Ponnusamy, "The Design and Implementation of a Parallel Unstructured Euler Solver Using Software Primitives", AIAA Paper 92-0562, 1992.

[8] V. Shankar, "A Gigaflop Performance Algorithm for Solving Maxwell's Equations of Electromagnetics", AIAA Paper 91-1578-CP, 1991.

[9] V. E. Benes "On Rearrangeable Three-stage Connecting Networks" *The Bell System Technical Journal*, vol. 41, pp 1481-1492, 1962

[10] C. P. Kruskal, L. Rudolph and M. Snir "The Power of Parallel Prefix" *IEEE Transactions on Computers*, Vol 34, pp 965-968

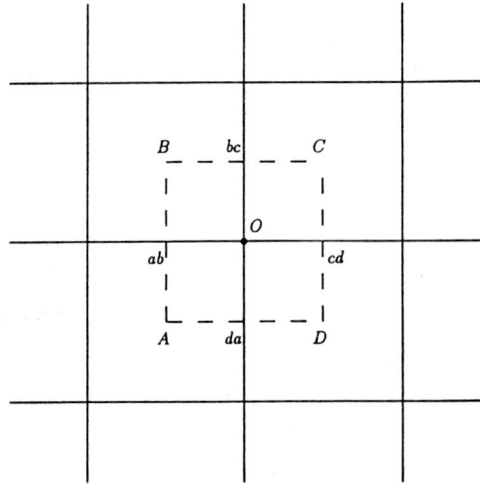

Figure 1: Primary cell for viscous terms discretization

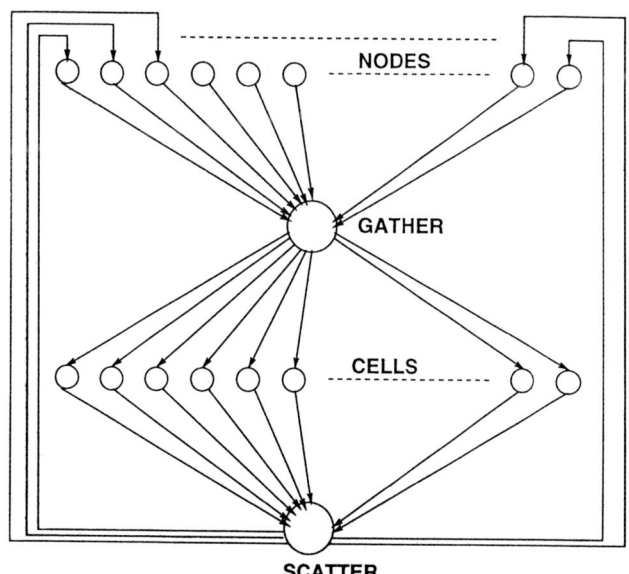

Figure 2: Data flow chart for solver

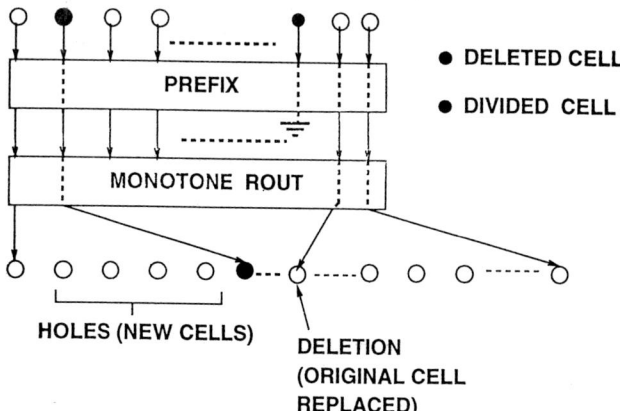

Figure 3: Data flow chart for adapter

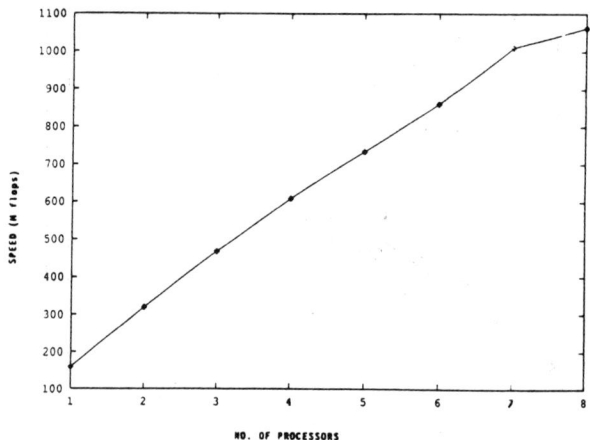

Figure 4: Parallel execution speed of Navier-Stokes Finite-Volume algorithm vs number of processors on the Cray-Y-MP8

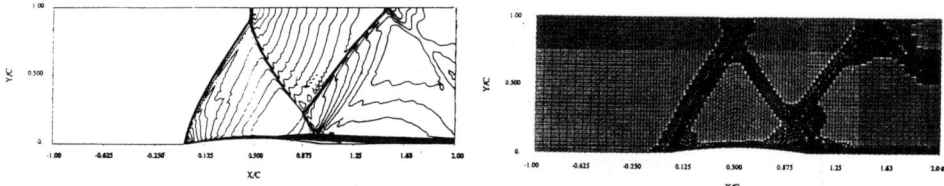

Figure 5: Parallel adaptive simulation of viscous flow through channel with a bumb

Figure 6: Parallel creation of two-level adapted for viscous flow through channel with a bump

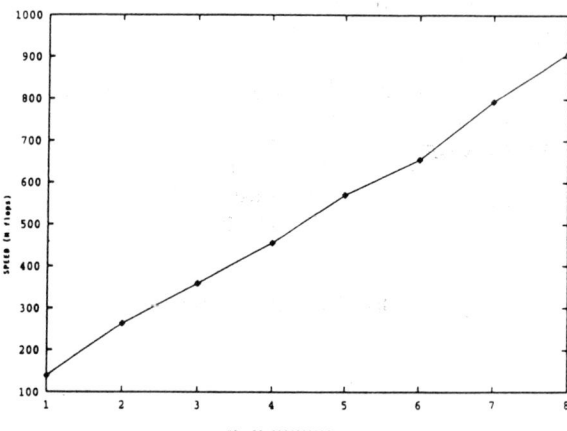

Figure 7: Parallel execution speed of combined solver/grid-adapter vs number of processors on the Cray-Y-MP8 (grid adaption applied every three time-steps).

Challenges posed for parallel processing on iPSC/860 by DNS schemes for supersonic flows

Foluso Ladeinde
Department of Mechanical Engineering, SUNY Stony Brook
Stony Brook, New York 11794-2300

Abstract

We present an implementation of a potentially useful numerical approach for studying the physics of turbulence in supersonic flows in systems with realistic geometries and boundary conditions. The numerical approach is based on the so-called ENO (essentially non-oscillatory) schemes. We discuss parallel implementation on the iPSC/860 Delta machine for arbitrary complex geometries. Other than the low Mflops per processor obtainable from this machine, the small RAM per processor and the absence of a virtual memory manager make the start-up process, which is I/O bound in our procedure, very cumbersome. The procedure presented here is not cheap and ways to speed up the calculations are suggested.

1. INTRODUCTION

The goal in our study is to produce a code which is sufficiently accurate for studying the physics of turbulence in supersonic flows. The type of calculation is sometimes referred to as the direct numerical simulation (DNS) of turbulence.

DNS poses many computational challenges, one of which is the need for highly-accurate calculations in space and time, including at discontinuities. This is a distinction from "engineering turbulence" calculations where accuracy is at best second-order and degrades to first-order at discontinuities. In our work we will address the accuracy problem for high-speed flows but stay within the low R_λ limitation of today's DNS.

Many numerical methods and schemes are capable of delivering the needed accuracy for DNS of incompressible or low Mach number (Ma) flows. However, most of these either break down completely or suffer accuracy degradation when applied to flow fields with strong discontinuities. One exception is the ENO scheme presented by Harten et al. (1987). The latter, or some variations of it, is suitable for DNS, as demonstrated in Osher and Shu (1988) and Shu et al. (1991). Although the procedures of Shu and his colleagues could conceptually be extended to handle more realistic systems implementation is difficult, and has not been attempted, at least to our knowledge.

The approach of Harten and Chakravarthy (1991) seems to be more appropriate for complex geometries and is of interest to us. Although the finite volume approach, in which their formulation was presented, is simpler and requires less programming skill, we have opted for the finite element method because of the flexibility with boundary conditions and for other advantages. We will discuss parallel implementation on iPSC/860 supercomputer for arbitrary complex three-dimensional geometries. The procedure here is not cheap, and our computing machine is inadequate, even for low R_λ's. Some suggestions to speed up calculations are given.

2. GOVERNING EQUATIONS

The equations, in non-dimensional form, and using Cartesian tensors, are

$$\rho_t + (\rho u_m)_{,m} = 0 \tag{1}$$

$$(\rho u_n)_t + (\rho u_n u_m)_{,m} = -(\gamma - 1)(\rho e)_{,n} + \frac{1}{Re}\tau_{nm,m} \tag{2}$$

$$E_t + [Eu_m + (\gamma - 1)\rho e u_m]_{,m} - \frac{1}{Re}(\tau_{nm}u_m)_{,n} - \frac{1}{(\gamma - 1)RePrM^2}(kT_{,m})_{,m} = 0 \tag{3}$$

Above, ρ, u_m, t, x_m, γ, e, and k are the density, mth component of velocity, time, mth coordinate direction, ratio of specific heats, internal energy, and thermal conductivity. We also have the total energy $E = \rho e + \frac{1}{2}\rho u_i u_i$. The parameters of the problem include the Reynold number, Re, the Prandtl number, Pr, and the Mach number, M. The absolute viscosity μ (which appears in equation (4) below), and k, are the remaining (non-dimensional) parameters. The viscous stress τ_{nm} is $\tau_{nm} = -\frac{2}{3}\mu u_{k,k}\delta_{nm} + (u_{n,m} + u_{m,n})$, and the pressure p, density ρ, and temperature T, are related by $p = (\gamma - 1)\rho e$, $e = \frac{1}{M^2\gamma(\gamma-1)}T$.

3. THE BASE SCHEMES

3.1 Spatial integration

The base scheme is Galerkin, but this is subsequently modified in the treatment of the convective terms, both to obtain nonlinear stability and in the implementation of the ENO procedure. Let V represent a real Hilbert Space and $B : V \times V \to \Re$ be a V-elliptic and continuous bilinear form. $f \in V$. For the basis $\{w_h^1, w_h^2, \ldots, w_h^N\}$ of a finite subspace $V_h \in V$, the Galerkin procedure for calculating $u_h \in V_h$ can be written as

$$\sum_{j=1}^{N} B(w_h^j, w_h^i)u_h^j = (f, w_h^i), \quad 1 \leq i \leq N. \tag{4}$$

The latter is a system of linear equations in N unknowns $(u_h^1, u_h^2, \ldots, u_h^N)$, and can be put in a matrix form.

In the present work the primary dependent variables are the density ρ, the mass flux ρu_n (where $n = 1, 2, 3$ are spatial coordinate directions), and total energy, E. For the system in equations (1) – (3), equation (4) can be written as follows.

$$M_{ij}\frac{\partial \rho^j}{\partial t} + S_{ij}^{1u}(\rho u)^j + S_{ij}^{1v}(\rho v)^j + S_{ij}^{1w}(\rho w)^j = F_i^1 \tag{5}$$

$$S_{ij}^{2\rho}\rho^j + M_{ij}\frac{\partial(\rho u)^j}{\partial t} + S_{ij}^{2u}(\rho u)^j + S_{ij}^{2v}(\rho v)^j + S_{ij}^{2w}(\rho w)^j = F_i^2 \tag{6}$$

$$S_{ij}^{3\rho}\rho^j + S_{ij}^{3u}(\rho u)^j + M_{ij}\frac{\partial(\rho v)^j}{\partial t} + S_{ij}^{3v}(\rho v)^j + S_{ij}^{3w}(\rho w)^j = F_i^3 \tag{7}$$

$$S_{ij}^{4\rho}\rho^j + S_{ij}^{4u}(\rho u)^j + S_{ij}^{4v}(\rho v)^j + M_{ij}\frac{\partial(\rho w)^j}{\partial t} + S_{ij}^{4w}(\rho w)^j = F_i^4 \tag{8}$$

$$S_{ij}^{5\rho}\rho^j + S_{ij}^{5u}(\rho u)^j + S_{ij}^{5v}(\rho v)^j + S_{ij}^{5w}(\rho w)^j + M_{ij}\frac{\partial E^j}{\partial t} + S_{ij}^{5e}E^j = F_i^5. \qquad (9)$$

(The matrices in (5)-(9) are defined in the appendix.) We are experimenting with various solution procedures for equations (5)-(9). The one reported here solves the equations in a segregated manner. In this case, to solve for $(\rho u)^j$, for example, only the second and third terms of (6) are retained on the left-hand side, with the other terms moved to the right. Contributions from elements within a subdomain are assembled and the resulting system of ordinary differential equations are integrated in time (section 3.2 below). Finally, the subspaces V_h of the various Hilbert spaces V are constructed by partitioning the subdomain (block) Ω (from domain decomposition) into Lagrange "elements". For $\{w_h^i\}$, we have experimented with the tensor products of both the linear and the cubic one-dimensional basis for the three-dimensional partitions of Ω (so that $N = 8$, 64 at the element level). The high-order basis functions are expected to give high accuracy and a convergence of spectral order, although this is still under investigation. (We sometimes use the words 'subdomain' and 'block' interchangeably even though we would have preferd to use them in physical and computational contexts, respectively.)

The following are some of the subtle points of the spatial integration procedure.

(a) Velocities and their gradients are expressed in terms of the primary dependent variables. For example, $u_{n,m}$ is written as $u_{n,m} = \frac{1}{\rho}(\rho u_n)_{,m} - \frac{1}{\rho^2}\rho_{,m}(\rho u_n)$. Similarly, the heat conduction term in the energy equation is written in terms of E^j and $(\rho u_n)^j$. This is a more expensive procedure, but the implicitness is expected to accelerate convergence. The procedure also removes the need to "smoothen" and interpolate secondary (derived) variables. Further, we have not used a group formulation as this has been found to introduce errors.

(b) It is not clear whether to write the term $(\gamma - 1)\rho e u_m$ in the energy equation as $(\gamma - 1)eu_m N_j \rho^j$, or as $(\gamma - 1)eN_j(\rho u_m)^j$. (This may not be trivial since, for example, we know from the finite volume method that the interpolation of $(uT)_{i+1/2}$ as $\frac{1}{2}((uT)_{i+1} + (uT)_i)$, in the energy equation written in terms of temperature T, is not acceptable for incompressible flows.) We found no consistent argument to prefer one form, and have, somewhat arbitrarily, chosen to work with the first form.

(c) For numerical integration, we have used $2 \times 2 \times 2$ Gauss quadrature points for the 8−node element, and $4 \times 4 \times 4$ for the 64−node element. These are sufficient, as they more than integrate the volumes of elements exactly. Further, we have not assumed as constant, coefficients (such as u in the convective terms) which otherwise could vary within elements. The approximation of a constant value speeds up computation, and will be examined in another study; but we remind the reader that such a procedure has been known to introduce aliasing errors.

3.2 Time integration

As is well-known, the use of both explicit and implicit schemes in one simulation is a relatively easy task in a domain partitioning approach, and this is exploited in the present study. In any case, high-order schemes are needed, of which explicit procedures are more commonplace compared to implicit. (See for example Shu et al., 1991 and Jameson et al., 1981 for high-order explicit schemes.) For an implicit integration we are using a base scheme which, initially, is a third-order, two-stage, Runge-Kutta (RK) procedure. The scheme is actually semi-implicit, in the language of ordinary differential equations. We say "initially" because the Lax-Wendroff method, which is one of the approaches we are testing for nonlinear stability, is potentially useful for increasing the accuracy beyond the third. Assuming an autonomous system $\dot{\mathbf{u}} = \mathbf{f}$, the RK procedure can be written as

$$\mathbf{u}^{l+1} = \mathbf{u}^l + \Delta t \sum_{i=1}^{s} b_i \mathbf{f}(\mathbf{G}_i) \tag{10}$$

where

$$\mathbf{G}_i = \mathbf{u}^l + \Delta t \sum_{j=1}^{s} a_{ij} \mathbf{f}(\mathbf{G}_j) \tag{11}$$

and l is the time level, Δt is the time step size, s is the stage of the RK scheme, and b_i, a_{ij}, and c_i (not shown) are the elements of the Butcher array (Lambert, 1991), which, for the present scheme, have the values $c_1 = \frac{3+\sqrt{3}}{6}$, $c_2 = \frac{3-\sqrt{3}}{6}$, $b_1 = b_2 = \frac{1}{2}$, $a_{11} = a_{22} = \frac{3+\sqrt{3}}{6}$, $a_{21} = -\frac{\sqrt{3}}{6}$, $a_{12} = 0$, and $s = 2$. Thus,

$$\mathbf{G}_1 = \mathbf{u}^l + \Delta t [a_{11} \mathbf{f}(\mathbf{G}_1)] \tag{12}$$

and

$$\mathbf{G}_2 = \mathbf{u}^l + \Delta t [a_{21} \mathbf{f}(\mathbf{G}_1) + a_{22} \mathbf{f}(\mathbf{G}_2)] \tag{13}$$

Each of the ODE's in (5) – (9) can be written as

$$M_{ij} \dot{u}_j + K_{ij} u_j = F_i, \tag{14}$$

where $\mathbf{M}$ is the usual mass matrix, $\mathbf{u}$ is the solution vector, consisting of the nodal values of a particular dependent variable, $\mathbf{K}$ is coefficient matrix for the variable, and $\mathbf{F}$ results from surface fluxes (for elements on the boundary), volumetric sources (where applicable), and from terms moved to the right as a result of the segregated solution procedure.

The equations for ρ^j and E^j are linear, with the segregated approach. However, $\mathbf{K} = \mathbf{K}(\mathbf{u})$ for the three momentum equations (ρu_n). Linearization with the Newton-Rhapson method results in the term $N'_{ij} = -(\rho u)_l \int \frac{N_{i,n} N_j N_l}{\rho_f N_f} d\mathbf{X}$, which has to be added to the coefficient matrix. (The N_i's are the basis functions.)

Substituting $\mathbf{f}$ from (14) into (12) and (13), allowing for Newton-Rhapson linearization, and using a residual formulation yields the following algebraic equations for $\mathbf{G}_1$, $\mathbf{G}_2$:

$$[M_{ij} + \alpha_1 K_{ij}(\mathbf{G}_{1,m}^{(l+1)/2}) + \alpha_1 \mathbf{N}_{ij}'(\mathbf{G}_{1,m}^{(l+1)/2})]\Delta G_{1,j,m}^{(l+1)/2} = M_{ij}u_j^l + \alpha_1 F_i(\mathbf{G}_1^{(l+1)/2})$$
$$- [M_{ij} + \alpha_1 K_{ij}(\mathbf{G}_{1,m}^{(l+1)/2})]G_{1,j,m}^{(l+1)/2} \quad (15)$$

$$[M_{ij} + \alpha_1 K_{ij}(\mathbf{G}_{2,m}^{l+1}) + \alpha_1 \mathbf{N}_{ij}'(\mathbf{G}_{2,m}^{l+1})]\Delta G_{2,j,m}^{l+1} = M_{ij}u_j^l +$$
$$\alpha_2[F_i(\mathbf{G}_1^{(l+1)/2}) - K_{ij}(\mathbf{G}_{1,m}^{(l+1)/2})G_{1,j}^{(l+1)/2}] + \alpha_1 F_i(\mathbf{G}_2^{l+1}) - [M_{ij} + \alpha_1 K_{ij}(\mathbf{G}_{2,m}^{l+1})]G_{2,j,m}^{l+1} \quad (16)$$

Above, $\alpha_1 = a_{11}\Delta t$, $\alpha_2 = a_{21}\Delta t$, m is the sub-iteration level within a time level, and i and j are indeces ranging from one to the number of equations. Also,

$$G_{1,j,m+1}^{(l+1)/2} = G_{1,j,m}^{(l+1)/2} + \Delta G_{1,j,m}^{(l+1)/2} \quad (17)$$

$$G_{2,j,m+1}^{l+1} = G_{2,j,m}^{l+1} + \Delta G_{2,j,m}^{l+1} \quad (18)$$

(Note that $\mathbf{G}_1$ must be solved to convergence before solving for $\mathbf{G}_2$.) The dependent variables can be obtained, at the end of the second stage, using

$$u_j^{l+1} = [1 + \lambda_1 - \lambda_2 - \lambda_3]u_j^l + [\beta_1 - \beta_2]G_{1,j} + \chi G_{2,j} \quad (19)$$

where

$$\lambda_1 = \frac{b_2 a_{21}}{a_{11}^2}, \quad \lambda_2 = \frac{b_1}{a_{11}}, \quad \lambda_3 = \frac{b_2}{a_{11}}, \quad \beta_1 = \frac{b_1}{a_{11}} \quad \beta_2 = \frac{b_2 a_{21}}{a_{11}^2}, \quad \chi = \frac{b_2}{a_{11}}$$

We have observed that (15) becomes the backward Euler scheme if α_1 is set to Δt, while (16) is the trapezoid rule if $\alpha_1 = \alpha_2 = \Delta t/2$, and $G_{1,j}^{(l+1)/2}$ is replaced by u_j^l. Thus the present RK scheme is actually an implicit predictor-corrector method, with a backward Euler predictor and a trapezoid rule corrector.

3.3 Nonlinear stability

Although numerous approaches to obtain nonlinear stability have been suggested it is not clear whether these are suitable for DNS. The exemption are procedures based on upwinding and flux-splitting, but only when these are interpolated to high order - third, fourth, etc; (Rai and Moin, 1991). Even these latter appraoches will degrade to first order in the vicinity of discontinuities, making them suspect for DNS of supersonic flows. In conjunction with the ENO procedure discussed in the next section, we are, at the moment experimenting with approaches based on Lax-Wendroff, upwind test functions (see for example Brueckner et al. in this volume), and schemes based on approximate Riemann solvers, in the sense of Roe (1986) and others.

4. THE ENO PROCEDURE

The ENO scheme can be traced back to the work of Godunov (1959) and the second-order scheme of van Leer (1978). Harten and his colleagues (1987) proposed an arbitrarily high-order procedure, which forms the basis of the ENO scheme. The work by Harten and Chakravarthy (1991) is more relevant to the present work. ENO is used in connection with the ability to obtain uniformly high-order accuracy, including at the vicinity of disontinuites. It is known that, with this method, oscillations of the order of the interpolation error cannot be ruled out in the smooth part of the flow.

We define $\bar{u}_j^l$ as the cell-average of the approximate solution $u(x,t^l)$. Given $\bar{u}_j^l = \{\bar{u}_j^l\}$ we compute $\bar{u}_j^{l+1}$ as

$$\bar{u}_j^{l+1} = A \cdot E(\tau) \cdot R(x; \bar{u}_j^l) \tag{20}$$

where A is the cell-average operator, $E(\tau)$ is solution operator for the evolution equation, and $R(x; \bar{u}_j^l) \equiv R(x; \bar{u})$ is an approximation to $u(x,t)$, obtained by a reconstruction from the cell-average values $\bar{u}_j^l$. In equation (21) above R is the only source of oscillation if E is exact or a TVD scheme is used. To minimize this oscillation, which occurs because we discretize across discontinuity, the information to construct R is taken from the smooth part of the flow.

The reconstruction $R(x; \bar{u})$ satisfies

$$R(x; \bar{u}) = u(x) + O(h^r), \tag{21}$$

where u is smooth, and we require for consistency, that

$$A(C_j)R(x; \bar{u}) = \bar{u}_j \tag{22}$$

We denote the reconstructed polynomial in element C_i by $R_i(x; \bar{u})$. Taylor series expansion of R_i about $\bar{x}_i$ gives

$$R_i(x; \bar{u}) = \sum_{k=0}^{r-1} \frac{1}{k!} \sum_{|l|=k} (x - \bar{x}_i)^l D_l; \quad x \in C_i \tag{23}$$

Above, we have used the multi-index notation, and it is easily shown that for $0 \leq |l| \leq r-1$,

$$D_0 = R_i(x_i; \bar{u}) = u(\bar{x}_i) + O(h^r); \quad \left(D_l = \frac{\partial^{|l|} u}{\partial x_1^{l_1} \partial x_2^{l_2} \ldots \partial x_s^{l_s}} + O(h^{r-|l|}) \right) \tag{24}$$

To obtain D_l needed for the reconstruction we average $R_i(x, \bar{u})$ over all elements in a stencil $J(i)$ associated with element C_i:

$$A(C_j)R_i = \bar{u}_j; \quad j \in J(i), \tag{25}$$

The matrix problem for D_l is

$$\sum_{k=0}^{r-1} \sum_{|l|=k} a_{j,l} D_l = \bar{u}_j; \quad j \in J(i) \tag{26}$$

where

$$a_{j,l} = \frac{1}{k!} A(C_j)(x - \bar{x}_i)^l = \frac{1}{k! |C_j|} \int_{C_j} (x - \bar{x}_i)^l dV \tag{27}$$

In the present work we consider $r = 4$. The equation for an element in the stencil becomes:

$$a_{j,(0,0,0)}D_{(0,0,0)} + a_{j,(0,0,1)}D_{(0,0,1)} + a_{j,(0,1,0)}D_{(0,1,0)} + a_{j,(1,0,0)}D_{(1,0,0)} + a_{j,(0,0,2)}D_{(0,0,2)} +$$
$$a_{j,(0,1,1)}D_{(0,1,1)} + a_{j,(0,2,0)}D_{(0,2,0)} + a_{j,(1,0,1)}D_{(1,0,1)} + a_{j,(1,1,0)}D_{(1,1,0)} + a_{j,(2,0,0)}D_{(2,0,0)} +$$
$$a_{j,(0,0,3)}D_{(0,0,3)} + a_{j,(0,1,2)}D_{(0,1,2)} + a_{j,(0,2,1)}D_{(0,2,1)} + a_{j,(0,3,0)}D_{(0,3,0)} + a_{j,(1,0,2)}D_{(1,0,2)} +$$
$$a_{j,(1,1,1)}D_{(1,1,1)} + a_{j,(1,2,0)}D_{(1,2,0)} + a_{j,(2,0,1)}D_{(2,0,1)} + a_{j,(2,1,0)}D_{(2,1,0)} + a_{j,(3,0,0)}D_{(3,0,0)} = \overline{u}_j$$

This is a 20×20 matrix problem for D_l. The Gauss-Legendre procedure for integration gives, for an arbitary complex geometry,

$$|C_j| = \int |J| \, d\xi d\eta d\zeta = \sum_i \sum_j \sum_k |J|(q_i, q_j, q_k) W_i W_j W_k$$
$$|C_j| \, \overline{x}_i = \sum_i \sum_j \sum_k \sum_m x_m N_m(q_i, q_j, q_k) |J|(q_i, q_j, q_k) W_i W_j W_k$$
$$|C_j| \, \overline{u}_j = \sum_i \sum_j \sum_k \sum_m u_m N_m(q_i, q_j, q_k) |J|(q_i, q_j, q_k) W_i W_j W_k$$
$$k! \, |C_j| \, a_{j,(l_1,l_2,l_3)} = \sum_i \sum_j \sum_k \left(\sum_m x_m N_m(q_i, q_j, q_k) - \overline{x}_i \right)^{l_1} \bullet$$
$$\left(\sum_m y_m N_m(q_i, q_j, q_k) - \overline{y}_i \right)^{l_2} \left(\sum_m z_m N_m(q_i, q_j, q_k) - \overline{z}_i \right)^{l_3} |J|(q_i, q_j, q_k) W_i W_j W_k$$

Above, the physical coordinates (x, y, z) have been transformed to the the computational coordinates (ξ, η, ζ), via the the Jacobian J, with determinant $|J|$. (q_i, q_j, q_k) are quadrature points of the Gauss-Legendre procedure, with associated weights (W_i, W_j, W_k). The N's are the finite element basis functions. Upon solution for D_l, we obtain $R(x; \overline{u}) \equiv R(q_\alpha; \overline{u})$, at the αth quadrature point of C_i using

$$R(q_\alpha; \overline{u}) = \sum_{k=0}^{r-1} \frac{1}{k!} \sum_{|l|=k} (x(q_\alpha) - \overline{x}_i)^{l_1} (y(q_\alpha) - \overline{y}_i)^{l_2} (z(q_\alpha) - \overline{z}_i)^{l_3} D_{(l_1,l_2,l_3)} \qquad (28)$$

In the present work we represent the convective terms by

$$-\sum_j \int N_{i,m} N_j R(x; \overline{u}_m) dV (\rho u_n)^j, \qquad (29)$$

where $R(x; \overline{u}_m)$ is a reconstruction from the element-average values, and

$$\overline{u}_m = \frac{1}{|C_j|} \int u_m dV \qquad (30)$$

The actual form of the convective terms will depend on the procedure used to control nonlinear instability. A non-Galerkin expression is obtained with the upwind procedures mentioned earlier in this paper.

5. PARALLEL IMPLEMENTATION

The host node is used for the start-up processes. The grids for the subdomains are pregenerated using standard finite element mappings. They are read into the solver and, for problems with no periodic boundary conditions or overlapping, are despatched to the processors. There is no global indexing of any array or data, which is possible because the grid is structured. If there is overlapping or if periodic boundary conditions are used there is, in the solver, a further processing of the grid before they are sent to the processors. We believe that some form of overlapping might be needed for cost-effective, time-accurate calculations. Moreover, overlapping results in a faster convergence and obviates the need to transfer grid data between processors. We allow between 0 to 3 elements in the overlapped region, in each of the (i, j, k) direction of a block. We interpolate dependent variables in the overlapped region in the manner discussed in Ladeinde (1992). The "overheads" associated with the overlapping procedure can be reduced by using larger subdomain sizes.

Periodic boundary conditions are of interest in theoretical turbulence, and a major accomplishment of our code is the way in which this is implemented. In specifics, we "wrap-around" periodic domain surfaces using the blocks containing those surfaces, and in a manner consistent with the overlapping approach. No grid data is transfered between two such coupled blocks (processors). With this approach the *coordinate* of a node in a block is *not* necessarily *unique* but depends on a "parent" block (that contained a particular element of the node) in the original grid. Thus, coordinates are stored element-by-element. Although the concept is easy to grasp, implementation is not. We plan to give a more systematic discussion of the procedure in another forum.

6. STENCIL SELECTION

From previous discussions we require that $R(x; \bar{u})$ be constructed from the smooth part of the flow, and in a manner that retains the conservative properties of the scheme. The stencil needed to reconstruct $R(x; \bar{u})$ for an element C_i is chosen in the following way (see Figure 1.)

(1.) Map out a stencil region consisting of $4 \times 4 \times 4$ elements. We select 4 candidate stencils from this region.
(2.) The first stencil is central, and made up of $3 \times 3 \times 3$ elements. In the interior of the domain, C_i is in the center of stencil. Three directionally-biased stencils with $4 \times 4 \times 2$, $4 \times 2 \times 4$, $2 \times 4 \times 4$ elements are also selected. All stencils for C_i must contain C_i, for conservation. Location of a discontinuity inside of C_i has not caused problems (Chakravarthy, personal communication). We bias stencils away from the (artificial) boundaries of blocks.
(3.) We then select 20 elements from each of the stencils, based on proximity to C_i, or $min_j | \bar{x}_i - \bar{x}_j |$.
(4.) We choose the stencil with minimum values of

$$\sigma_r \approx \sum_{|l|=r-1} | \frac{\partial^l u}{\partial x^l}(\bar{x}_i) | \quad . \tag{31}$$

We give preference to the central stencil, in the manner presented by Harten and Chakravarthy (1991) or Shu (1991).

7. SAMPLE TESTS AND THE COST OF ENO

Our experience with the ENO scheme is preliminary and efforts are currently concentrated on code validation and a cost-effective implementation. The way to validate the code will be to solve equations whose exact solutions are known. For this purpose we consider the following hyperbolic system describing three-dimensional wave.

$$u_t + grad\ u = 0; \quad x \in D \subset \Re^3,$$

where $D \in [-1,1] \times [-1,1] \times [-1,1] \times [0,\infty)$. The initial condition is $u(x,y,z,0) = sin2\pi x sin2\pi y sin2\pi z$, and periodic condition is $u(-1,y,z,t) = u(1,y,z,t)$, $u(x,-1,z,t) = u(x,1,z,t)$, $u(x,y,-1,t) = u(x,y,1,t)$.

The exact solution is $u(x,y,z,t) = sin2\pi(x-t)sin2\pi(y-t)sin2\pi(z-t)$.

The numerical solution (at a time during simulation) for a processor handling the subdomain $(x,y,z) \in [.3333, 1.0] \times [.3333, 1.0] \times [.3333, 1.0]$ is shown in Figure 2. For this test there were effectively 27 processors (!), each solving the same size of problem, and having one element in the overlapped region. The number of points in a subdomain is 3375 after stripping-off the overlap. The agreement between the numerical and exact solution, which we have found to improve dramatically with grid refinement, is only fairly good for this grid, but we estimate that grid points in excess of 80^3 will probably be needed for a perfect agreement. The audience will realize that this is a fairly demanding calculation in terms of the number of grid points because the domain spans two periods of the solution in each of the x, y, z coordinate directions.

Although the ENO approach presented here shows a lot of promise in terms of accuracy the usefulness is currently undermined by the high costs, and some features of the iPSC/860 machine are not very helpful in this regard. As an explanation we present floating point operation counts per element in Table 1.

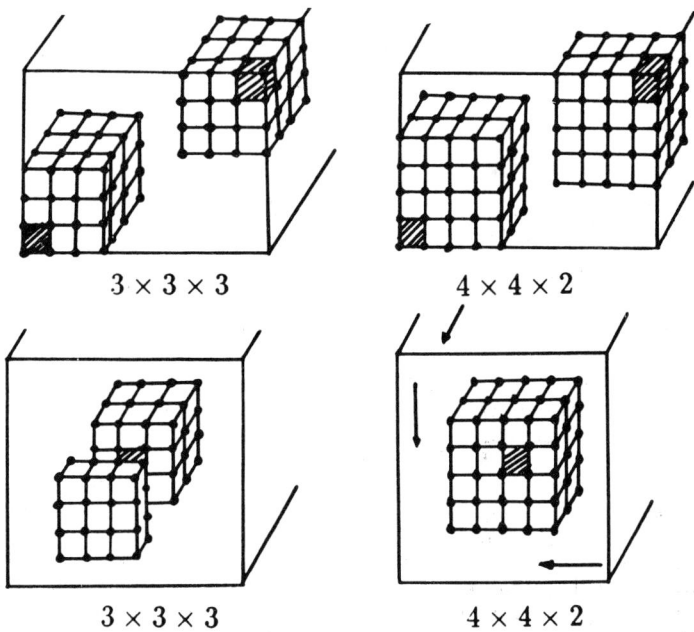

Figure 1. Sample stencils for ENO.

Figure 2. Computed three-dimensional wave. Contour levels: -0.98(0.164)0.82

Table 1. Estimate of floating point operation counts per element associated with the ENO procedure presented in this paper.

ENO Calculation	Number of operations
Calculating $\bar{\rho}$, $\bar{u}$, $\bar{v}$, $\bar{w}$, $\bar{x}$, $\bar{y}$, $\bar{z}$, $\mid C_j \mid$	3,000
Calculating $a_{j,l}$	200,000
Inverting A and solving for D_l	40,000
Calculating $R(x; \bar{u}_m)$	12,000

To generate the table we merely count the number of "visible" additions, subtractions, multiplications, etc. in the code. From the table we can deduce that a procesor, handling a subdomain of 5000, 8-node elements, which is typical for us, with an effective Mflops of 5 will require approximately 35 hours for the ENO part of a simulation that requires 500 time steps. This is simply unaffordable.

To address the foregoing problem we need to re-examine the way we implement the ENO procedure. We also have to accept that a faster machine is needed. There are other problems with the Delta machine, other than the low Mflops: RAM is only 16 Megabytes, which we find very limiting during start-up, as the latter is I/O bound. Many of the "pre-processing" calculations such as overlapping and, in particular, our "wrapping-around" procedure for periodic boundary conditions pose additional challenge for this Intel machine, as data cannot be stored in-core. To make matters worse the system has no virtual memory manager, so that the disk cannot be accessed for "emergency" memory space. The start-up process is expensive and ways to speed it up are needed. In fact, we have done some simulations where the cost of start-up and the actual calculations are comparable! Finally, after the start-up process we have to live with the fact that the *cfs* file system in this machine is sequential, and does not allow a direct-access of files, which will make out-of-core procedures expensive.

8. REDUCTION OF CPU TIME

The following modifications are sugested for possible reductions in the cost of ENO. Although we anticipate problems with these suggestions we are not certain that they will actually occur.

(1.) Avoid numerical integration, and related "overheads", for example by constructing from point-values of $u(x)$, using $a_{j,l} = \frac{1}{k!}(\bar{x}_i - \bar{x}_j)^{l_1}(\bar{y}_i - \bar{y}_j)^{l_2}(\bar{z}_i - \bar{z}_j)^{l_3}$. Cost of computing $a_{j,l}$ will be reduced to 19,000 operations from 200,000. However, there is a catch in our procedure where most of the logic is element-based, and node-based calculations will have to be dublicated for all elements containing a node (500,000 operations), or we store the node information, which may also not be cheap.

(2.) Use $J(i)$ for all elements $C_j \in J(i)$. The danger with this is that a discontinuity in C_i might cause problems for other elements in $J(i)$!

(3.) Use $R_i(x; \bar{u})$ for all $C_j \in J(i)$. This might cause conservation problems.

(4.) Lump A in $AD_l = \bar{u}$. This is attractive for the point-values approach where most of the cost goes to inverting A. This might degrade accuracy.

(5.) Retain cell-average approach, but store (factored) $a_{j,l}$ for the 4 stencils. This will cost $\approx 64MB$ in a block (or 400 entries of $a_{j,l} \times 5,000$ elements in a block $\times 8$ bytes per word). This is attractive and could lead to a much faster calculation. However, the data will most probably be stored in a disk, which is not comforting for systems such as the one used here where I/O performance is rather poor.

9. CONCLUSION

Most of the available CFD codes are not sufficiently accurate for calculations meant for studying the physics of turbulence in supersonic flows. The high-order ENO procedure is an exception, and a variation of it has been validated for DNS, although for rather simple geometries and boundary conditions. Implementation for more realistic systems is presented here; it is not cheap, and it might be worthwhile to pursue the suggestions in this paper. Further, a bigger machine (per processor) than the one used here is called for, even for the low R_λ values of today's DNS.

10. APPENDIX

The vectors and matrices appearing in equations (5) – (9) are defined below.

$M_{ij} = \int N_i N_j d\mathbf{X}$

$S_{ij}^{1n} = -\int N_{i,n} N_j d\mathbf{X},$

where $n = u, v, w$ (left-hand side), or $1, 2, 3$ (right-hand side)

$F_i^1 = -\int N_i(\rho u_l)\eta_l dS$

$F_i^n = -\int N_i(\rho u_n)\eta_n dS - (\gamma - 1)\int N_i \rho e \eta_n dS,$

where $n = 2, 3, 4$ (left-hand side), or $1, 2, 3$ (right-hand side)

$F_i^5 = \int N_i[Eu_m + (\gamma - 1)\rho e u_m]\eta_m dS + \frac{1}{Re}\int N_i \tau_{nm} u_m \eta_n dS$

$\quad + \frac{1}{(\gamma-1)RePrM^2}\int N_i kT_{,l}\eta_l dS$

$S_{ij}^{n\rho} = -(\gamma - 1)\int N_{i,n} N_j e_l N_l dS,$

where $n = 1, 2, 3$

$S_{ij}^{5\rho} = -(\gamma - 1)\int N_{i,m} N_j e_l N_l u_{m\lambda} N_\lambda dS$

$S_{ij}^{2u} = -\int N_{i,m} N_j u_{mq} N_q d\mathbf{X} + \frac{1}{Re}\int N_{i,m}\mu[r_1 N_{j,m} - r_2 N_{k,m}\rho_k N_j]d\mathbf{X}$

$\quad + \frac{1}{Re}\int N_{i,1}\mu[r_1 N_{j,1} - r_2 N_{k,1}\rho_k N_j]d\mathbf{X} - \frac{2}{3}\frac{1}{Re}\int N_{i,1}\mu[r_1 N_{j,1} - r_2 N_{k,1}\rho_k N_j]d\mathbf{X}$

$S_{ij}^{2v} = \frac{1}{Re}\int N_{i,2}\mu[r_1 N_{j,1} - r_2 N_{k,1}\rho_k N_j]d\mathbf{X}$

$\quad - \frac{2}{3}\frac{1}{Re}\int N_{i,1}\mu[r_1 N_{j,2} - r_2 N_{k,2}\rho_k N_j]d\mathbf{X}$

$S_{ij}^{2w} = \frac{1}{Re}\int N_{i,3}\mu[r_1 N_{j,1} - r_2 N_{k,1}\rho_k N_j]d\mathbf{X}$

$\quad - \frac{2}{3}\frac{1}{Re}\int N_{i,1}\mu[r_1 N_{j,3} - r_2 N_{k,3}\rho_k N_j]d\mathbf{X}$

$S_{ij}^{3u} = \frac{1}{Re}\int N_{i,1}\mu[r_1 N_{j,2} - r_2 N_{k,2}\rho_k N_j]d\mathbf{X}$

$\quad - \frac{2}{3}\frac{1}{Re}\int N_{i,2}\mu[r_1 N_{j,1} - r_2 N_{k,1}\rho_k N_j]d\mathbf{X}$

$S_{ij}^{3v} = -\int N_{i,m} N_j u_{mq} N_q d\mathbf{X} + \frac{1}{Re}\int N_{i,m}\mu[r_1 N_{j,m} - r_2 N_{k,m}\rho_k N_j]d\mathbf{X}$

$\quad + \frac{1}{Re}\int N_{i,2}\mu[r_1 N_{j,2} - r_2 N_{k,2}\rho_k N_j]d\mathbf{X} - \frac{2}{3}\frac{1}{Re}\int N_{i,2}\mu[r_1 N_{j,2} - r_2 N_{k,2}\rho_k N_j]d\mathbf{X}$

$$S_{ij}^{3w} = \tfrac{1}{Re} \int N_{i,3}\mu[r_1 N_{j,2} - r_2 N_{k,2}\rho_k N_j]d\mathbf{X}$$

$$-\tfrac{2}{3}\tfrac{1}{Re} \int N_{i,2}\mu[r_1 N_{j,3} - r_2 N_{k,3}\rho_k N_j]d\mathbf{X}$$

$$S_{ij}^{4u} = \tfrac{1}{Re} \int N_{i,1}\mu[r_1 N_{j,3} - r_2 N_{k,3}\rho_k N_j]d\mathbf{X}$$

$$-\tfrac{2}{3}\tfrac{1}{Re} \int N_{i,3}\mu[r_1 N_{j,1} - r_2 N_{k,1}\rho_k N_j]d\mathbf{X}$$

$$S_{ij}^{4v} = \tfrac{1}{Re} \int N_{i,2}\mu[r_1 N_{j,3} - r_2 N_{k,3}\rho_k N_j]d\mathbf{X}$$

$$-\tfrac{2}{3}\tfrac{1}{Re} \int N_{i,3}\mu[r_1 N_{j,2} - r_2 N_{k,2}\rho_k N_j]d\mathbf{X}$$

$$S_{ij}^{4w} = -\int N_{i,m}N_j u_{mq}N_q d\mathbf{X} + \tfrac{1}{Re} \int N_{i,m}\mu[r_1 N_{j,m} - r_2 N_{k,m}\rho_k N_j]d\mathbf{X}$$

$$+\tfrac{1}{Re} \int N_{i,3}\mu[r_1 N_{j,3} - r_2 N_{k,3}\rho_k N_j]d\mathbf{X} - \tfrac{2}{3}\tfrac{1}{Re} \int N_{i,3}\mu[r_1 N_{j,3} - r_2 N_{k,3}\rho_k N_j]d\mathbf{X}$$

$$S_{ij}^{5u} = \tfrac{1}{Re} \int N_{i,1}\mu u_{mk} N_k [r_1 N_{j,m} - r_2 N_{l,m}\rho_l N_j]d\mathbf{X}$$

$$+\tfrac{1}{Re} \int N_{i,n}\mu u_{1k} N_k [r_1 N_{j,n} - r_2 N_{l,n}\rho_l N_j]d\mathbf{X}$$

$$-\tfrac{2}{3}\tfrac{1}{Re} \int N_{i,n}\mu u_{mk} N_k \delta_{nm} [r_1 N_{j,1} - r_2 N_{l,1}\rho_l N_j]d\mathbf{X}$$

$$-\tfrac{\gamma}{2RePr} \int N_{i,m} k\{\, r_1 u_{1\lambda} N_\lambda N_{j,m} + N_j[\, r_1 N_{\lambda,m} u_{1\lambda} - r_2 u_{1\lambda} N_\lambda N_{l,m}\rho_l \,]\,\}d\mathbf{X}$$

$$S_{ij}^{5v} = \tfrac{1}{Re} \int N_{i,2}\mu u_{mk} N_k [r_1 N_{j,m} - r_2 N_{l,m}\rho_l N_j]d\mathbf{X}$$

$$+\tfrac{1}{Re} \int N_{i,n}\mu u_{2k} N_k [r_1 N_{j,n} - r_2 N_{l,n}\rho_l N_j]d\mathbf{X}$$

$$-\tfrac{2}{3}\tfrac{1}{Re} \int N_{i,n}\mu u_{mk} N_k \delta_{nm} [r_1 N_{j,2} - r_2 N_{l,2}\rho_l N_j]d\mathbf{X}$$

$$-\tfrac{\gamma}{2RePr} \int N_{i,m} k\{\, r_1 u_{2\lambda} N_\lambda N_{j,m} + N_j[\, r_1 N_{\lambda,m} u_{2\lambda} - r_2 u_{2\lambda} N_\lambda N_{l,m}\rho_l \,]\,\}d\mathbf{X}$$

$$S_{ij}^{5w} = \tfrac{1}{Re} \int N_{i,3}\mu u_{mk} N_k [r_1 N_{j,m} - r_2 N_{l,m}\rho_l N_j]d\mathbf{X}$$

$$+\tfrac{1}{Re} \int N_{i,n}\mu u_{3k} N_k [r_1 N_{j,n} - r_2 N_{l,n}\rho_l N_j]d\mathbf{X}$$

$$-\tfrac{2}{3}\tfrac{1}{Re} \int N_{i,n}\mu u_{mk} N_k \delta_{nm} [r_1 N_{j,3} - r_2 N_{l,3}\rho_l N_j]d\mathbf{X}$$

$$-\tfrac{\gamma}{2RePr} \int N_{i,m} k\{\, r_1 u_{3\lambda} N_\lambda N_{j,m} + N_j[\, r_1 N_{\lambda,m} u_{3\lambda} - r_2 u_{3\lambda} N_\lambda N_{l,m}\rho_l \,]\,\}d\mathbf{X}$$

$$S_{ij}^{5e} = -\int N_{i,m} N_j u_{mq} N_q d\mathbf{X} + \tfrac{\gamma}{RePr} \int N_{i,m} k[r_1 N_{j,m} - r_2 N_{l,m}\rho_l N_j]d\mathbf{X}+$$

11. REFERENCES

GODUNOV, S. K. 1959. Matematicheskii Sbornik **47**, pp. 271-290.
HARTEN, A., ENGQUIST, B., OSHER, S. & CHAKRAVARTHY, S. 1987. J. Comp. Phys. **71**, pp. 231-303.
HARTEN, A., & CHAKRAVARTHY, S. 1991. ICASE Report 91-76.
JAMESON, A., SCHMIDT, W., & TURKEL, E. 1981. AIAA Paper 81-1259, June 1981.
LADEINDE, F. 1992. 13*th* Symposium on Turbulence, September 21-23, 1992, Rolla, Missouri.
LAMBERT, J. D. 1991."Numerical Methods for Ordinary Differential Systems: *The Initial Value Problem*". Publ. John Wiley & Sons, Chichester.
OSHER, S. & SHU, C.-W. 1988. Proc. of the Intl. Workshop in Compressible Turbulence, Princeton, Springer-verlag.
RAI, M. M. & MOIN, P. 1991. J. Comp. Phys. **96**, pp. 15.
ROE, P. L. 1986. Annual Review of Fluid Mechanics, **18**, pp. 337.
SHU, C.-W., ERLEBACHER, G., ZANG, T. A., WHITAKER, D., & OSHER, S. 1991. ICASE Report 91-38.
VAN LEER, B. 1979. J. Comp. Phys. **32**, pp. 101-136.

Block-Structured Multigrid for the Navier-Stokes Equations: Experiences and Scalability Questions

J. Linden, G. Lonsdale, H. Ritzdorf, A. Schüller

Gesellschaft für Mathematik und Datenverarbeitung mbH, P.O. Box 1316, D-5205 St. Augustin 1, Germany

Abstract

This paper summarizes investigations concerning the algorithmic scalability of multigrid methods for partial differential equations on MIMD distributed memory systems. It is shown that even multigrid methods which are distinguished by h-independent convergence rates are not scalable in a rigorous sense. We develop their parallel asymptotic computational complexity for different types of multigrid cycles and analyze their critical components with respect to scalability. Experimental results for two Navier-Stokes test problems presented in the last section of this paper show, however, that the theoretically predicted dependency of the combined numerical and parallel efficiencies of multigrid methods on the number of processors employed is in fact very weak. This leads to the conclusion that multigrid is also with respect to scalability an appropriate candidate for solving partial differential equations on massively parallel machines.

1 INTRODUCTION

The effective use of highly parallel machines introduces new aspects for algorithmic development. In particular, on MIMD systems with distributed memory, on which we concentrate in this paper, load balancing, data locality and the suitability of the problem granularity for the target architecture are crucial for efficient execution.

A feature which is becoming increasingly important is scalability. In fact, the scalability of the architectures which are beginning to appear is one of the most important and attractive properties from the user's point of view. In principle, machines of the same architectural type can be built with only a few processors or with hundreds or thousands. The same machine can, therefore, be used for program and algorithmic developments or for testing purposes and – in a scaled–up version – for production runs.

To exploit this hardware scalability, however, a corresponding scalability property of the algorithm to be employed is necessary: the efficiency of the algorithm must not deteriorate (at least not substantially) as the number of processors utilized increases. Scalability of an algorithm can be defined by considering its performance for either

- a fixed total problem size or
- a fixed problem size per processor.

With the first definition, the parallel speed-up is limited by the inverse of the sequential portion of the algorithm, reflecting Amdahl's law. Thus, the parallel efficiency will tend to zero as the number of processors increases. The second definition provides a more informative approach and is in fact more relevant for distributed memory multiprocessors since the available total memory increases with the number of processors. Scalability in this sense would ensure that twice as many processors solve a problem of double the size within the same time.

With this view of assessing scalability and restricting attention to the use of iterative solution methods for grid-oriented problems, "optimal" scalability is equivalent to requiring: convergence rates which are independent of both the grid size and the number of processors employed; a computational cost per iteration and per grid-point which is $\mathcal{O}(1)$ and independent of the number of processors.

With these aims in mind, a natural candidate algorithm is multigrid, which is explicitly designed to achieve grid-independent convergence. With parallelization by grid-partitioning, this essential property is maintained. This paper investigates the scalability of such parallel multigrid iterations. While we will initially consider basic multigrid attributes which affect its parallelization, we will then restrict our attention to the use of multigrid for time-independent, incompressible flows on block-structured grids. The choice of block-structured grids to provide geometrical flexibility while maintaining logically rectangular structures for efficient implementation, is one which arises naturally as an extension of finite-volume handling of simple domains and is frequently employed for complex flow simulations. The application of the parallelization approach of grid-partitioning to block-structured grids, whereby the partition is defined by the block-structure, is straightforward, though technically more complex due to the increased variety of possible communications requirements.

In Section 2 we discuss some basic results on the theoretical complexity of parallel multigrid iterations, showing that they are not scalable in the above rigorous sense. More precisely, we show that the parallelization strategy introduces restrictions on the handling of coarse grid levels which result in the cost of the algorithm being dependent on the number of processors. It is, in fact, quite unlikely that an optimally scalable algorithm can be found, so that in practice a weaker form of scalability will have to be accepted. Of interest then is the form of the dependence on the number of processors utilized. In Section 3 for the incompressible Navier–Stokes and Boussinesq equations we, therefore, present experimental results on the dependency of computing times on the number of processors on Suprenum and the Intel iPSC/860. These results were obtained with the parallel block-structured multigrid solver L_iSS [10]. They confirm that a careful treatment of the coarse grid problems yields practical scalability at least within a wide range of processor numbers. Due to space limitations we do not describe any details concerning multigrid, grid partitioning and their use with block-structured grids. The interested reader is referred to [1, 4, 5, 6, 7, 8, 9, 11, 13].

2 ASYMPTOTIC COMPLEXITY ANALYSIS

In order to analyze the effect of parallelization by grid partitioning on the computational cost of a multigrid algorithm, we will consider a linear, 2-D equation and will assume that *each processor has $N = (2^l + 1)^2$ grid-points on the finest level*, giving l levels when standard coarsening is used, and that p processors are employed. Computational cost will be given in terms of work units, with one work unit representing the computational work/time required for the relaxation of N grid points. This analysis concentrates solely on the computational costs and ignores the communications costs.

Grid partitioning exploits the locality of the multigrid components, leaving the sequence of operations within the multigrid cycle and its structure unchanged from the sequential case. That is, *except* for the definition of the coarsest level to be used in the hierarchy of grids: the coarsest grid possible whereby each processor remains active has $\mathcal{O}(p)$ points (i.e. $\mathcal{O}(1)$ points per processor) at least. For ease of explanation we will refer to this grid as the 'global coarsest grid'. The effect of this restriction on the form of a multigrid W-cycle is illustrated in Figure 1. The increased computational cost arises from the need for a significant reduction of the residuals on the global coarsest grid, in order to ensure fast convergence of the multigrid iteration.

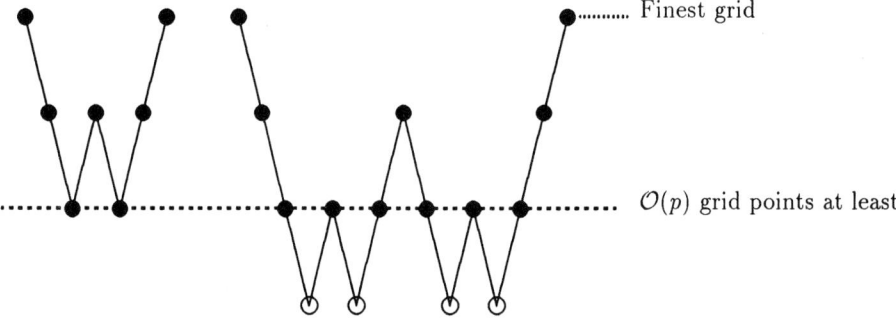

Figure 1: W-cycles with and without the restriction to the global coarsest grid.

2.1 Coarse grid relaxation

One possibility is simply to perform relaxation on the global coarsest grid. In order to maintain the overall multigrid convergence, the coarse grid error reduction must be below the expected fine grid convergence rate, ρ. Under certain simplifying assumptions, ρ can be predicted by local mode analysis: either smoothing or two-level analysis (see [1]). Since the relaxation is to be performed in parallel it must have a Jacobi-character, so that an asymptotic convergence rate of $1 - \mathcal{O}(p^{-1})$ can be assumed. With such a convergence rate, the reduction of residuals by a factor, $\varepsilon \leq \rho$, requires $\mathcal{O}(p)$ relaxations, implying a cost of $\mathcal{O}(\frac{p}{N})$ work units per processor. (It should be noted that, dependent on the size of the residual reduction required, the number of relaxation sweeps needed in practice may not reflect the asymptotic behaviour. Thus, the above cost is a worst-case estimate.)

However, in multilevel cycling the global coarsest grid may be "visited" more than once, depending on the cycle type. For example, while a V-cycle visits the coarse grid only once, a W-cycle over l levels involves 2^{l-2} coarse grid visits. This offers an opportunity to reduce the accuracy criterion for each of these visits. The underlying multigrid philosophy is that the reduction of low frequency error modes is achieved effectively only on coarse levels, while all high frequency errors are rapidly reduced on the fine grid. If couplings between error modes introduced by inter-grid transfers are assumed not to be significant, and if the multigrid components are assumed not to re-introduce low frequency error components, then the reduction of the lowest frequency modes may be distributed over the several visits to the coarsest grid. This suggests that for each visit to the global coarsest grid, a residual reduction of only $\varepsilon^{\frac{1}{\kappa}}$ is required, where κ denotes the number of coarse grid visits. This approach is supported by the rigorous multilevel cycle convergence estimate given in [13, Chapter 4] which is based on norms of the two-level operators.

In order to illustrate the feasibility of this approach, Table 1 presents convergence rates for Poisson's equation on the unit square, with 128×128 intervals on the finest grid but where grid coarsening was terminated so that only 3 or 4 levels were used. The relaxation method used was lexicographic point Gauss-Seidel. On the coarsest grid, relaxation was performed until a residual reduction by a factor ϵ was achieved *per visit*. The cycle used one relaxation sweep before and after coarse grid correction, so that the aymptotic two-grid convergence rate predicted by two-level local mode analysis is 0.193. Thus, the choice $\varepsilon \leq 0.18$ should be sufficient to deliver the expected rate. From Table 1 it is clear that choosing $\epsilon \leq \varepsilon^{\frac{1}{\kappa}}$ does indeed deliver the expected convergence, whereas a greater value of ϵ leads to a significant deterioration.

Table 1:
Calculated asymptotic convergence factors per multigrid cycle.

	V-cycle		W-cycle	
ϵ	3 grids	4 grids	3 grids	4 grids
0.1	0.187	0.189	0.186	0.186
0.18	0.187	0.193	0.186	0.186
$\sqrt{0.18}$	0.422	0.405	0.186	0.186
$\sqrt[4]{0.18}$	0.645	0.616	0.416	0.186
0.8	0.789	0.775	0.623	0.365
	$\kappa = 1$	$\kappa = 1$	$\kappa = 2$	$\kappa = 4$

With the residual reduction to $\varepsilon^{\frac{1}{\kappa}}$, the cost of the relaxations per visit to the global coarsest grid is then $\frac{1}{\kappa}\mathcal{O}(\frac{p}{N})$ work units per processor.

2.2 Continued Coarsening

Assuming that the number of processors, p, is of the form $p = (2^k)^2$, for some k, then an alternative to employing relaxation on the global coarsest grid is to simply continue grid-coarsening, but to leave processors idle during the part of the solution process when their grid points are not included in the coarser levels. For the parallel application, the time on each coarser level introduced is equal to the time taken by those processors possessing grid points on that level (the remaining processors simply waiting to become active again). Thus, relaxation on the newly introduced coarser levels remain as expensive as relaxations on the global coarsest grid. The total number of relaxations performed on the global coarsest level and on coarser levels is cycle-type dependent: $\mathcal{O}(\log_2 p)$ for V-cycles, $\mathcal{O}(\sqrt{p})$ for W-cycles. The corresponding computational costs of the residual reduction on the global coarsest grid (per visit) are then: for V-cycles $\mathcal{O}(\frac{\log_2 p}{N})$ work units per processor, for W-cycles $\mathcal{O}(\frac{\sqrt{p}}{N})$ work units per processor. A similar analysis for 2- and 3-D cases can be found in [11].

2.3 Summary of costs

The previous sections dealt with the computational costs of approaches for the reduction of the residuals on the global coarsest grid. These costs can then be used to give the computational costs of various cycle-types. We will concentrate here only on V- and W-cycles. For both, the computational cost on all levels above the global coarsest grid is $\mathcal{O}(1)$ work units per processor. The number of visits to the global coarsest grid, κ, is given by: $\kappa = 1$ for V-cycles; $\kappa = \mathcal{O}(\sqrt{N})$ for W-cycles. Table 2 gives the computational cost of one multigrid cycle using both V- and W-cycles and the three approaches for the global coarsest grid.

Table 2:
Work units per processor for one multigrid cycle.

Method on Global Coarsest Grid	Relaxation : reduction ε	Relaxation : reduction $\varepsilon^{\frac{1}{\kappa}}$	Continued Coarsening
V-cycle	$\mathcal{O}(1) + \mathcal{O}(\frac{p}{N})$	$\mathcal{O}(1) + \mathcal{O}(\frac{p}{N})$	$\mathcal{O}(1) + \mathcal{O}(\frac{\log_2 p}{N})$
W-cycle	$\mathcal{O}(1) + \mathcal{O}(\frac{p}{\sqrt{N}})$	$\mathcal{O}(1) + \mathcal{O}(\frac{p}{N})$	$\mathcal{O}(1) + \mathcal{O}(\sqrt{\frac{p}{N}})$

It should be emphasized that the terms depending on p are divided by N or $\sqrt{N}$. It may thus be expected that the dependence on p is quite weak and may not be observed in practical computations on typical parallel machines as they are available today with a limited number of processors. This will also be seen in Section 3 for a suitable treatment of the problem on the global coarsest grid.

A similar analysis can be performed with respect to the amount of communication required. If the problem on the global coarsest grid is solved by relaxation, similar asymptotic results are obtained since the number of communication steps per relaxation is constant for each process.

The continued coarsening strategy considered above keeps as many processors busy as possible on the coarse grids. With respect to communication, however, this may not be the most efficient approach. Instead, it may pay off to agglomerate the problem on the global coarsest mesh and on coarser meshes stepwise to fewer processors and finally to only one process. This would reduce the communication cost while the arithmetic cost in the above sense would increase. An example of this approach can be found in [5].

3 EXPERIMENTS WITH NAVIER-STOKES APPLICATIONS

The method and the program used for the numerical experiments below are all based on the use of general block-structured grids for the Navier-Stokes equations, which imposes particular restrictions on the achievement of efficient multigrid performance. Both the use of V-cycles and the continued grid coarsening are problematic for block-structured Navier-Stokes applications although the asymptotic performance estimates given in Table 2 would suggest their employment.

A stable first-order discretization of the Navier-Stokes equations necessarily introduces 'artificial viscosity'-like terms which are meshsize-dependent. This results in a degradation in the coarse grid correction. A local-mode analysis of this situation (for model problems) predicts asymptotic two-grid convergence rates approaching 0.5, which has, depending on the cycle-type, a significant consequence for the overall multigrid convergence rates as the number of levels increases; the multigrid convergence rates when using V-cycles rapidly approach unity. The speed at which the convergence rate approaches unity is much lower for W-cycles, making them an acceptable compromise - the use of more coarse grid intensive cycle types would drastically increase the computational costs. Detailed analyses of such coarse grid correction problems, together with practical examples, can be found in [2, 12].

In general, block-structuring introduces certain limitations with respect to agglomeration or other continued coarsening approaches. In complicated geometries, the global coarsest grid may, for example, represent the coarsest definition of the underlying geometry so that a continued coarsening may be impossible. For the numerical experiments presented in this section we, therefore, did not consider any continued coarsening strategies although, for the particular examples considered, this would have been possible. The test problems taken involve geometries which can be defined in terms of a single block, thus also allowing performance to be measured on a single processor.

The multigrid method used is briefly characterized by the following components:

- W-cycles,
- coarse grid solution by relaxation on the global coarsest grid,
- standard coarsening,
- block-wise alternating line collective relaxations with additional block boundary smoothing and updating of overlap regions after every partial relaxation step (cf. [8, 9]),
- full weighting in interior residual transfers, injection at boundaries,
- bilinear interpolation of corrections.

To compute suitable initial approximations on the finest grid, we use the full multigrid method, starting from the global coarsest grid, with only one W-cycle on all intermediate grids. The iteration on the finest grid is terminated when the residuals in the maximum norm are below 5×10^{-4}. Depending on the block configuration, the numbers of iterations required to achieve this criterion differ slightly. It should be noted here that, due to the block-wise line relaxation, the multigrid method actually depends (but only weakly) on the particular block configuration.

The times given below correspond to the total solution time for the complete multigrid application as described above, neglecting initialization and I/O costs.

3.1 Test cases

Two test problems will be considered:

(A) incompressible Navier-Stokes equations for flow at Reynolds number 1000 through a channel with a half-circular bump (Fig. 2),

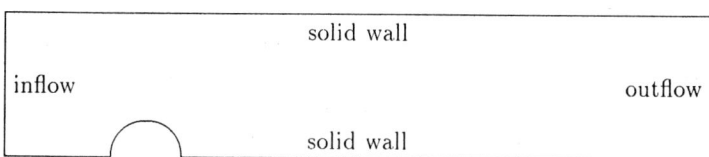

Figure 2: Boundary conditions for the circular bump problem.

(B) Boussinesq equations for the free convection in a square cavity at Rayleigh number 30000 (Fig. 3).

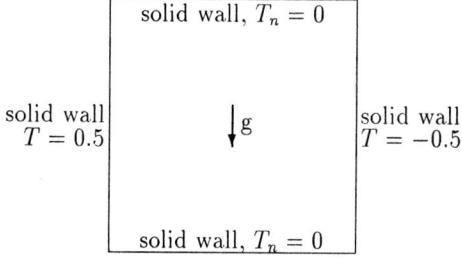

Figure 3: Boundary conditions for free convection problem and direction of gravity.

In both cases the discretization was based on the flux difference splitting method for the equations in terms of the primitive variables as described in [3].

3.2 Results

In order to show the dominating influence of the coarse grid calculations with respect to scalability, two different termination criteria for the relaxation on the global coarsest meshes were used. With the first criterion, the coarse grid relaxation stops after a residual reduction of .01 (but after at most 100 relaxation sweeps) each time the global coarsest grid is visited. With the second criterion, only ten relaxations per visit are performed. In the figures we refer to these criteria as Criteria (1) and (2).

The solution accuracy enforced by Criterion (1) is much more stringent than is really needed. For both test cases, we can expect multigrid convergence factors of between 0.3 and 0.5 per W-cycle. Following our discussion in Section 2, an accuracy of roughly $0.3^{1/\kappa}$ should be sufficient, where κ denotes the number of visits to the global coarsest grid. However, Criterion (1) was nevertheless employed in order to clearly demonstrate the asymptotic influence of the coarse grid relaxation.

The second criterion represents the other extreme with a fixed amount of work on the global coarsest grid, but in fact still yields sufficient coarse grid accuracy in the above sense for the processor numbers considered here. It is therefore expected that with this second criterion, the increase of computing times will be much weaker than with the first criterion for increasing numbers of processors. This is confirmed by the experimental results in Figures 4 and 5 for Test Problem (A) obtained on Suprenum and the Intel iPSC/860. Within the range of available processors (up to 32 on the Intel machine and up to 192 on Suprenum), Criterion (2) actually leads to a scalable algorithm while for Criterion (1) the computing times increase. It should be noted that, due to a bug in the compiler, the tests on Suprenum had to be performed in scalar mode, i.e. without use of the vector units, which led to the huge computing times.

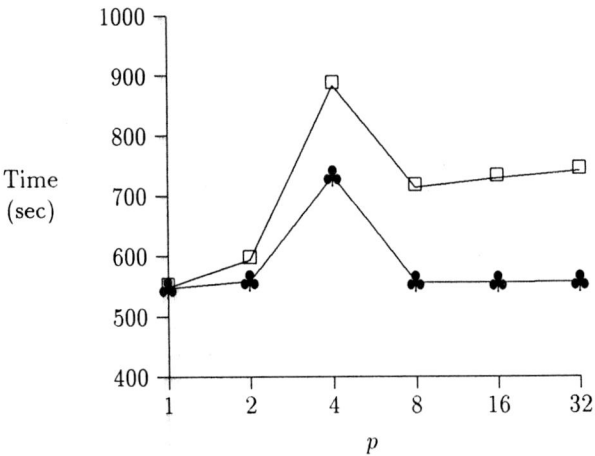

Figure 4: Problem (A); Intel iPSC/860, 129×65 grid points per processor;
□: coarse grid termination (1); ♣: coarse grid termination (2).

The results for Test Problem (B) show the same asymptotic tendency. They are presented in Figures 6 and 7.

The deviations for 4 and 128 processors in (A) result from the fact that the number of multigrid iterations for achieving the prescribed accuracy on the finest grid are slightly different. This number of iterations together with the number of grids used and the

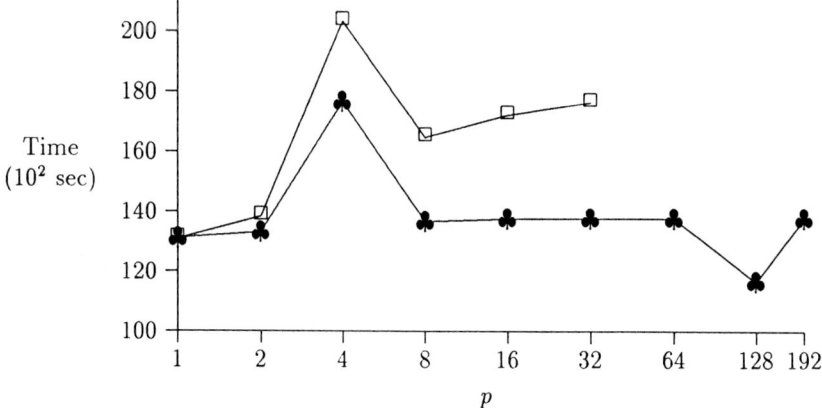

Figure 5: Problem (A); Suprenum; 129 × 65 grid points per processor;
□: coarse grid termination (1); ♣: coarse grid termination (2).

Table 3:
Parallel efficiencies for Problem (A); 129 × 65 grid points per processor; 6 grids; Criterion (2).

p	1	2 × 1	2 × 2	4 × 2	4 × 4	8 × 4	8 × 8	16 × 8	16 × 12
W-Cycles	6	6	8	6	6	6	6	5	6
Residuals (10^{-4})	3.1	4.6	1.7	4.1	4.2	2.1	3.4	4.5	2.2
Suprenum	100	97	93	91	89	89	88	88	88
Intel iPSC/860	100	97	93	92	90	90	–	–	–

achieved parallel efficiencies are shown in Table 3 (using the formula

$$E := \frac{\sum_{i=1}^{p} a_i}{p \max_i (a_i + c_i)}$$

where a_i denotes the CPU time for arithmetic computations in processor i and c_i the corresponding total communication time including idle time as used in [8, 9]).

The parallel efficiencies remain essentially constant for both test problems, (A) and (B), if 16 or more processors are used (Tables 3 and 4). For the tests using fewer processors, the overhead produced by the communication varies significantly: tests on one processor do not require any communication; in the tests using 2 (4,8,16) processors, the processors have to communicate with 1 (2,3,4) neighbouring processors. Correspondingly, the parallel efficiencies decrease somewhat until the full communication complexity (communication with four neighbouring processors) is reached for 16 = 4 × 4 processors.

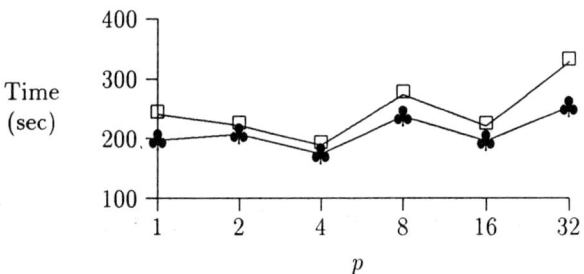

Figure 6: Problem (B); Intel iPSC/860, 65 × 65 grid points per processor;
□: coarse grid termination (1); ♣: coarse grid termination (2).

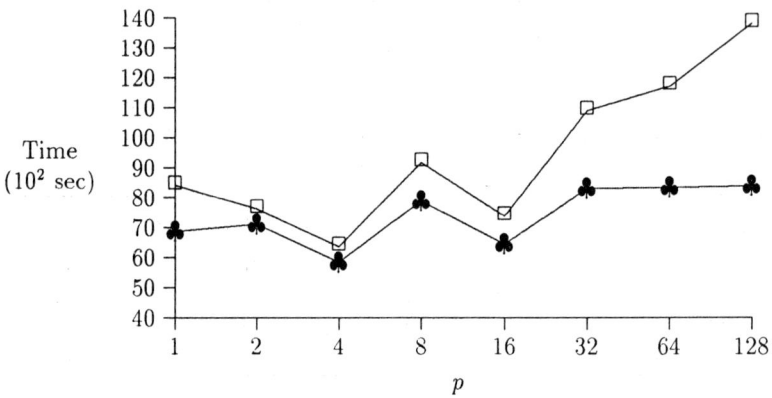

Figure 7: Problem (B); Suprenum; 65 × 65 grid points per processor;
□: coarse grid termination (1); ♣: coarse grid termination (2).

Table 4:
Parallel efficiencies for Problem (B); 65 × 65 grid points per processor; Criterion (2).

p	1	2 × 1	2 × 2	4 × 2	4 × 4	8 × 4	8 × 8	16 × 8
W-Cycles	4	4	3	4	3	4	4	4
Grids	5	5	6	6	6	6	6	6
Residuals (10^{-4})	0.6	2.9	2.4	2.5	4.8	2.3	2.1	3.2
Suprenum	100	97	91	88	83	83	83	83
Intel iPSC/860	100	96	90	86	81	81	–	–

For Test Problem (B) the parallel efficiencies are slightly worse than for (A). This is because the ratio of arithmetic and communication work is smaller since we have exploited the fact that the underlying grid for (B) is square Cartesian while for (A) the grid is curvilinear and non-orthogonal.

4 CONCLUSIONS

It has been shown that multigrid methods, although not being rigorously scalable in an asymptotic sense, can be used efficiently and without loss of performance on parallel systems with up to two hundred processors. In geometrically simple cases, in which coarse grid agglomeration is possible, an improved parallel multigrid performance can be expected. With or without agglomeration, multigrid proves to be well suited for solving partial differential equations on parallel machines at least within the range of processor numbers indicated above, combining high numerical and parallel efficiency.

5 ACKNOWLEDGMENTS

We thank our colleagues K. Stüben and V. Mikulinsky for fruitful discussions. We would also like to thank the Zentralinstitut für Angewandte Mathematik of the Research Centre Jülich and the Theoretical and Computational Science Division, S.E.R.C. Daresbury Laboratory, for providing the computing time on the Intel iPSC/860.

6 REFERENCES

[1] A. Brandt, *Multigrid techniques: 1984 guide with applications to fluid dynamics* (GMD-Studie Nr. 85, St. Augustin, 1984).

[2] A. Brandt and I. Yavneh, Accelerated multigrid convergence and high-Reynolds recirculating flows, Weizmann Institute Report, Rehovot, 1991.

[3] E. Dick and J. Linden, A multigrid method for steady incompressible Navier-Stokes equations based on flux-difference splitting, *Int. j. numer. methods fluids* **14** (1992) 1311-1323.

[4] R. Hempel and H. Ritzdorf, The GMD communications subroutine library for grid-oriented problems, Arbeitspapiere der GMD Nr. 589, St. Augustin, 1991.

[5] R. Hempel and A. Schüller, *Experiments with parallel multigrid using the SUPRENUM communications library* (GMD-Studie Nr. 141, St. Augustin, 1988).

[6] J. Linden, B. Steckel and K. Stüben, Parallel multigrid solution of the Navier-Stokes equations on general 2D domains, *Parallel Computing* **7** (1988) 461-475.

[7] J. Linden and K. Stüben, Multigrid methods: An overview with emphasis on grid generation processes, in: J. Häuser and C. Taylor, eds., *Numerical grid generation in Computational Fluid Dynamics* (Pineridge Press, Swansea, 1986) 483-509.

[8] G. Lonsdale and A. Schüller, Parallel and vector aspects of a multigrid Navier-Stokes solver, Arbeitspapiere der GMD Nr. 550, GMD, St. Augustin, 1991.

[9] G. Lonsdale and A. Schüller, Maintaining multigrid and parallel efficiencies for the Navier-Stokes equations, in: K.G. Reinsch et al., eds., *Parallel Computational Fluid Dynamics '91* (Elsevier Science Publishers B.V., Amsterdam, 1992) 271-284.

[10] G. Lonsdale and K. Stüben, The L_iSS package, Arbeitspapiere der GMD Nr. 524, GMD, St. Augustin, 1991.

[11] O.A. McBryan, P.O. Frederickson, J. Linden, A. Schüller, K. Solchenbach, K. Stüben, C.-A. Thole and U. Trottenberg, Multigrid methods on parallel computers - a survey of recent developments, *Impact Comput. in Sci. Engrg.* **3** (1991) 1-75.

[12] A. Schüller, *Mehrgitterverfahren für Schalenprobleme*, GMD-Bericht Nr. 171 (Oldenbourg Verlag, München, 1988).

[13] K. Stüben and U. Trottenberg, Multigrid methods: Fundamental algorithms, model problem analysis and applications, in: W. Hackbusch and U. Trottenberg, eds., *Multigrid methods,* Lecture Notes in Mathematics Vol. 960 (Springer, Berlin, 1982).

PARALLEL ALGORITHMS FOR GAS DYNAMICS

Lyle N. Long

The Pennsylvania State University, University Park, PA 16802, USA
(814) 865-1172   LNL@ECL.PSU.EDU

ABSTRACT

Algorithm and performance issues relevant to simulating a wide range of gas dynamic phenomenon on massively parallel computers are discussed. Particular attention is given to the various types of Connection Machine computers. Several different algorithms and codes are discussed. One of the main points of this paper is that communication speed is just as important as computation speed on massively parallel computers. We should strive for teracomputers with teraflops, terabytes/sec., terabytes of RAM, and terabytes of disk space -- not simply teraflop computers. Our goals should be to drastically reduce wall-clock time and increase the size of problems that can be solved, not necessarily maximize megaflop rates.

INTRODUCTION

This paper describes several different algorithms that were implemented on various Connection Machine computers [1-3]. All of these codes were designed to simulate gas dynamics on massively parallel computers. The languages used were *Lisp and CM-Fortran.

When massively parallel computers first appeared, there were two main classes [4]: single instruction multiple data (SIMD) and multiple instruction multiple data (MIMD). The SIMD/MIMD debate is not as relevant as it once was, since the hardware on all parallel computers is becoming more similar. There is still, however, a distinct difference between the data-parallel and message-passing approaches, but this is a programming or software issue. SIMD and MIMD refer to hardware differences. While most of the algorithms described below were implemented on data-parallel computers, they could have been implemented on message-passing machines. The early SIMD computers (e.g. CM-1 and DAP) used bit-serial processors. These processors are not cost effective for floating point operations, since the number of operations they must perform scale like N^2, where N is the number of significant bits (e.g. 32, 64, or 128).

The data-parallel and message-passing approaches both have their advantages and disadvantages, as shown in Table 1. Some people believe it is easier to parallelize dusty-deck programs using the message-passing approach. While this may be true for a 32 processor system, it may not be true for a 10,000 processor system. Proper scalability is crucial for massively parallel systems. It will be difficult to design a teraflop-class computer in the near future with individual processor speeds above 100-200 megaflops each, so thousands of processors will be required. Scalability and programming complexity are important issues for parallel processors. Data parallel computers typically have more of the communications tasks performed by the compiler. When using thousands of processors, it may be unreasonable to assume applications programmers will want to (or be able to) manage the communications.

Table 1. Data Parallel Processing vs. Message Passing

APPROACH	ADVANTAGES	DISADVANTAGES
DATA PARALLEL	- Simple code for some problems - Compiler performs domain decomposition and communications	- Codes may have to be rewritten - Inefficient branching
MESSAGE PASSING	- Larger portions of existing Fortran-77 codes may be useful	- User responsible for all data transfer between processors and synchronization

The rapid advances made in parallel computing are well illustrated by comparing Figure 1 of Reference [5] with a similar figure presented several years ago in Reference [6]. Just a few years ago most massively parallel computers were clustered around the 1 gigaflop performance level. The current designs have the capability of achieving 100-1000 gigaflops. Table 1 summarizes the various Connection Machine computers. The CM-5 [3] has both SIMD and MIMD features and permits both data parallel and message passing approaches. While coarse-grain parallel machines offer cost effective solutions to some numerical problems, the emphasis in this paper will be on algorithms for large scale computers capable of solving grand challenge problems [7].

Table 2. The History of Connection Machine Computers.

Type	Clock Speed (Mhz)	Processor	Number of Processors	Network	Maximum Memory (GBytes)	Peak Speed (Gflops)	Approx. Size (ft.) W x L x H	Date
CM-1	4	TMC	8K - 16K	Hypercube	0.1	0.001?	6 x 6 x 6	1985
CM-2a	7	Weitek/TMC	128 - 256	Hypercube	1	4	2 x 2 x 4	1988
CM-2	7	Weitek/TMC	512 - 2048	Hypercube	8	28	6 x 6 x 6	1987
CM-200a	10	Weitek/TMC	128 - 256	Hypercube	1	5	2 x 2 x 4	1989
CM-200	10	Weitek/TMC	512 - 2048	Hypercube	8	40	6 x 6 x 6	1989
CM-5	32	Sparc/Vector	32 - 16K	Fat Tree	524	2100	30 x 10 x 7 (for 1K)	1991

GAS DYNAMIC MODELS

The applications discussed in this paper will be limited to gas dynamics. Liquids can actually be more difficult to solve and are not as well understood theoretically as gases. The various regimes of gas dynamics can be represented by a three-dimensional graph in Mach, Reynolds, and frequency space, as shown in Figure 1.

It is instructive to consider the various limits of this graph. For example, the plane defined by Mach=0 is the region governed by the incompressible Navier-Stokes equations. The plane represented by Reynolds=∞ corresponds to the Euler equations. All of steady-state gas dynamics is

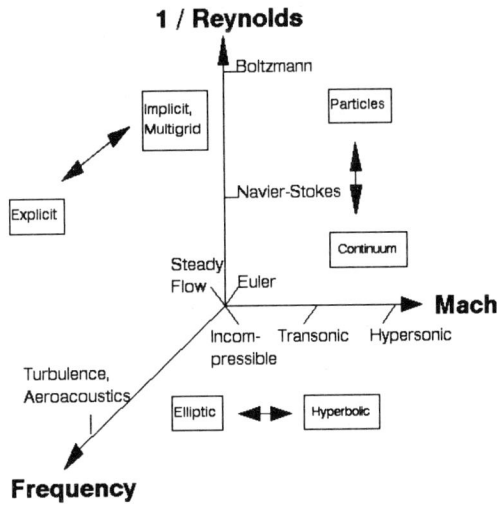

Figure 1. The Range of Gas Dynamic Phenomenon

included in the Mach-Reynolds plane where frequency=0. At the extremes of each of the axes, three different regimes exist: hypersonic, rarefied, and unsteady gas dynamics. Aeroacoustics and turbulence are examples of highly unsteady flows. Traditionally, CFD has been most concerned with steady-state gas dynamics. Even turbulent flows are often recast in terms of time-averaged quantities. This approach (i.e. Reynolds averaging) is the most widely-used method, but has clear limitations in terms of its validity and its usefulness.

Another reason for showing Figure 1, is to illustrate the wide variety of algorithms and equations required to model gas dynamics. There is no one algorithm that can effectively model this entire range of gas dynamic flows. Multi-grid and implicit algorithms, while effective for steady-state flows, are not as effective for unsteady problems where one is interested in time-accurate phase and amplitude predictions. Similarly, in terms of Reynolds number, the continuum approach works well at higher Reynolds numbers (or smaller Knudsen numbers), but at low Reynolds numbers the particle approach may be more effective. This occurs because the Navier-Stokes equations become invalid and the Boltzmann equation is too difficult to solve. In the aerospace community rarefied gas dynamics and high Mach number flow often occur together; however, this is not always true. There are also instances of very low Reynolds number flows that have low Mach numbers (such as in microdevices or microrobots).

The third axis illustrates the effect of compressibility. For very low Mach numbers the governing partial differential equations are essentially elliptic, while at higher Mach numbers one encounters the hyperbolic nature of gas dynamics. At very low Mach numbers, many algorithms become ineffective due to the stiffness of the equations, so people resort to pseudo-compressibility techniques or pressure relaxation techniques [8]. At the higher Mach numbers strong shock waves can form and upwind or Riemann solvers become more useful. Dissociation, ionization, and recombination can also occur in this regime.

In each of the above regimes the computer power required is enormous, even when the algorithm is specially designed for that regime. Therefore, at the present time it does not seem feasible that one algorithm or model will be capable of simulating all types of gas dynamic flows. Hybrid schemes may be especially useful to pursue.

Some of the more difficult gas dynamic problems occur at the outer portions of the three axis, that is, rarefied gas dynamics, hypersonic flow, and turbulence/aeroacoustics. While significant progress has been made in the areas of hypersonics and rarefied gas dynamics, turbulence is still one of the most elusive problems in physics. Massively parallel computers, however, could provide a factor of 1000 speed-up over existing vector computers. This speed will allow us to use far more capable algorithms for modeling turbulence, such as large-eddy simulation (LES) and direct numerical simulation (DNS). The Reynolds-averaged Navier-Stokes (RANS) approach has probably reached the limit of its usefulness.

In order to show the potential of massively parallel computers, one can investigate the performance required to simulate the various scales of turbulent flow. For a boundary layer on a flat plate (assuming 1 percent turbulence levels), the three main length scales of turbulence vary with Reynolds number approximately according to the following relations [9]:

$$\delta \sim 0.37 \, L \, Re_L^{-1/5} \qquad \lambda \sim 24 \, L \, Re_L^{-3/5} \qquad \eta \sim 24 \, L \, Re_L^{-4/5}$$

where L is the distance along the plate. The terms δ, λ, and η refer to the boundary layer thickness, Taylor microscale, and Kolmogorov scale, where $\delta > \lambda > \eta$. The three types of turbulence simulation techniques: Reynolds-averaging, LES, and DNS can be considered to vary in terms of which turbulent scales are computed and which are modelled. The Reynolds-averaged approach (e.g. algebraic or 2 equation models) essentially models all scales and computes none of them. On the other hand, DNS models none of the scales and computes all of them. LES is in between these two approaches. For the present discussion, consider LES to model all scales smaller than the Taylor microscale, and compute scales down to the size of the Taylor microscale.

In order to illustrate the varying amounts of computer power required for RANS, LES, and DNS, imagine a uniform computational grid over a volume V. The number of nodes in the grid is then simply $N = V / \Delta x^3$ where Δx is the node spacing. If one considers the RANS approach to be feasible on present computers and the computer power required is proportional to the number of nodes required, one can estimate from the above formulae the relative increase in computer power required for LES and DNS. On can show that

$$\frac{N_\lambda}{N_\delta} = \frac{\delta^3}{\lambda^3} \sim 10^{-6} \, Re_L^{6/5} \qquad \frac{N_\eta}{N_\delta} = \frac{\delta^3}{\eta^3} \sim 10^{-6} \, Re_L^{9/5}$$

At a large Reynolds number, say 10^7, one can show that $N_\lambda \sim 10^3 \, N_\delta$ and $N_\eta \sim 10^7 \, N_\delta$.

This is quite encouraging, since it indicates that with a factor of 1000 improvement in computer power, one should be able to use LES for the same flow that we can use RANS on current vector supercomputers. It also indicates that DNS will not be feasible for high Reynolds number flows for quite some time.

ALGORITHM AND HARDWARE MATCHING

This section will discuss the importance of matching the algorithm to the computer hardware and software. Algorithms designed to run effectively on vector computers are not necessarily the best ones for massively parallel computers. Likewise, algorithms that run well on 32 processors may not scale well to thousands of processors.

The main difference between parallel and vector computers is the memory access and communications requirements. On a parallel computer (especially massively parallel), communication and computation are both important. For this reason, the goal of a "teraflop" computer by the mid 1990's may be misleading unless we understand that to mean a computer that has computational speed, communication bandwidth, RAM memory, and mass storage all thousands of times larger than previous computers. Not only are teraflops required, one may also need (or desire) terabytes/second bandwidth, terabytes of fast memory, and terabytes of mass storage. The experience to date on massively parallel computers indicates that as one solves gas dynamic problems with more complicated physics or geometries, the communication speed is a more and more important factor.

The ratio of communication speed to processing speed dramatically effects the algorithm choice on parallel computers. For example, implicit algorithms will not be effective on computers with poor communication performance. While it is easy to determine which floating point chips (and their peak speed) are used in the various parallel computers, it is not as easy to determine how well their communication networks work. While the new benchmarking codes give an indication of how well the parallel computers perform on more realistic problems, it would be useful to have a measure of performance that includes the communication speed, just as floating point operations per second denote computational speed. Ideally one would like to know the peak bandwidth for several types of message routing, but even the ratio of peak communication to peak computation speeds would be useful. Ideally, one would like the computer to be able to communicate and compute simultaneously.

As an example of an algorithm that has low communications requirements, one could consider the 3-D heat equation:

$$\frac{\partial f}{\partial t} = \nu \left(\frac{\partial^2 f}{\partial x^2} + \frac{\partial^2 f}{\partial y^2} + \frac{\partial^2 f}{\partial z^2} \right)$$

Using a simple explicit, centered-difference scheme on a nonuniform, cartesian grid would yield :

$$f_{ijk}^{n+1} = (1-a_1) f_{ijk}^n + a_2 (f_{i+1\,jk}^n + f_{i-1\,jk}^n) + a_3 (f_{i\,j+1\,k}^n + f_{i\,j-1\,k}^n) + a_4 (f_{ij\,k+1}^n + f_{ij\,k-1}^n)$$
where

$$a_1 = 2 \nu \Delta t \left(\frac{1}{\Delta x^2} + \frac{1}{\Delta y^2} + \frac{1}{\Delta z^2}\right), \quad a_2 = \frac{\nu \Delta t}{\Delta x^2}, \quad a_3 = \frac{\nu \Delta t}{\Delta y^2}, \quad \text{and} \quad a_4 = \frac{\nu \Delta t}{\Delta z^2}.$$

For one processor assigned to every node, this algorithm has eleven computations and six communications per node per iteration. As more and more nodes are assigned to each processor, however, less processor-to-processor communication is required since the surface to volume ratio of the subgrid changes. The ratio of computations to communications scales linearly with the virtual processing (VP) ratio, which is essentially the number of nodes per processor. Computers should be able to exploit this fact, whether they are SIMD or MIMD.

If one had a 1000 x 1000 x 1000 grid (10^9 nodes) and 10^4 processors, there would be 10^5 nodes per processor. This implies roughly a 46 x 46 x 46 grid on each processor. This also means each subgrid would have 10^4 surface nodes. At every time step one would have roughly 10^6 computations and 10^4 interprocessor communications per processor. If each processor is capable of 100 mflops, the computations could be performed in 0.01 seconds. In order for the communications to be performed in the same time, using 64-bit precision, one would need a communication speed of roughly 10 MBytes/second per processor (this includes time required for data to be stored in memory). So, for this example, the ratio of communication (megabytes/second) to computation (megaflops) is roughly 0.1. Virtually all other algorithms will require more communications than this algorithm. Implicit schemes, unstructured grids, particle methods, or higher-order spatial accuracy would all significantly increase the communications requirements. So the above example could be considered a minimum. For good performance on a variety of algorithms, one would like to have this ratio be closer to unity.

It should also be pointed out that this algorithm requires one to store 6 floating point numbers (f^{n+1}, f^n, a_1, a_2, a_3, and a_4) per node. If each processor has 10^5 nodes, one would need 5 megabytes per processor. Algorithms for solving the Navier-Stokes equations would require roughly 10 times more memory. Notice also that, even for this simple problem, a grid with 10^9 nodes would mean that at <u>each time step</u> storing the solution (f) would require 8 gigabytes of disk space. Clearly any massively parallel system must be well balanced.

One could also implement an implicit scheme for the heat equation, but this would have a much higher communications to computations ratio. For example, an alternating direction implicit (ADI) scheme could be used [8]. This would require the solution of 3 tridiagonal systems (of order 1000) at each time step. To solve these one could use parallel cyclic reduction. Or one could use more conventional tridiagonal solvers, but this has not been very successful on massively parallel computers. For ADI you need to sweep through the 3-D field of data in each of the three directions individually, which usually means only one of the three tridiagonal solves will be efficient. That is, one will require almost no communication while the other two will involve enormous amounts of communication. To avoid this, one could transpose the system 3 times to arrange the data properly [10], but the transpose also requires large amounts of communication. The ADI approach is just not well suited for massively parallel computers. This is somewhat unfortunate since it is a widely used approach for CFD.

In an effort to achieve convergence characteristics similar to implicit schemes, while maintaining the low communications cost of explicit schemes, a family of an unconditionally stable explicit schemes were developed in Reference [11]. These schemes may be very effective on massively parallel computers.

On vector computers most well-designed gas dynamics algorithms require computer time roughly according to the following formula:

$$\text{CPU Time} = \left[\frac{\text{Computations/Node/Iteration}}{\text{flops}} \right] \cdot \text{Nodes} \cdot \text{Iterations}$$

which assumes the algorithm is O(N) where N is the number of nodes or iterations. Nodes and iterations refer to the number of spatial nodes used in the computational grid and the number of time steps (or iterations used), respectively. These are usually problem dependent, but the choice of algorithm can also dramatically effect them. For example, a multi-grid scheme may only require a few hundred iterations while an explicit scheme may require thousands. The quantity "Computations/Node/Iteration" is more or less algorithm dependent, while flops are machine dependent.

As an example, consider an algorithm that requires 2000 computations per node per iteration and a problem that requires 1000 iterations and 10^5 nodes. In addition, assume the computer is capable of sustaining 100 mflops. The CPU time would then be roughly 2000 seconds. On vector computers the key parameter is usually the megaflop rate.

On the other hand, a first approximation to algorithm performance on massively parallel computers would be significantly more complicated. A simple model that illustrates a few key parameters in massively parallel algorithms is :

$$\text{CPU Time} = \left[\left(\frac{\text{B.C. Computations}_1}{\text{flops/Proc.}} \right) \cdot \frac{\text{B.C. Nodes}}{\text{B.C. Proc.}} + \left(\frac{\text{Computations}_2}{\text{flops/Proc.}} + \frac{\text{Communications}}{\text{Bytes / Sec./Proc.}} \right) \cdot \frac{\text{Nodes}}{\text{Proc.}} \right] \text{Iterations}$$

The ultimate goal is to minimize the CPU time, <u>not</u> maximize the flops. Typically, the CPU time varies linearly with number of nodes and iterations. One of the key tests of parallel algorithms and computer codes is whether they will run twice as fast with twice as many processors or whether they will take only twice as much CPU time if you double the size of the problem. Ideally one should be able to double the problem size and number of processors simultaneously and the CPU time should remain the same.

The first term represents the time spent in boundary conditions, serial code, synchronization, or load balancing. The second term accounts for the bulk of the computations and is analogous to the term for vector computers. The last term represents the time spent in interprocessor communication. Notice that this simple performance model shows that one could achieve the peak speed of a SIMD computer only if there were zero communications and boundary conditions.

As an example, consider a computer with 10^4 processors, 100 mflops/processor, and 10 mBytes/seconds/processor. Also imagine a problem that requires 10^9 interior nodes, 10^3 iterations, and 10^6 Boundary condition nodes. If the algorithm required 100 boundary condition computations/B.C. node/iteration, 2000 interior point computations / node / iteration, and 4000 Bytes / node / iteration, this example would give :

$$\text{CPU Time} = \left[\left(\frac{100}{10^8}\right)\frac{10^6}{3000} + \left(\frac{2000}{10^8} + \frac{4000}{10^7}\right)\frac{10^9}{10^4}\right] 10^3 \approx 42{,}000 \text{ seconds}$$

Notice that if the communications speed was on the same order as the computational speed, say 100 mBytes/sec., the CPU time would be only 6000 seconds.

This simple model assumes that the CPU time comes from essentially three parts: boundary condition computations, interior point computations, and communications. If one were able to perform communication and boundary condition calculations simultaneously with interior point calculations, then the CPU time would simply be the time to perform the slowest one. The above three components are still relevant however. This model would be quite accurate for a SIMD machine, such as a CM-2, an AMT DAP, or a MasPar. Since it is difficult, however, to achieve synchronous communication and computation on massively parallel computers (even MIMD machines), the above may be valid on a range of computers.

ALGORITHMS IMPLEMENTED ON THE CONNECTION MACHINE

This section describes some of the gas dynamics algorithms that have been implemented on the Connection Machine by the author and his colleagues. The algorithms include techniques to solve the Euler, Navier-Stokes, and Boltzmann equations (including particle-based techniques). Some of the experience gained from these implementations will be described here. In particular, the CPU time requirements for each code will be described and the relative fractions of time spent in the three portions of the algorithm will be described.

The first techniques developed were for the 3-D Euler/Navier-Stokes equations for structured and unstructured grids [6 and 12]. A finite-volume Runge-Kutta scheme was used with implicit and explicit residual smoothing. These codes used primarily an adaptive dissipation scheme developed by Jameson et al [13]. An algebraic renormalization group (RNG) turbulence model was also implemented. These early codes were written in *Lisp. More recently, 3-D Navier-Stokes codes have been written in CM-Fortran for structured [14] and unstructured grids. These also used a finite-volume, Runge-Kutta scheme. The unstructured solver, however, uses Roe's approximate Riemann solver [15] instead of an adaptive dissipation scheme.

The CM-Fortran structured-grid solver, used nonreflecting far-field (Riemann invariant) boundary conditions. These boundary conditions are a classic example of what not to do on

SIMD computers. They require a large number of floating point operations (mainly to compute exponentials) and most of the processors are idle while they are being computed. Depending on the compiler options chosen and the problem being solved, these boundary conditions could require up to 50% of the CPU time, which is clearly inappropriate. On the other hand, the time spent in communication compared to interior point calculations was roughly balanced. A good discussion of how to implement boundary conditions in the same manner as interior point calculations is included in Reference [16], which showed a finite difference scheme on the Connection Machine could achieve 14 gigaflops.

The unstructured-grid solver that has been implemented is more accurately called an "arbitrary-grid" solver, since it can handle grids that have cells with arbitrary numbers of faces. This is analogous to a finite element solver that can handle different types of elements in the same grid. The main types of elements anticipated are tetrahedral and hexahedral cells. Hexahedral cells are useful in boundary layer regions, while in the far-field tetrahedral cells are adequate. This solver also uses a finite volume, Runge-Kutta time-marching scheme, but it uses Roe's approximate Riemann Solver. Even though this is an unstructured-grid solver, it runs at over 1 gigaflop on full CM-2 and should be extremely well-suited to the CM-5. It is also well suited to adaptive grid refinement. The grids are currently being generated quickly and easily using SDRC's IDEAS package.

The other major advantage of this scheme is that the boundary conditions are implemented in essentially the same manner as the interior points. This is possible due to the use of the Riemann (upwind) solver. Since every cell face flux calculation accurately accounts for the characteristics, one can just have imaginary "ghost" cells at the boundaries. These ghost cells can be permanently loaded with the free-stream flow properties and the Riemann solver will properly compute the flux between these ghost cells and the first layer of interior cells. The boundary points require virtually no CPU time. They consist of nothing more than assignment statements nested within a WHERE statement (with no calculations performed inside the WHERE). Roughly 30 and 70 percent of the CPU time is spent in calculations and communications, respectively.

In addition to the continuum codes, two CM-Fortran codes for rarefied gas dynamics have been written for the Connection Machines. One of these codes [5] was an implementation of the well-known Direct simulation Monte Carlo scheme [17], which is a particle-based method. This was a 1-D implementation that solved the supersonic Rayleigh problem. The DSMC algorithm is the most widely-used and validated algorithm in rarefied gas dynamics. It is used internationally and routinely by government labs, industry, and universities. The simulated molecules are moved and collided randomly through a grid of cells. The calculations involved in both the motions and the collisions are relatively simple. The most difficult and expensive portions are the sorting and the collisions. No partial differential equations are involved.

A 1-D version of this algorithm was implemented on the Connection Machine. The code demonstrated that the original DSMC algorithm is scalable on a data-parallel computer. A more capable 3-D unstructured-grid version is planned for the future. Performance on a 32K processor CM-2 was comparable to a 1-Processor Cray YMP. This is, however, a communication dominated algorithm. The CPU breakdown shows 0, 10, and 90 percent of the time is spent in boundary conditions, computations, and communications, respectively. A 128 processor CM-5 (without vector units) ran seven times faster than a 4K CM-2.

Bartell et al [18] have also implemented DSMC on a massively parallel nCUBE computer. Their implementation seems to be more efficient (having achieved 30 times the performance of a Cray YMP/1 on 512 processors). This seems to be due to the fact that they are performing domain decomposition and consequently reducing the communications required. They are also moving the data associated with each molecule from processor to processor, instead of just changing a pointer and sorting. This would be more efficient, especially if the collision routine is called often, since the communication is performed once and the collision routine would require no communications.

The other rarefied gas dynamics code developed [19] used a new algorithm to solve a model form of the Boltzmann equation known as the Bhatnagar-Gross-Krook (BGK) equation. This BGK code was used to simulate shock wave structure and the supersonic Rayleigh problem. This algorithm was essentially designed for the Connection Machine computer. It uses a Runge-Kutta time marching scheme with upstream differencing. The integrals in velocity space are computed using Gaussian quadrature and a least squares technique is used to enforce of conservation. Initially, 343 points were used in velocity space (all stored "on chip"). Quadrature and conservation was accomplished using no communication. A rough CPU breakdown shows 0, 90, and 10 percent of the time was spent in boundary conditions, computation, and communication. On an 8K CM-200 this code ran at approximately 1 gigaflop and on a 128 processor CM-5 it ran at 6 gigaflops.

CONCLUSIONS

The wide range of possible gas dynamic problems cannot all be solved with a single algorithm. The hyperbolic, parabolic, elliptic, and even particle nature of gas dynamics is just too complex. Every computer is better suited to some algorithms than others. Thinking machines CM-Fortran (with CMSSL) is capable of solving a wide variety of problems. The data parallel approach is quite natural for most scientific problems. A good balance between communications and computations is required to solve real engineering applications (complex physics and complex geometries), not just teraflops.

ACKNOWLEDGMENTS

The support of the National Science Foundation (ASC-9009998) is gratefully acknowledged.

REFERENCES

1. Hillis, W.D.; "The Connection Machine," Scientific American, Vol. 256, June, 1987.

2. "CM-200 Technical Summary," Thinking Machines Corp. Technical Report HA87-4, Camb., Mass., June, 1991.

3. CM-5 Technical Summary, Thinking Machines Corp., Cambridge, MA, Oct., 1991.

4. Trew, A. and Wilson, G.; Past, Present, Parallel, Springer-Verlag, London, 1991.

5. Wong B.C. and L.N. Long; "A Data-Parallel Implementation of the DSMC Method on the Connection Machine," Computing Systems in Engineering, to appear, 1992.

6. Long, L.N., Khan, M.M.S., and Sharp, H.T.; "A Massively Parallel Euler/Navier-Stokes Method for the Connection Machine," AIAA Jnl., Vol. 29, No. 3, Mar, 1991.

7. Grand Callenges 1993: High Performance Computing and Communications, Office of Science and Technology Policy, Washington, D.C., 1992

8. Hirsch, C.; Numerical Computation of Internal and External Flows, Wiley, NY, 1990

9. Tennekes, H. and Lumley, J.L.; A First Course in Turbulence, MIT, Camb., 1989.

10. Chyszewski, T.; Private Communication, Aug., 1992

11. Richardson, J.L., Ferrell, R.C., and Long, L.N.; "Unconditionally Stable Explicit Algorithms for Nonlinear Fluid Dynamics Problems," J. Comp. Phys., to appear, 1992.

12. Long, L.N.; "A Three-Dimensional Navier-Stokes Code for the Connection Machine," in Scientific Applications on the Connection Machine, World-Scientific, 1989.

13. Jameson, A., Schmidt, W., and Turkel, E.; "Numerical Solutions of the Euler Equations by Finite Volume Methods Using Runge-Kutta Time-Stepping Schemes," AIAA Paper No. 81-1259, June, 1981.

14. Long, L.N.; "3-D Navier-Stokes in Parallel Fortran," in Proceedings of the 10th AIAA CFD Conference, Honolulu, Hawaii, June, 1991.

15. Roe, P.L.; "Approximate Riemann Solvers, Parameter Vector, and Difference Schemes," Jnl. of Comp. Physics, Vol. 43, 1981, pp. 357-372.

16. Myczkowski, J., Bromley, M., and McCowan, D.; "Extremely Fast Finite Difference Techniques for the Connection Machine," AIAA Paper No. 91-0436, Jan., 1991.

17. Bird, G.A.; Molecular Gas Dynamics, Clarendon Press, Oxford, 1976.

18. Bartell, T. J., "DSMC Simulation of Ionized Rarefied Flows on Large MIMD Supercomputers," 18th Rarefied Gas Dynamics Symposium, Vancouver, July, 1992.

19. Long, L.N., M. Kamon, T.S. Chyczewski, and J. Myczkowski; "A Deterministic Parallel Algorithm to Solve a Model Boltzmann Equation (BGK)," Computing Systems in Engineering, to appear, 1992.

Evaluation of Different Approaches to Parallel Processing for CFD

C. de Nicola[a], G. De Pietro[b] and L. Paparone[b]

[a]Gasdynamics Institute, University of Naples
Piazzale Tecchio 80, Napoli, Italy
Email: NIDODAPI@ICNUCEVM

[b]Parallel Information System Research Institute, National Research Council
Via P.Castellino 111, 80128, Napoli, Italy
Email: PINO@AREANA.NA.CNR.IT
Email: LUIGI@IRSIP.NA.CNR.IT

1. INTRODUCTION

Parallel processing is the present challenge to attain high performance computing; the recent technological growth in the hardware and software fields makes this challenge more realistic, by creating many expectations in the scientific community.

Computational Fluid Dynamics represents one of the research fields for which computing power requirements cannot be met by the present generation of supercomputers; since the accuracy level of the numerical flow simulation depends on the complexity of the mathematical model adopted and on the degree of resolution required (for example, for flows over a complete aircraft it is necessary at least one million of grid points or more), a large use of CFD for aerospace industry applications is strongly related to the development of new computing systems [1].

The variety of parallel architectures today available, resulting from different approaches adopted to exploit parallelism (e.g., data parallelism, algorithmic parallelism) urges the user to make a preliminary analysis regarding which computational model (SIMD or MIMD) may be more suited to parallelize his code and on which machine to implement it.

In this paper experimental results related to the implementation of CFD codes on parallel machines will be presented, in order to give a global overview on advantages and drawbacks of different parallel approaches. The present analysis has been carried out by using two different architectures, the Connection Machine CM 200 (a SIMD computer with 8K processors and 64-bit FPUs) and a Meiko Computing Surface (a MIMD computer with 256 T800 transputers); for the performance evaluation the 1-D inviscid Burgers equation and a family of Eulerian codes have been used.

2. MATHEMATICAL MODELS

The so-called inviscid Burgers equation can be written in conservative form as follows:

$$\frac{\partial u}{\partial t} + \frac{\partial}{\partial x}(u^2/2) = 0 \tag{1}$$

Because of its nonlinear and hyperbolic nature, the Burgers equation is a simple and meaningful model for Fluid Dynamics equations, and it can be used as starting point for testing the effectiveness of numerical schemes and methods on parallel machines [2].

The 3-D Euler equations, which describe the unsteady, inviscid, compressible and rotational flows, can be written, in conservative integral form as follows:

$$\iint f + \int \mathbf{n} \cdot \mathbf{H} = 0 \tag{2}$$

This integral relation express the consevation of mass, momentum and energy for any region of the flow domain. The variables w is the vector of the unknowns

$f = (\rho, \rho u, \rho v, \rho w, \rho E)^T$

where ρ is the density, u,v and w are velocity components along x,y and z axes respectively, and E is total energy per unit mass. The function H is given by

$\mathbf{H}(f) = [E(f), F(f), G(f)]$

where:

$E(f) = (\rho u, \rho u^2+p, \rho uv, \rho uw, \rho uH)^T$

$F(f) = (\rho v, \rho uv, \rho v^2+p, \rho vw, \rho vH)^T$

$G(f) = (\rho w, \rho uw, \rho vw, \rho w^2+p, \rho wH)^T$

Here p is pressure and H enthalpy; these are defined by

$p = (\gamma - 1) \rho [E - (u^2 + v^2)/2]$

$H = E + p/\rho$

where γ is the ratio of specific heats.

It has to be reminded that nowadays Euler equations represent a fundamental tool for aerodynamic design of complete aircraft configuration [3], and could be suited for industrial applications, when coupled with appropriate pre and post-processing tools and an efficient computing environment.

All the mathematical models used for testing the parallel machines have been numerically solved by using a finite volume central scheme with blended second and fourth order explicit, self adaptive artificial viscosity, and a multistage time stepping procedure [4].

3. PARALLEL IMPLEMENTATION STRATEGIES

The implementation of parallel codes can be obtained by designing new parallel algorithms or by means of the parallelization of existing sequential codes. If, on one hand, there are no methods to design new parallel algorithms (rather some general criteria in order to obtain a good load balancing and to minimize communication costs can be followed), on the other hand some techniques for the parallelization of existing sequential codes are available; for example, domain decomposition, odd-even reduction, coloring schemes, recursive doubling represent some of the techniques used to parallelize a wide class of existing codes and algorithms. Clearly, the fundamental differences existing between SIMD and MIMD architecture imply that, in general, there is no a parallelization strategy which is well suited for both architectures.

The most used approach for the parallelization of sequential codes is the exploitation of data parallelism by means of domain decomposition. For SIMD architectures this is obtained by changing the sequential code by using vector notations; clearly, the efforts due to this work depend on the algorithm structure and on the "vectorization" level of the original code. On MIMD machine, the user has to explicitly handle the decomposition, that is he has to properly subdivide the domain into several subdomains which have to be assigned to the available processors; moreover, the user has to explicitly manage the interprocessor communications for data exchange.

For the implementation of the Eulerian codes, a domain decomposition technique has been used [5]; in particular the computational domain has been divided along one, two or three dimension of the grid reference frame. In this case each processor basically performs the same algorithm on its own subdomain, and any special handling of the interface conditions between subdomains is involved. Instead, for the implementation of the code for the solution of the inviscid Burgers equation the Multiblock technique has been used [6], providing a subdivision of the computational domain in a fixed number of blocks, according to proper topological rules. The block decomposition introduces artificial boundary faces for each block, on which additional conditions need to be specified. This technique is well suited for MIMD architectures; in fact the calculation on each block can proceed independently until flux continuity conditions have to be satisfied at each time step. However, it should be noted that in this case, because the number of blocks and their computational workload come from geometrical decomposition of the computational domain carried out independently from the particular computer system used, the problem of assigning blocks to processors is the well known mapping problem.

4. RESULTS

Table 1. shows the performance, on parallel architectures, of the 1-D multiblock Burgers code, with an additional comparison with executions time on Cray Y-MP. These results refer to a non continuous solution of the eq. (1), with 256 blocks, each with 1024 cells.

Tab. 1
Execution time for multiblock Burgers code

	Cray Y-MP	Meiko	CM200
Time (sec)	16.89	25.40 (256 procs)	3.89

The best results have been obtained on CM 200; it should be pointed out the different "class" of the machines used in terms of peek performance, ranging from 380 MFlops for the Meiko to 5 GFlops for the CM 200.

Further tests on CM 200 have been performed by using a 2-D Euler code, applied to the prediction of lifting airfoil flows. The implementation has essentially required a code vectorization, so the main work performed on the sequential code was the modification of a number of subroutines by using vector notations and some specific instruction supported by the CM compiler for data layout.

The performance are very encouraging, as can be seen in Tab. 2; the results on Cray Y-MP are obtained by using the vectorization facilities of the compiler. However, it should be noted that an improvement can be achieved on both CM 200 and Cray Y-MP properly changing the code structure according to the hardware characteristics of each machine.

Tab. 2
Execution time for 2-D Euler code

	Cray Y-MP	CM 200
Time (sec)	68.95	21.92

The weight of the scalar computation on CM 200 is exploited by analyzing the per cent evaluation of the time spent on the Front-End and on the parallel machine, shown in Tab. 3: by considering that the most of the computation is performed on parallel machine, it should be concluded taht scalar operations on the Front-End make the parallel execution less efficient.

Tab. 3
Percentage of the time spent on CM 200

FE cpu (user)	CM (total)
63 %	37 %

The 3-D Euler code has been tested on Meiko Computing Surface applied to a wing transonic flow; Tab. 4 shows a CPU time comparison between Cray Y-MP and Meiko. In this case greater effort was spent in the implementation of the parallel code, because of the explicit handling of the interprocessor communications and of the subdivision of the computational workload among processors.

Tab. 4
CPU time for 3-D Euler code

	Cray Y-MP	Meiko
Time (sec)	17.95	136.55 (192 procs)

This results are referred to a 3-D domain decomposition, where the computational domain was divided along the three axes of the grid reference frame; in this case the highest volume to surface ratio is reached, which means that, for the same computational workload, a minor amount of data needs to be exchanged among processors respect to other decompositions. However, this does not imply, in general, that the communication time decreases; in fact, because an higher number of communication is required, the time spent for the communications set-up increases. It should be pointed out that the grid size was chosen in such a way that the tests were performed in optimal load balancing conditions. In Fig. 1 the Speed-up curve for the present case is shown.

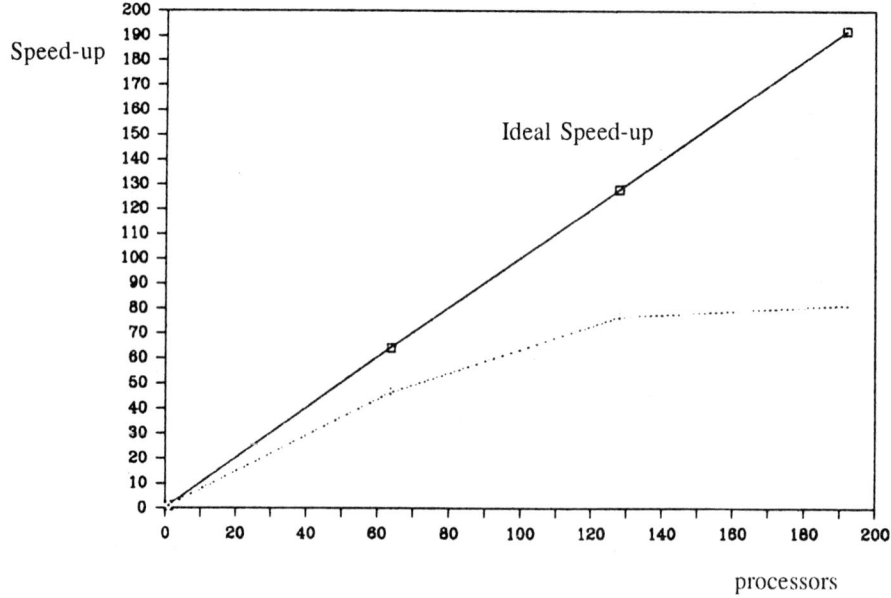

Fig. 1: Speed up curve of Euler 3-D code

5. CONCLUSION

On the basis of the experience made using MIMD and SIMD parallel machines, the following considerations can be made:

- the greatest problem for CFD parallel processing on MIMD architecture is due to efficiency of communications tools;
- the explicit handling of both communications and data partitioning among processors gives to MIMD approaches an high degree of versatility;
- SIMD arcitectures are very easy to use, especially regarding the implementation of sequential codes, but in general they do not offer an high degree of versatility as MIMD architectures.

ACKNOLEDGEMENTS

The authors are indebted with Dr. Vittorio Puoti for his contribution in performing some tests.

6. REFERENCES

[1] C. de Nicola, G. De Pietro, V. Puoti, P. Schiano, R. Vaccaro, "Parallel Processing for Typical CFD Problems", XI AIDAA Conference, October 14-18, Forli', 1991.

[2] C. de Nicola, L. Paparone, R. Tognaccini, "An Investigation on Multiblock Structured Algorithms for Parallel CFD Applications by a Solution of 1-D Burgers Equation", XI AIDAA Conference, October 14-18, Forli', 1991.

[3] A. Amendola, R. Tognaccini, J. W. Boerstoel, A. Kassies, "Validation of a Multiblock Euler Flow Solver with Propeller-Slipstream Flows", AGARD CP 437, Lisbon, May 1988.

[4] A. Jameson, "Numerical Solution of the Euler Equations for Compressible Inviscid Fluids", MAE Report 1643.

[5] C. de Nicola, G. De Pietro, V. Puoti, "Solving Euler Equations on Massively Parallel Computers", Parallel CFD '91 Conference, Stuttgart, June 10-12, 1991.

[6] H. C. Chen, N. J. Yu, "Development of a General Multiblock Flow Solver for Complex Aircraft Configurations", 8th GAMM Conference on Numerical Methods in Fluid

Compressible Vortex Reconnection on the Connection Machine

R.B. Pelz, T. Scheidegger, N.J. Zabusky, O.N. Boratav

Department of Mechanical and Aerospace Engineering, Rutgers University, Piscataway, NJ 08855-0909, U.S.A.

ABSTRACT

An efficient Connection Machine (SIMD) algorithm for solving the Navier Stokes equations for a three-dimensional, unsteady, compressible fluid flow is presented. A Fourier pseudospectral method is used for spatial discretization in a 3-torus domain, and a low-storage, 3rd Order Runge-Kutta scheme is used for time-marching. With initial conditions being orthogonal offset vortex tubes, we find that the effect of compressibility is to inhibit vortex reconnection and change the geometry of the region of antiparallel alignment of the vortex tubes.

1 Introduction

Vortex reconnection in incompressible flows has been the subject of numerous experimental and numerical studies [1] [2] [3] [4] [5]. Loosely put, it is the rapid collapse of antiparallel vortex flux tubes, dissipation and breaking of vortex lines in each tube, and subsequent connection of the vortex lines in one tube with those in the other. This change in topology of vortex lines can only occur in a viscous fluid.

Vortex reconnection has been connected with intermittency in turbulence, finite-time singularities in Euler equations and noise in jets [6] [7] [8]. It has been proposed that magnetic reconnection, which is similar in many ways, occurs in the solar corona and in the Earth's magnetosphere [9].

The large range of scales in this problem makes it a computationally intensive problem.

Whereas high resolution is required to image the steep gradients produced during the collapse, nonlocal contributions to the rate of strain which causes the collapse must also be resolved. It is also a finite-time event which can be modeled as a scattering problem with an in-going and out-going state [10].

The problem of vortex reconnection in compressible fluids has received little attention. To the best of our knowledge, the only other work in this field is by Kerr, Virk and Hussain [11]. Recently there have been numerous works in compressible turbulence, which use similar schemes but on serial computers [12] [13] [14] [15] [16].

Because the problem has a simple geometry but is inherently four-dimensional, pseudospectral methods are employed. [18] They have an accuracy greater than any power of the grid size for smooth problems and hence minimize the resolution requirements.

The Connection Machine CM-2 is used for our studies. It has been shown by Pelz [19] [20] that the efficiency of pseudospectral methods for incompressible flows can be high on a number of multiprocessors despite the global nature of the basis functions. We shall present here what we believe to be the most efficient algorithm for compressible flow simulations on the CM.

This paper is the first in a series of papers that concern vortex reconnection in a compressible fluid. It is focused more on the issues of CM algorithm and implementation, which is the theme of the "Parallel CFD" conferences. Latter papers will deal more with the physics.

In section 2 we present the CM algorithm we use for integrating the Navier-Stokes equations. In section 3 the CM implementation of the 3-D FFT, the heart of the algorithm, is discussed. In section 4 we talk about the initial conditions and in section 5 we present some preliminary results.

2 CM Algorithm for Compressible Flows

In this section we present the algorithm which is used to integrate numerically the Navier-Stokes equations for a compressible viscous fluid. Boundary conditions are periodic in all three space directions (3-torus). The spatial discretization scheme iss Fourier pseudospectral. Initial conditions and results will be discussed in subsequent sections.

We use the equations of conservation of mass, momentum and energy in conservative form. The independent variables are x, y, z in a right-handed Cartesian coordinate system and time t. The dependent variables are the density ρ, the momentum vector ρu, ρv and ρw where u, v and w are the components of the velocity vector, and the total energy per unit volume E_T.

The continuity equation can be written:

$$\frac{\partial \rho}{\partial t} + \frac{\partial}{\partial x}(\rho u) + \frac{\partial}{\partial y}(\rho v) + \frac{\partial}{\partial z}(\rho w) = 0. \qquad (1)$$

The equation for the momentum conservation equation is:

$$\frac{\partial}{\partial t}(\rho u) + \frac{\partial}{\partial x} F_{xx} + \frac{\partial}{\partial y} F_{xy} + \frac{\partial}{\partial z} F_{xz} = 0,$$

$$\frac{\partial}{\partial t}(\rho v) + \frac{\partial}{\partial x} F_{yx} + \frac{\partial}{\partial y} F_{yy} + \frac{\partial}{\partial z} F_{yz} = 0,$$

$$\frac{\partial}{\partial t}(\rho w) + \frac{\partial}{\partial x} F_{zx} + \frac{\partial}{\partial y} F_{zy} + \frac{\partial}{\partial z} F_{zz} = 0. \qquad (2)$$

The energy equation is:

$$\frac{\partial}{\partial t}(E_T) + \frac{\partial}{\partial x} E_x + \frac{\partial}{\partial y} E_y + \frac{\partial}{\partial z} E_z = k\Delta\Theta. \qquad (3)$$

where k is the coefficient of thermal conductivity (which is taken to be a constant), Θ is the temperature and the symmetric tensor $\mathbf{F}$ and vector $\mathbf{E}$ are defined as follows:

$$\begin{aligned} F_{xx} &= \rho u^2 - T_{xx}, & F_{xy} &= \rho u v - T_{xy} \\ F_{xz} &= \rho u w - T_{xz}, & F_{yy} &= \rho v^2 - T_{yy} \\ F_{yz} &= \rho v w - T_{yz}. & F_{zz} &= \rho w^2 - T_{zz} \end{aligned} \qquad (4)$$

and

$$\begin{aligned} E_x &= u(E_T - T_{xx}) - v T_{xy} - w T_{xz} \\ E_y &= -u T_{yx} + v(E_T - T_{yy}) - w T_{yz} \\ E_z &= -u T_{zx} - v T_{zy} + u(E_T - T_{zz}). \end{aligned} \qquad (5)$$

The stress tensor $\mathbf{T}$ is written

$$T_{xx} = -P - \frac{2}{3}\mu D + 2\mu S_{xx}, \quad T_{yy} = -P - \frac{2}{3}\mu D + 2\mu S_{yy},$$

$$T_{zz} = -P - \frac{2}{3}\mu D + 2\mu S_{zz}, \quad T_{xy} = 2\mu S_{xy}, \quad T_{xz} = 2\mu S_{xz}, \quad T_{yz} = 2\mu S_{yz}, \qquad (6)$$

where P is the pressure, D is the divergence of the velocity, μ is the coefficient of viscosity (taken to be constant) and $\mathbf{S}$ is the rate of strain tensor. A Stokes fluid is assumed. The rate of strain tensor is written

$$S_{xx} = \frac{\partial u}{\partial x}, \quad S_{yy} = \frac{\partial v}{\partial y}, \quad S_{zz} = \frac{\partial w}{\partial z},$$

$$S_{xy} = \frac{1}{2}\left(\frac{\partial u}{\partial y} + \frac{\partial v}{\partial x}\right), \quad S_{xz} = \frac{1}{2}\left(\frac{\partial u}{\partial z} + \frac{\partial w}{\partial x}\right), \quad S_{yz} = \frac{1}{2}\left(\frac{\partial v}{\partial z} + \frac{\partial w}{\partial y}\right). \qquad (7)$$

We also assume that we have a calorically perfect fluid with equations of state

$$P = (\gamma - 1)e, \ P = \rho R \Theta, \tag{8}$$

where γ is the ratio of specific heats, e is the internal energy per unit volume and R is the universal gas constant. Note that $E_T = e + \rho u^2/2$. The reason for this rather verbose presentation of the equations of motion is so that the algorithm can be explained easily.

Each of the dependent variables is expanded in a finite Fourier series. Using the orthogonality property of the trigonometric series, the equations for the Fourier coefficients can be written. The derivative of a dependent variable can be found analytically. The nonlinear terms become convolutions on the coefficients. For fast evaluation of these terms, an inverse discrete Fourier transform (DFT) is first used to find the physical (grid point) representation of the dependent variables involved, then the nonlinear term is evaluated, then a forward DFT is used to transform the product into Fourier space. This method is called the pseudospectral or collocation method.

The rules for the pseudospectral method are to perform derivatives on Fouier space and nonlinear operations in physical space. FFTs are used to changes bases.

We now give the algorithm. It is also presented in Table I. We begin timestep n with the dependent variables $\hat{\rho}$, $\hat{\rho u}$, $\hat{\rho v}$, $\hat{\rho w}$, $\hat{E}_T$ in Fourier space. The hat above denote the Fourier coefficients. If n = 0, The initial conditions are supplied. Five 3-d FFTs transform the dependent variables into the physical space representation. The velocity components are computed and the pressure is found from the equation of state. These new variables are stored in 4 work arrays. To calculate the rate of strain tensor **S**, the velocity components are transformed into Fourier space. The off-diagonal terms are first computed and stored in three additional work arrays. The diagonal terms are computed and overwrite the velocity components. Six FFTs take **S** into physical space. The divergence is found from the trace and stored in a final work array. The next step is to compute the stress tensor using P, D and **S**.

The next few stages are related to the energy equation. We overwrite the pressure with the temperature found from the equation of state. Upon transforming it to Fourier space, we find the Laplacian. Using E_T and the stress tensor, E_x is computed and overwrites D. E_x is then transformed and differentiated with respect to x and subtracted from $\widehat{\Delta \Theta}$ and stored in the place of $\widehat{\Delta \Theta}$. A similar procedure is done for E_y and E_z. Finally $\partial \hat{E}_T / \partial t$ is found.

The tensor **F** is then created using the dependent variables and the stress tensor, which **F** overwrites. Now the five dependent variables and the **F** tensor are transformed back into Fourier space. The time derivatives of the dependent variables are found by systematic differentiation of $\hat{\mathbf{F}}$ and the momenta. The dependent variables can now be updated.

The procedure makes up one stage of the 3th-order, low-storage Runge-Kutta timestepping scheme [21]. Three stages are executed for each timestep.

For one stage, 29 3-d FFTs and eight work arrays in addition to the five dependent variables are required. If the mesh has n^3 points. The operation count is $O(n^3 log n)$ for the FFTs and $O(n^3)$ for all other operations. The time to execute the FFTs makes up about 80 % of the total time. For n = 128 the code ran at about 45 second per timestep on a

16k CM-2. Using $5n \log n$ as the complexity of the FFT, and only using the FFT operation count, a MFLOPS rate of over 200 was obtained.

A spectral filter of the form $\sigma(k) = 1$ if $|k| \leq k_c$ and for $|k| > k_c$

$$\sigma(k) = exp[-\alpha(|k| - k_c)^4] \qquad (9)$$

where k is the wavenumber $(-n/2 \leq k \leq n/2)$, $k_c = n/3$ and $\alpha = 3$, is applied to the Fourier coefficients of the five primary variables at each time step. This exponential filter is a good compromise between overall accuracy and representation of discontinuities [18]. It is needed in the parameter range of our calculations to smooth and stabilize the solution. Its application to all variables, including the density ρ and energy E_T, makes it partially nonphysical. At some wavenumbers internal energy is drained and kinetic energy is increased. But filtering the velocity components only has proven to be insufficient to stabilize the solution.

No additional technique is used to remove aliasing errors. The standard 2/3-truncation rule for quadratic nonlinearities becomes a 1/2-rule for cubic terms, which is very costly.

3 The CM FFT

The CMSSL routine "detailed FFT" was employed for multidimensional Fourier transforms of complex sequences. [22] To lower communication time, the array layout was SEND and the arrays were left in bit-reversed order in physical space.

For the forward 3-d discrete transform an n^3 real array, the first transform, say in the x direction, has as input n^2 real sequences of length n. It returns n^2 length n conjugate symmetric sequences of which only n/2 complex elements are stored. The remaining y and z stages of the transform have as input $n^2/2$ complex sequences of length n. Thus, the supplied FFT routine for complex sequences is used for the y and z stages, but would be wasteful in both memory and operations if it were used on the x stage.

To do the x transform efficiently, we use the algorithm of Cooley, Lewis and Welch [23] to do the transform of a length n real sequence with a transform of a length n/2 complex sequence and some post-processing. For more information on parallel implementation see Pelz [19].

We should mention here that the post-processing of the sequence involves the ith and (n-i)th elements of the intermediate complex sequence, i = 1,...n/2. Bring these two elements together when the sequence is distributed across a hypercube in a send ordering is the worst possible communication operation. These elements are on opposite sides of the hypercube. If the NEWS (BRGC) ordering is used, bringing elements i and n-i together is a nearest neighbor communication. This ordering, however, increases the communication distance of processors in the FFT from nearest to next nearest neighbors. In the end, the time for both orderings is about the same.

If the VP ratio is greater than n, the best approach to the FFT of the real sequence is to layout the x direction as serial. Then if we write the postprocessing loops in Fortran 77, we can be assured that no communication calls will be issued. If the VP ratio is less than n, a high layout weight of the x direction will minimize communication time.

One very disturbing feature of CM Fortran is that once an array is laid out with a certain dimension, it must remain that way even in subroutines. The input array to the FFT is overwritten with the output. Thus, working with real arrays, which must be dimensioned as complex, requires real and aimag routines to be used to select odd and even indexed elements of the x direction.

4 Initial Conditions

In this section we present a discussion of the initial flow field for vortex reconnection. The centerlines of two vortex tubes are offset from and orthogonal to each other. The center lines are located $(\pi, 3\pi/4, z)$ and $(\pi, y, 5\pi/4)$, $0 \leq y, z \leq 2\pi$. Our goal is to explore the evolution of two such vortices as a prototypical interaction in turbulence. The standard initial condition for vortex reconnection in incompressible flow is antiparallel vortex tubes perturbed by the most unstable eigenmode of the Crow instability analysis [17]. Probablistically, however, a more likely interaction in turbulence is when the vortices are orthogonal. This condition is unstable and needs no perturbation to initiate the reconnection process.

The vorticity distribution in each tube is taken to be a constant $\hat{\omega}$ for $\eta \leq 0$, zero for $\eta \geq 1$, and for $0 \leq \eta \leq 1$

$$\omega = \hat{\omega} \cdot (1 - exp(-2.56\eta^{-1}exp(1/(\eta - 1)))) \tag{10}$$

where $\eta = (r - r_0)/(r_1 - r_0)$. r refers to the distance from the centerline and r_0 and r_1 are the inner and outer radii. When isolated it is a steady solution the Euler equations. This vortex distribution is placed around the vortex centerlines lines given above. The total vorticity field is found from simple superposition.

The velocity is found by solving the Poisson equation

$$\Delta \vec{v} = \nabla \times \vec{\omega}. \tag{11}$$

The mean of each component is set to zero.

The pressure is initialized by assuming a polytropic relation

$$P = const \cdot \rho^\gamma \tag{12}$$

Together with the fact that the velocity field given above is divergence free, this allows us to compute the pressure from the Poisson equation

$$\Delta P^{(\gamma-1)/\gamma} = -const^{(-1/\gamma)}\frac{\gamma-1}{\gamma}\nabla\cdot(\vec{v}\cdot\nabla\vec{v}) \qquad (13)$$

This is actually an equation for the temperature. The integration constant is found by adjusting the mean temperature such that a desired Mach number is realized. Note that the resulting density field is nonuniform despite the divergence free velocity. The initial density fluctuations are varied by choosing different polytropic exponents.

5 Results

In this section we present some preliminary results of compressible vortex reconnection. A more detailed description of the physics will be found in forthcoming articles.

For the results shown here, we kept the Reynolds number constant at 1300, and varied the Mach number. The Reynolds number is defined as Γ/ν, and the Mach number is defined as $v(r=r_0)/a_0$.

In Figure 1 we show isosurfaces of density for three times and two Mach numbers. The first row of figures is for M=0.25 at times 4.3, 4.8 and 5.3; the last row is for M=0.86 at times 4.7, 7.9 and 5.3. The low density marks the core region of the tubes. The evolution in the low Mach number case is very similar to the incompressible case. As the Mach number increases the isosurfaces indicate that the region with antiparallel alignment of the tubes increases in length. In the low-Mach-number case the tubes pinch at one point along the antiparallel region and form an "X" point. Later in the evolution, the X point is deformed so that it appears to be more like a "double Y" point. For higher Mach numbers, the X point seems never to appear, and the orientation is like the double Y immediately.

The reconnection process changes as Mach number increases. One way to quantify the difference is to look at vortex lines. We find integral lines through the vorticity field which start in the plane $y = 0$ and have a magnitude $75 \pm 2.5\%$ of the maximum value in this plane. By plotting the vortex lines at successive times, we can define the beginning of the reconnection process when one vortex line spans both tubes, that is, when a vortex line that enters the domain at $y = 0$, exits it at $z = 2\pi$. When all the vortex lines have reconnected, we say that the reconnection is complete. Figure 2 shows vortex lines for the same Mach numbers as in the previous figure. The times for the first row are as above, and for the last row they are 4.9, 6.7 and 8.6. Eventually all lines are reconnected.

In Figures 3 and 4, we show the time when reconnection starts and how long it takes (according to the definition above) as a function of initial Mach number. As we can see the timescale gets longer as Mach number increases. The rapid increase in reconnection times seems to coincide with the first appearance of a supersonic zone between the antiparallel vortices.

6 Summary

From our simulations of orthogonal vortex reconnection of a compressible fluid, we have observed an increase in the timescales of reconnection when the initial Mach number is increased. We believe that this phenomenon is related to compressibility effects in the antiparallel region.

To the best of our knowledge, this is the largest Fourier pseudospectral simulation of compressible flows ever done. This fact and the fact that the CM-2 provides a relatively friendly environment, suggests that this machine can be used quite easily as a research tool for fluid dynamics research.

The work was supported by a grant from the Office of Naval Research N00014-89-J-1320. The computations were done on the CM-2 at the Pittsburgh Supercomputer Center and the Naval Research Laboratory in Washington DC.

References

[1] Schatzle, P.R., (1987) "An Experimental Study of Fusion of Vortex Rings," Ph.D. Thesis, Stanford University.

[2] Melander, M.V. & Hussain, F., (1988) "Cut and Connect of two antiparallel vortex tubes," Center for Turbulence Research Proceedings, 257.

[3] Zabusky, N.J. & Melander, M.V., (1989) "Three-dimensional vortex tube reconnection: morphology for orthogonally-offset tubes," *Physica D*, **37**, 555.

[4] Boratav, O.N., Pelz, R.B. & Zabusky, N.J. (1992) "Reconnection in orthogonally interacting vortex tubes: Direct numerical simulations and quantifications," *Phys Fluids A*, **4**, 581.

[5] Meiron, D.I., Shelley, M.J. & Orszag, S.A., (1992) "Dynamical aspects of vortex reconnection," to appear in *J. Fluids Mech.*.

[6] Frisch, U. & Orszag, S.A., (1990) *Physics Today*, 24.

[7] Hussain, F., & Melander, M.V. (1991) in *Lumley Symposium: Recent Developments in Turbulence*, (Springer, New York)

[8] Kerr, R. (1992) submitted to *Phys Fluids A*.

[9] Lau, Y-T. & Finn, J.M. (1990) *Astrophys J.* **350** 672.

[10] Zabusky, N.J., Boratav, O.N., Pelz, R.B., Gao, M. Silver, D. & Cooper, S.P. (1991) "Emergence of Coherent Patterns of Vortex Stretching during Reconnection: A Scattering Paradigm," *Phys Rev. Lett.*, **67**, 2469.

[11] Kerr, R.M., Virk, D. & Hussain, F. (1990) "Effects of incompressible and compressible vortex reconnection," in *Topological Fluid Mechanics*, Eds Moffatt & Tsinober, Cambridge Univ Press, 500.

[12] Feiereisen, W.J., Reynolds, W.C. & Ferziger, J.H. (1981) Rep. TF-13, Thesis, Stanford University

[13] Passot, T., and Pouquet, A., (1987). Numerical simulation of compressible homogeneous flows in the turbulent regime, *J. Fluid Mech.* **181**, 441.

[14] Kida, S., and Orszag, S. A., (1990). Energy and spectral dynamics in forced compressible turbulence, *J. Sci. Comp.* **5**, 85.

[15] Sandham, N. D., and Reynolds, W. C., (1991). Three-dimensional simulations of large eddies in the compressible mixing layer, *J. Fluid Mech.* **224**, 133.

[16] Sarkar, S., Erlebacher, G., Hussaini, M. Y., and Kreiss, H. O., (1991). The analysis and modeling of dilatational terms in compressible turbulence, *J. Fluid Mech.* **227**, 473.

[17] Crow, S.C. (1970) "Stability theory for a pair of trainling vortices," *AIAA J.* **8**, 2172.

[18] Canuto, C., Hussaini, M.Y., Quarteroni, A. & Zang, T.A., 1987 *Spectral Methods in Fluid Dynamics* (Springer, Berlin).

[19] Pelz, R.B. 1991 "Parallel Fouier spectral methods on ensemble architectures," *Comp. Meth. Appl. Mech. Engr.*, **89**, 529.

[20] Pelz, R.B. (1990) "The parallel Fourier pseudospectral method," *J. Comp. Phys.*, **52**.

[21] Williamson, J.H., (1980) Low-Storage Runge-Kutta Scheme, *J. Comp. Phys.*, **35**, 48.

[22] *CMSSL for CM Fortran* (1991) Thinking Machines Corporation.

[23] Cooley, J.W., Lewis, P.A.W. & Welch, P.D., (1970) *J. Sound and Vibration*, **12**, 315.

$\hat{\rho}$	$\hat{\rho u}$	$\hat{\rho v}$	$\hat{\rho w}$	$\hat{E_T}$							
ρ	ρu	ρv	ρw	E_T							
					u	v	w				P
					$\hat{u}$	$\hat{v}$	$\hat{w}$				
					$\hat{S_{xx}}$	$\hat{S_{yy}}$	$\hat{S_{zz}}$	$\hat{S_{xy}}$	$\hat{S_{xz}}$	$\hat{S_{yz}}$	
					S_{xx}	S_{yy}	S_{zz}	S_{xy}	S_{xz}	S_{yz}	D
					T_{xx}	T_{yy}	T_{zz}	T_{xy}	T_{xz}	T_{yz}	
											Θ
											$\hat{\Theta}$
											$\Delta\Theta$
										E_x	
										$\hat{E_x}$	
											$-\frac{\partial}{\partial x}\hat{E_x}$
										E_y	
										$\hat{E_y}$	
											$-\frac{\partial}{\partial y}\hat{E_y}$
										E_z	
										$\hat{E_z}$	
											$-\frac{\partial}{\partial z}\hat{E_z}$
											$=\frac{\partial}{\partial t}(\hat{E_T})$
					F_{xx}	F_{yy}	F_{zz}	F_{xy}	F_{xz}	F_{yz}	
$\hat{\rho}$	$\hat{\rho u}$	$\hat{\rho v}$	$\hat{\rho w}$	$\hat{E_T}$	$\hat{F_{xx}}$	$\hat{F_{yy}}$	$\hat{F_{zz}}$	$\hat{F_{xy}}$	$\hat{F_{xz}}$	$\hat{F_{yz}}$	
					$\frac{\partial}{\partial x}\hat{F_{xx}}$	$\frac{\partial}{\partial y}\hat{F_{yy}}$	$\frac{\partial}{\partial z}\hat{F_{zz}}$				$\frac{\partial}{\partial x}(\hat{\rho u})$
					$+\frac{\partial}{\partial y}\hat{F_{xy}}$	$+\frac{\partial}{\partial x}\hat{F_{xy}}$	$+\frac{\partial}{\partial x}\hat{F_{xz}}$				$+\frac{\partial}{\partial y}(\hat{\rho v})$
					$+\frac{\partial}{\partial z}\hat{F_{xz}}$	$+\frac{\partial}{\partial z}\hat{F_{yz}}$	$+\frac{\partial}{\partial y}\hat{F_{yz}}$				$+\frac{\partial}{\partial z}(\hat{\rho w})$
					$=\frac{\partial}{\partial t}(\hat{\rho u})$	$=\frac{\partial}{\partial t}(\hat{\rho v})$	$=\frac{\partial}{\partial t}(\hat{\rho w})$				$=\frac{\partial\hat{\rho}}{\partial t}$

Table 1: Timestepping scheme

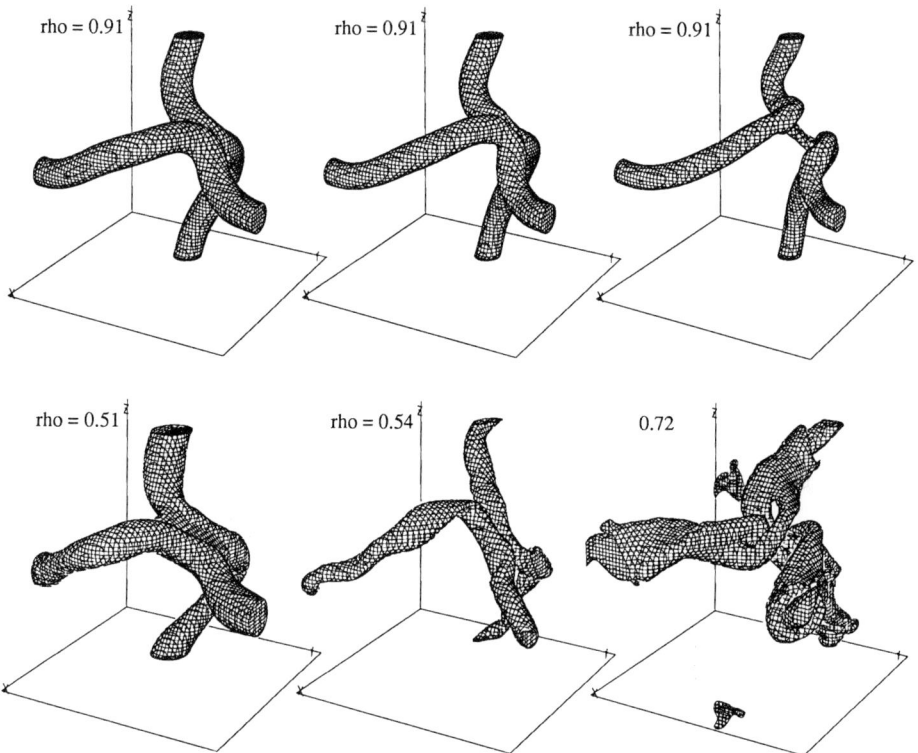

Figure 1: Density Isosurfaces: top M=.29, t=4.3, 4.8, 5.3; bottom M=.86, t=4.7, 7.9, 11.1

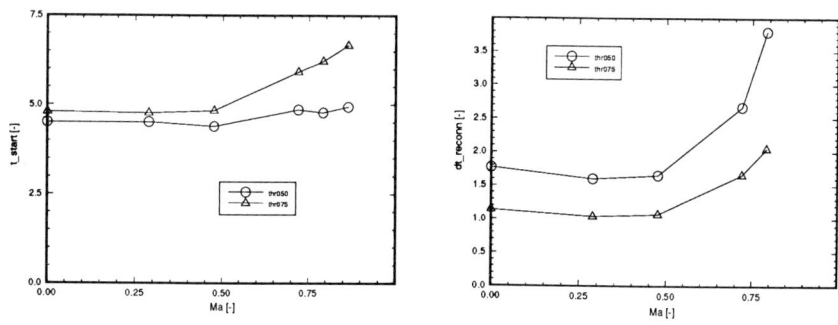

Figure 3: Time when reconnection begins   Figure 4: Time for reconnection

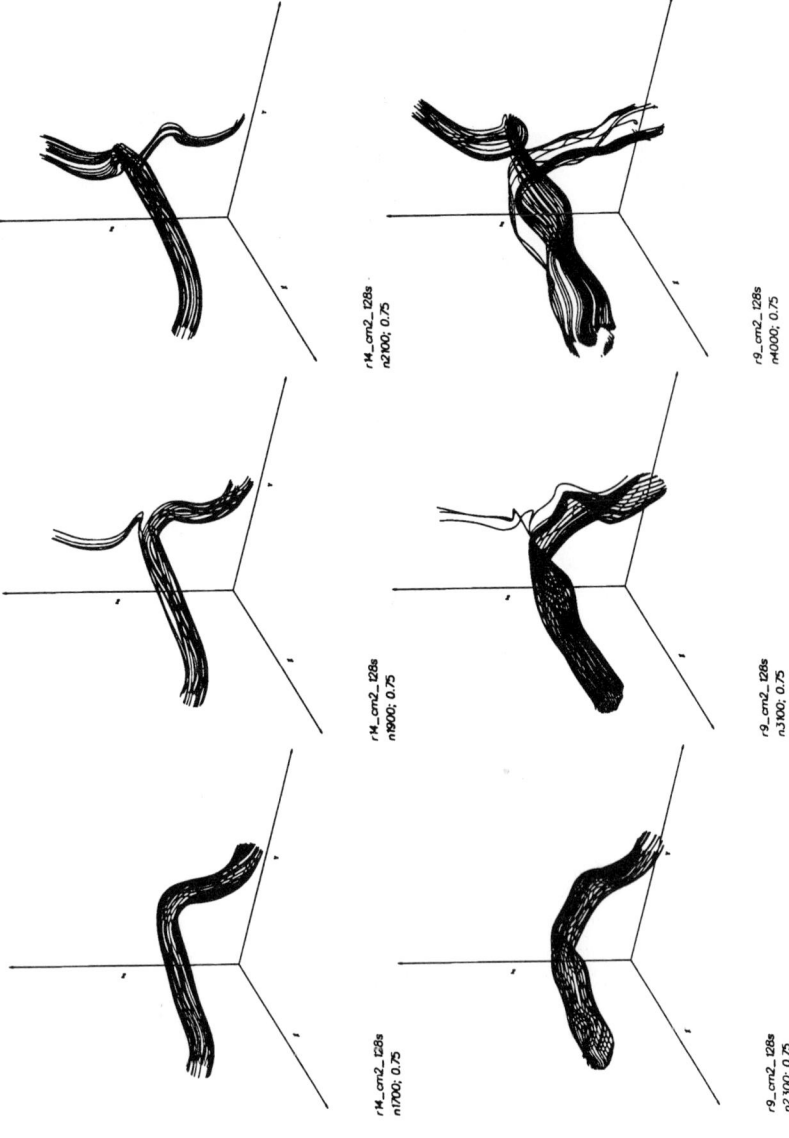

Figure 2: Vorticity Lines: top M=.29, t=4.3, 4.8, 5.3; bottom M=.86, t=4.9, 6.7, 8.6

Numerical Simulation of Complex Fluid Flows on MIMD Computers

M. Perić, M. Schäfer and E. Schreck

Lehrstuhl für Strömungsmechanik, Universität Erlangen-Nürnberg,
Cauerstr. 4, D-8520 Erlangen, Germany

Abstract

The paper presents a parallelization strategy for finite volume solution methods which employ block-structured, non-orthogonal grids. The solution method used here is fully implicit and employes second order central-difference approximations, SIMPLE algorithm for pressure-velocity coupling, a multigrid outer iteration scheme and various inner iteration solvers. The parallelization is based on domain decomposition approach, which is found best suited for use with block-structured grids and MIMD computers. Special attention is paid to the analysis of parallel, numerical and load balancing efficiency. Applications are presented for two-dimensional problems, showing good efficiencies. The possibilities for improving the parallel efficiency are also analysed.

1 Introduction

Calculation of flow in complex geometries requires the use of either block-structured or unstructured grids. Except for extremely complex configurations, the block-structured approach has the advantage of allowing the use of efficient solvers developed for single-block structured grids.

When solving steady flow problems (either laminar or Reynolds-averaged turbulent flows), implicit methods are usually used. They involve two iteration levels, which are here called *outer* and *inner* iterations. In the inner iterations, large linear equation systems are solved, whose coefficient matrix is sparse and within each block has a diagonal structure. The outer iteration loop provides for the update of the coefficient and source matrices in order to take into account the non-linearity and coupling of the equations for the individual variables.

The outer iterations are explicit in nature, since the new quantities are calculated using variable values from the previous iteration. This part of the solution algorithm is therefore easily parallelized. On the other hand, inner iterations - except for the Jacobi and the so called "red-black" Gauß-Seidel iteration methods - are implicit and pose special requirements to be performed in parallel. For example, ILU solvers can be parallelized in two dimensions if a structured grid is subdivided into stripes (Bastian and Horton, 1989). However, this approach can not be used in case of complex block-structured grids.

Recently, the present authors presented parallel implicit solution method based on domain decomposition technique, using ILU solver after Stone (1968) for the inner iterations and a full-approximation multigrid scheme for the outer iterations (Schreck and Perić, 1991; Perić et al, 1991). This approach extends in a straightforward manner to the block-structured grids. The present paper describes the extended method and concentrates especially on the analysis of the numerical efficiency.

2 Description of Solution Method

Solution method used in this study is described in detail by Demirdžić and Perić (1990), so only a summary of main features will be given here. The method is of finite volume type and uses non-orthogonal boundary-fitted grids with a colocated arrangement of variables. Figure 1 shows a typical control volume (CV). The working variables are the cartesian velocity components, pressure and temperature. The continuity equation is used to obtain a pressure-correction equation according to the SIMPLE algorithm (Patankar and Spalding, 1972). Second order discretization is used for all terms (central differences, linear interpolation). The part of diffusion fluxes which arises from grid non-orthogonality is treated explicitly. The convection fluxes are treated using the so called "deferred correction" approach: only the part which corresponds to the first order upwind discretization is treated implicitly, while the difference between the central differencing and upwind fluxes is treated explicitly. The effect of non-orthogonality is also treated explicitly in the pressure-correction equation. In the first step, the pressure-correction equation is solved with these terms excluded (which suffices if the grid is not severely non-orthogonal). In the second step the non-orthogonal contribution is evaluated using pressure correction calculated in the first step, and the second pressure-correction equation is solved. It has the same coefficient matrix as the first one but a different source term.

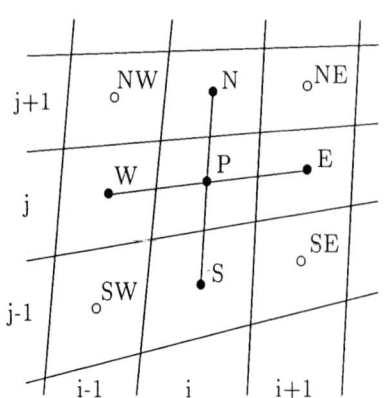

Fig. 1: A typical control volume and a computational molecule

Equations for the cartesian velocity components U and V, pressure correction P' and temperature T are discretized and solved one after another. Linear algebraic equation systems are solved iteratively using one of the following solvers: Gauß-Seidel method (GS), ILU-decomposition (SIP) after Stone (1968) or preconditioned conjugate gradient (PCG) method (Hestens and Stiefel, 1952). Inner iterations are stopped either after reducing the absolute sum of residuals over all CVs by a factor of five to ten, or after a prescribed number of iterations has been performed. Outer iterations are performed to take into account the non-linearity, coupling of variables and effects of grid non-orthogonality; this is why the linear equations need not be solved more accurately.

Computation is stopped when at the begining of an outer iteration the sum of absolute residuals over all CVs in all equations becomes 4 to 5 orders of magnitude smaller than the initial values. This corresponds roughly to a 4–5 digit accuracy. For steady flow calculations considered in this study, under-relaxation is used to improve the convergence of outer iterations. Typical values of under-relaxation factors range between 0.5 and 0.8 for the velocity components and temperature. The fraction of pressure correction added to pressure after solving the pressure-correction equation is typically equal to $1 - \alpha_u$, where α_u is the under-relaxation factor for velocities, which leads to nearly-optimum convergence rate. A flow diagram of the outer and inner iterations (for the SIP solver) is shown in Fig. 2.

The number of outer iterations increases linearly with the number of CVs, leading to the quadratic increase in computing time. For this reason a multigrid method is implemented, which keeps the number of outer iterations approximately independent of the number of CVs. The method is based on the so called "full approximation scheme" (FAS). It is implemented in the so called "full multigrid" (FMG) fashion. The solution is first obtained on the coarsest grid using the method described above, cf. Fig. 2. This solution provides initial fields for the calculation on the next finer grid, where the multigrid method using V-cycles is activated. The procedure is continued until the finest grid is reached. The coarse grids are subsets of the finest grid; each coarse grid CV is made of four CVs of the next finer grid. The equations solved on the coarse grids within a multigrid cycle contain an additional source term which describes the current solution and the residuals of the next finer grid. The multigrid method used in this study is described in detail in Hortmann et al (1990). It should be noted that the multigrid method is applied only to the outer iteration loop; inner iterations are performed with one of the above described solvers irrespective of the grid fineness. This is due to the fact that the linear equations need not be solved accurately, so only a few inner iterations are necessary. Only for the pressure-correction equation would implementation of a *multigrid solver* bring reduction of computing time at fine grids, because it converges more slowly and may require higher accuracy, especially in case of unsteady flows. Any of the above described solvers could be used as a *smoother* within such a multigrid solver.

3 Parallelization Strategy and Efficiency

3.1 Domain Decomposition Technique

Domain decomposition technique is the basis of the parallelization strategy used in the present study. It is completely analogous to the block-structuring of grid in complex geometries. The number of blocks is dictated by the geometry of the solution domain, and the number of CVs within each block may vary substantially. However, we may create more blocks than the geometry requires. The aim is to have as many subdomains of approximately the same size as there are processors, and assigning each subdomain to one processor. If some blocks are much smaller than the others, more than one block may form a subdomain assigned to one processor.

The subdomains do not overlap, i. e. each processor calculates only variable values for CVs within its subdomain. However, each processor uses some variable values which are calculated by other processors to which neighbour subdomains are assigned. This requires, in case of MIMD computers, an overlap of storage: each processor stores data from one or more layers of CVs along its boundary belonging to neighbour subdomains. This data is exchanged between processors, typically each time it is updated.

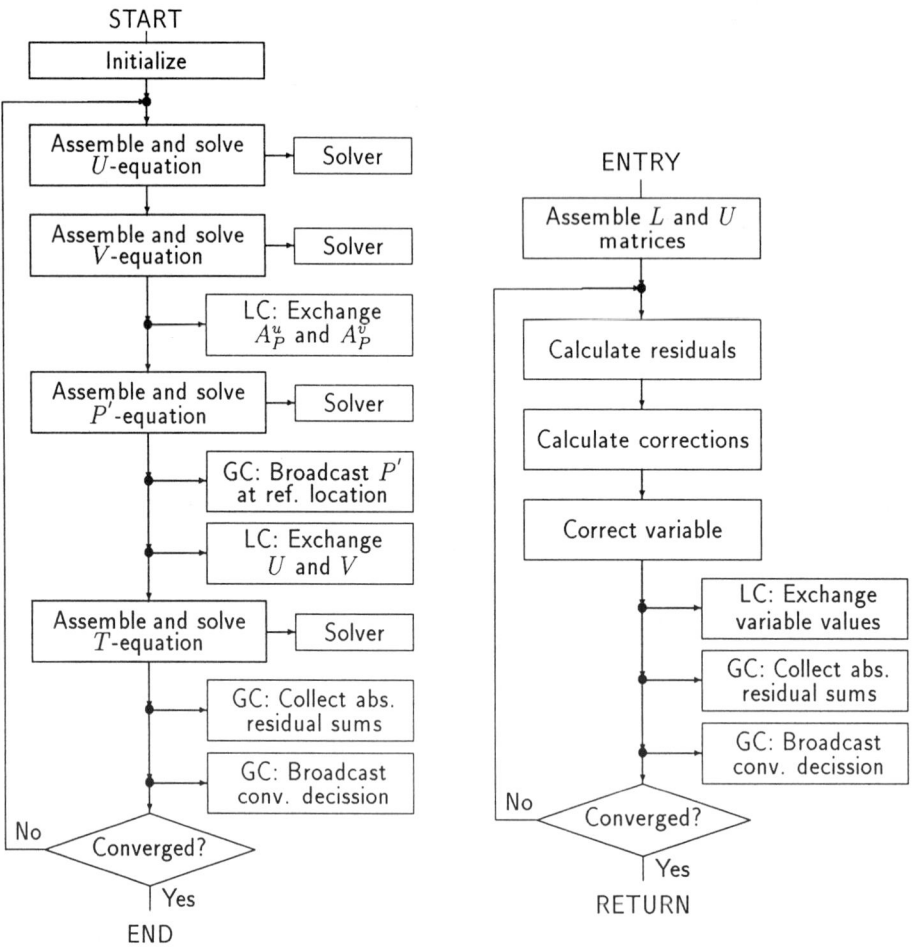

Fig. 2: Flow chart of the outer (left) and the inner (right) iteration loop (SIP solver)

The grid may be calculated on one processor and then partitioned into subdomains which are assigned to individual processors; however, in case of large problems we may specify for each processor its subdomain boundaries and generate the grid locally. The grid coordinates at subdomain boundaries must match, since the CV faces are common to two subdomains. In two-dimensional applications the subdomains have a shape of logical rectangles. Ideally, each subdomain has four neighbours; however, in case of complex geometries this may not be always achievable, so one side of one subdomain may be shared with two or more neighbour subdomains. This increases communication overhead for processors assigned to such subdomains. An example will be shown later.

The solution strategy for block-structured grids will be described by considering a quadrilateral solution domain subdivided into four subdomains (blocks) as shown in Fig. 3. Discretized partial differential equation for a variable ϕ leads at each CV to an algebraic equation of the form

$$A_P \phi_P + A_E \phi_E + A_W \phi_W + A_N \phi_N + A_S \phi_S = Q_P, \tag{1}$$

where the indices E, W, N, S and P represent the nodes of the computational molecule, cf. Fig. 1. The coefficients A arise from implicit parts of the convection and diffusion fluxes, and Q is the source term (which includes all explicitly treated terms from the discretized equation). For the whole solution domain there results an equation system which can be written in matrix form as

$$[A]\{\phi\} = \{Q\}, \tag{2}$$

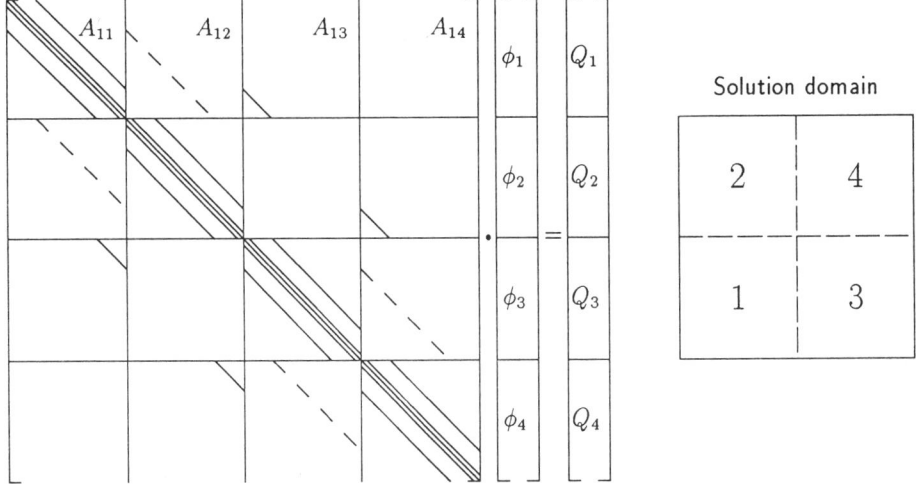

Fig. 3: Matrix structure for a quadrilateral domain subdivided into four subdomains

where $\{\phi\}$ is the column matrix containing variable values at CV centers ordered in a certain way, $\{Q\}$ is the corresponding column matrix containing the source term and $[A]$ is the coefficient matrix. If the nodes are odered within each block by starting at the southwest corner, proceeding northwards along the first column of CVs and then eastwards to the last column, than in case of a five point computational molecule the coefficient matrix $[A]$ has the structure shown in Fig. 3. The matrix $[A]$ consists of 4×4 block matrices. The diagonal block matrices $[A_{ii}]$ are the main matrices of the subdomains and have the same form as the matrix $[A]$ would have if the whole domain was considered as one block. The off-diagonal block matrices describe coupling of subdomains. For example, matrix $[A_{12}]$ of Fig. 3 describes coupling of block 1 with block 2 through the coefficient A_N in CVs next to the north boundary of block 1. The block matrix $[A_{13}]$ in Fig. 3 describes coupling of block 1 with block 3 through the coefficient A_E in CVs along east boundary of block 1. The block matrix $[A_{14}]$ has no non-zero entries, since blocks 1 and 4 do not have common interfaces.

On a single processor, one could adopt several possible iterative solution strategies. In a parallel environment, the following iteration scheme is the simplest choice:

$$[M_i]\{\phi_i^m\} = \{Q_i\} - [M_i - A_{ii}]\{\phi_i^{m-1}\} - [A_{ij}]\{\phi_j^{m-1}\} \quad (j \neq i; \text{ summation on } j) , (3)$$

where $[M_i]$ is the iteration matrix in block i, m is the iteration counter and the index j defines neighbour blocks. In case of the GS solver, the matrix $[M_i]$ is the lower triangular matrix of $[A_{ii}]$, whereas in case of the SIP solver, it is the product of the lower and upper triangular matrices $[L]$ and $[U]$. The iteration matrix need not be the same in each block; for example, one can use GS for small and SIP for large blocks.

This way of iteratively solving the equation systems for the solution domain as a whole is adopted in this study. Each subdomain is within one inner iteration treated as if it were an independent solution domain; reference to nodes inside other subdomains is treated explicitly. The addition of explicit parts due to the coupling matrices shown above usually does not slow the convergence down significantly if the variable values along interfaces are updated after each *inner* iteration. This communication option is denoted here as the *EI mode*.

Another option is to exchange the interface variable values between processors only after each outer iteration. This obviously decouples the subdomains within inner iterations completely and is bound to increase the number of outer iterations; usually, heavier under-relaxation has also to be employed to ensure convergence. This communication option is denoted here es the *EO mode*.

3.2 Efficiency of Parallel Implementation

The effectiveness of parallel computing can be characterized by the total efficiency, defined as the ratio of computing time on one processor using the most efficient serial algorithm, T_s, and the n-fold computing time using parallelized algorithm and n processors, T_n:

$$E_n^{tot} = \frac{T_s}{nT_n} = E_n^{par} E_n^{num} E_n^{lb} . \qquad (4)$$

Schreck and Perić (1991) have shown that the total efficiency can be expressed as a product of three factors termed *parallel*, *numerical* and *load balancing* efficiency. These factors describe: *(i)* the increase in the number of floating point operations per grid node required to reach solution of the same accuracy when the number of subdomains is increased, *(ii)* the increase of elapsed time for a parallel computation due to communication between processors during which computation can not take place, and *(iii)* idle time of some processors due to uneven load. Communication can be further split into *local* and *global*; the former describes exchange of interface information between logical neighbours, and the latter gathering of some information (e. g. level of residuals) from all processors to the "master" and broadcasting of some information (e. g. decission on convergence) from the master to all other processors. The difference is that the local communication runs in parallel, i. e. all processors are — except for the effect of unequal size of interfaces — involved in communication simultaneously. In case of global communication, only certain number of processors is involved in communication at any time between begin and end of gathering or scattering of information. This is why the global communication is the limiting factor for massive parallelization, unless communication and computation are allowed to take place simultaneously.

Schreck and Perić (1991) have shown that — for a chosen algorithm and communication pattern — the parallel efficiency can be predicted as a function of the computer parameters like latency time (μs), communication speed (MB/s), calculation speed (Mflops), grid size and number of processors used. This allows the choice of communication pattern (EI or EO mode, prescribed number or convergence check for inner iterations etc.) which leads to the highest expected total efficiency for the given computer. Comparisons of predicted and measured parallel efficiencies for various grid sizes, number of processors and different computers showed good agreement between predicted and measured values.

As noted before, the parallel efficiency will increase substantially if the computation and communication can take place simultaneously. Most new generation parallel computers offer this posibility. In Fig. 2 the local (LC) and global (GC) communications within the present solution algorithm are indicated. The solution algorithm can be rearranged to allow local communication to take place while doing calculations for the inner region, and performing calculations for CVs along interfaces at the end of each sequence of operations. Only in case of coarse grids (which are always encountered in multigrid methods) may computation have to be halted until communication finishes. Global communication can also be overlayed with computation. For example, collection of residual levels and broadcasting of decission on convergence can be allowed to take time of one whole outer iteration: the convergence decission can be based on the residual level of the previous outer iteration and the extrapolated convergence rate. The level of pressure correction at the reference location may also be taken from the previous iteration. This is possible since at convergence the level of pressure correction will be zero everywhere.

One of the major factors affecting the effectivness of parallelization is the numerical efficiency. Unfortunately, it can not be predicted. One can expect that the numerical efficiency will be high for the EI mode (it may even exceed 100%!) and low for the EO mode. However, it may still be advantageous to use the EO mode on computers with large latency time and fast computation, since the gain in parallel efficiency may compensate the loss of numerical efficiency, cf. Eq. (4).

The load balancing effects arise from the fact that in complex geometries the subdomains may not have the same size or the same boundary surface. Furthermore, complex boundary conditions on external boundaries — which exist only for some subdomains — may also cause delays. Algorithms for automatic grid partitioning usually optimize the size and shape of subdomains and the communication pattern, since many parallel computers communicate faster between directly connected than between remote processors.

While both parallel and load balancing efficiencies can be analysed theorethically as a function of the most important parameters (cf. Schreck and Perić, 1991), numerical efficiency can only be determined by numerical experiments. In the next section some examples and analysis of numerical efficiency are presented.

4 Applications and Analysis of Performance

Three test cases were chosen in order to analyse the effect of domain decomposition on numerical efficiency. Two cases involve a non-rectangular solution domain which can also be treated as a single block; the third one requires the use of block-structured (globally unstructured) grids.

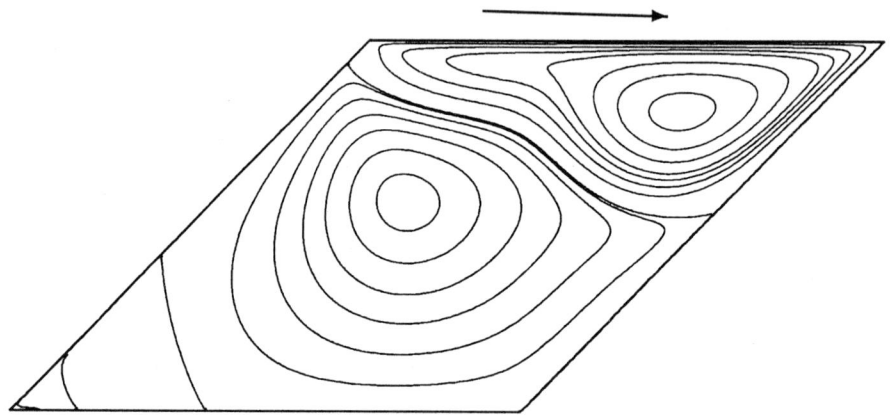

Fig. 4: Geometry of inclined cavity and streamlines for lid driven flow at $Re = 1000$

The first test case is the inclined lid driven cavity; figure 4 shows the geometry and predicted streamlines. All sides are of the same area and the inclination angle of side walls is $45°$. For the present study the Reynolds number was set to 1000. Accurate reference solution for this test case was presented by Demirdžić et al, 1992. Table 1 shows the numbers of outer iterations and the total number of work units (1 work unit = 1 GS iteration on a given grid) for the single grid method and various grids and numbers of subdomains, using EI communication. The calculations were stopped when the absolute sum of residuals over all CVs at the begining of an outer iteration had fallen 4 orders of magnitude for all equations. These results show that the numerical efficiency of the single grid method is not significantly affected by the domain decomposition, as the numbers of outer iterations and work units are only for 100 processors and SIP solver appreciably higher than for one processor. This is expected since the rate of convergence is slow – the large number of outer iterations allows for loose coupling of subdomains.

On the other hand, the multigrid method is much more efficient, as shown in Table 2 which presents results for the same problem using FMG procedure and variable number of inner iterations (residuals reduced one order of magnitude). The number of outer iterations actually reduces as the grid is refined, since FMG provides a more accurate initial field for finer grids. One would normally expect that the domain decomposition would affect the multigrid method more strongly than the single grid method, as it requires on a 160 × 160 CV grid only 23 outer iterations to reduce the residual levels 4 orders of magnitude. However, calculations show that the numbers of iterations remain almost the same for 1, 4, 25 and 100 processors. The total number of work units increases substantially with increasing number of subdomains only on coarse grids (the coarsest grid used here was 10 × 10 CV).

Table 1: Single grid calculations for lid driven cavity flow, prescribed number of inner iterations

Grid	Solver	Number of outer iterations (total work units)			
		1 Proc.	2 × 2 Proc.	5 × 5 Proc.	10 × 10 Proc.
40 × 40CV	GS	316 (19276)	319 (19459)	329 (20069)	321 (19581)
80 × 80CV	GS	835 (50935)	839 (51179)	854 (52094)	879 (53619)
40 × 40CV	SIP	282 (18894)	295 (19765)	288 (20448)	316 (22436)
80 × 80CV	SIP	545 (36515)	540 (36180)	548 (38908)	611 (43594)

Table 2: FMG calculations for lid driven cavity flow with a variable number of inner iterations

Grid	Solver	Number of outer iterations (total work units)			
		1 Proc.	2 × 2 Proc.	5 × 5 Proc.	10 × 10 Proc.
80 × 80CV	GS	48 (4660)	48 (4662)	48 (4680)	50 (5009)
160 × 160CV	GS	23 (2426)	23 (2427)	24 (2640)	23 (2442)
80 × 80CV	SIP	50 (4923)	50 (5248)	50 (6011)	50 (6939)
160 × 160CV	SIP	24 (2763)	24 (2932)	24 (2832)	24 (3273)

For the GS solver, using prescribed number of inner iterations turned out more efficient for both lid and buoyancy driven cavity flows: only 4 inner iterations were required for U, V and T and 6 for P' on the finest grids to achieve the optimum result. The number of inner iterations needed to be increased only for coarse grids and large numbers of processors. However, when the SIP solver is used, one processor needs only 1 inner iteration for U, V and T and up to 6 for P' for optimum results. In case of more than one subdomain, one has to do more than 1 inner iteration for all equations in order to couple the subdomains. As the number of subdomains increases, the number of inner iterations also has to be increased. For 100 subdomains the same number of inner iterations as with GS solver is necessary to achieve convergence in approximately the same number of outer iterations.

The above means that the SIP solver results in a reduction of numerical efficiency due to the necessity to increase the number of inner iterations when more than one subdomain is involved. The GS solver, on the other hand, needs more than one inner iteration anyway, so in a parallel implementation it has the parallel efficiency of almost 100%. This remains valid for the strict definition of the numerical efficiency, cf. Eq. (4), which requires the comparison to be made with the most efficient single processor solver. Since the first SIP iteration involves about the same number of floating point operations as 4 GS iterations and any subsequent inner iteration in the SIP solver takes about twice as long as a GS iteration, the computational effort for optimum convergence on one processor is about 10% lower for GS than for SIP. It should be noted, however, that the GS solver can only be used efficiently in conjunction with the FMG solution algorithm; for a single grid algorithm it is too inefficient and requires substantially more inner and outer iterations for all but the coarsest grids, cf. Table 1.

Another example analysed in this study is the buoyancy driven flow in the same inclined cavity as before. Figure 5 shows predicted streamlines for the Rayleigh number of 10^6. This is also one of the bench-mark test cases presented by Demirdžić et al (1992). Table 3 shows results for both lid and buoyancy driven flow using prescribed number of inner iterations and GS solver, and requiring reduction of absolute residual sums five orders of magnitude. For most of grids the numbers of outer iterations are the same using 1, 4, 25 and 100 processors, and the numerical efficiencies are close to 100%. The same number of outer iterations results when using SIP solver, but the numerical efficiency is lower for reasons discussed above.

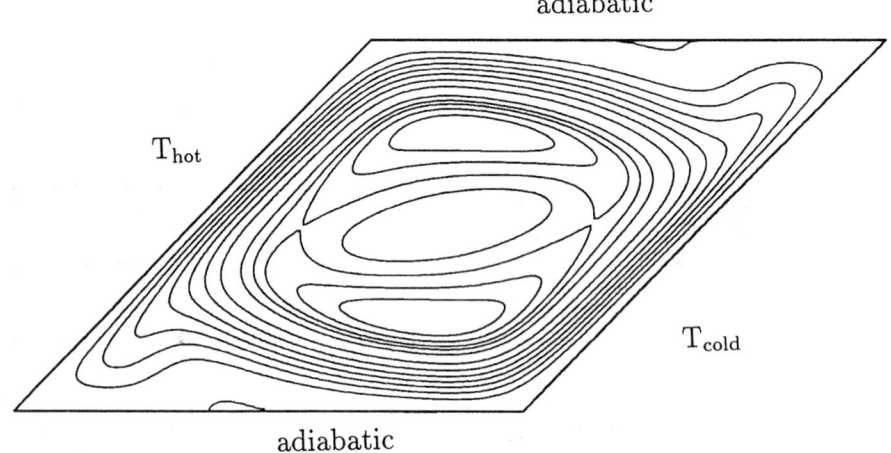

Fig. 5: Streamlines for buoyancy driven flow in inclined cavity at $Ra = 10^6$, $Pr = 0.1$

The reason for good performance of FMG under domain decomposition lies in the nature of multigrid methods. On the finest grid, only the high frequency error components are smoothed. These error components are of local character, so nodes near subdomain interface require only the information from few CVs across the interface. For this purpose, few inner iterations with exchange of variable values after each are sufficient. The low frequency errors spread accross the whole solution domain; they are eliminated on the coarsest grids, where the CVs are so large that one exchange of variable values carries information far across the domain. By doing usually few more inner iterations on the coarsest grid, the result matches single block performance.

In EO mode, the FMG algorithm offers good numerical efficiency with up to 10 processors; thereafter under-relaxation factors have to be reduced to achieve convergence and the numerical efficiency deteriorates sharply. The single grid algorithm is less sensitive to the change of communication mode, but the trend is the same.

The above results were all obtained on a workstation, simulating parallel processing, since we were only interested in numerical efficiency. Table 4 shows run times from computations on a Meiko Computing Surface which uses T800 transputers with 0.45 Mflops computing speed, 22 μs latency time and 1.4 MB/s data transfer rate. Fine grids could not be run on one processor due to memory limit (4 MB per processor), so the total efficiency was estimated by scaling run time on a workstation. With the relatively low latency, high efficiency is achieved for fine grids.

Table 3: FMG calculations for lid (LD) and buoyancy (BD) driven cavity flows using prescribed number of inner iterations and GS solver

Grid	Flow	Number of outer iterations (total work units)			
		1 Proc.	2 × 2 Proc.	5 × 5 Proc.	10 × 10 Proc.
80 × 80CV	LD	73 (6244)	73 (6244)	75 (6514)	75 (6828)
160 × 160CV	LD	36 (3376)	36 (3376)	36 (3376)	36 (3445)
320 × 320CV	LD	27 (2183)	27 (2183)	27 (2183)	27 (2196)
80 × 80CV	BD	45 (4798)	45 (4798)	50 (5331)	45 (5080)
160 × 160CV	BD	24 (2786)	24 (2786)	24 (2786)	24 (2855)
320 × 320CV	BD	22 (2177)	22 (2177)	22 (2177)	22 (2192)

Table 4: FMG calculations for lid (LD) and buoyancy (BD) driven cavity flows using prescribed number of inner iterations and GS solver with 25 Processors (5 × 5)

Flow	Run time (estimated total efficiency)			
	40 × 40 CV	80 × 80 CV	160 × 160 CV	320 × 320 CV
LD	28.7 s (54.4%)	60.8 s (71.0%)	150.7 s (81.1%)	437.0 s (89.1%)
BD	27.8 s (53.4%)	52.3 s (64.8%)	102.0 s (80.4%)	292.1 s (88.5%)

The third example involves laminar flow at low Reynolds number (10 based on channel width) in a complex branching duct. The geometry requires the use of at least 5 blocks if the grid is made of straight lines. Calculations were made using Parsytec MC3 computer based on T805 transputer, with 55 μs latency time, 0.5 Mflops computing speed and 1.5 MB/s data transfer rate (using PARIX communication software). The subdivision of the global grid into subdomains was performed using an automatic load balancing algorithm. Figure 6 shows the coarsest grid consisting of 376 CV subdivided into 40 subdomains.

For this geometry and a prescribed number of grid lines in each channel branch, it is impossible to create blocks of equal size, so the load balancing efficiency comes into play. The load balancing could have been improved by increasing or decreasing the number of CVs in some regions, which is probably what one would do in a real application. It was also impossible to keep the number of neighbours equal for each block; moreover, some blocks have three neighbours on one side, so that more complex communication patterns than in the previous examples had to be used.

This example illustrates the difficulties in parallel computation for complex geometries. For large problems it may not be possible to generate the whole grid on one processor; on the other hand, too many input data is required if each processor is to generate its own grid. The best way is to split the geometry into blocks which require minimum of input data for grid generation and where each block can still be handled by one processor or a front-end computer. Each such block can then be further subivided into subdomains automatically, according to the number of processors to be used and trying to optimize the load balancing. In this example 7 blocks were used to generate the grid, cf. Fig. 6.

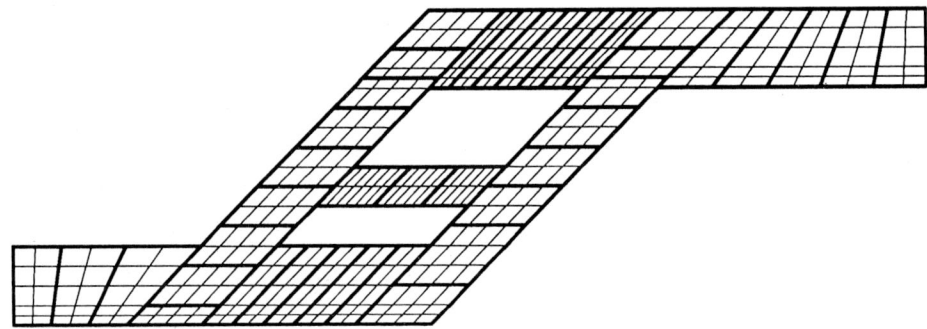

Fig. 6: Geometry and coarsest grid for the branching channel flow showing 40 subdomains

Table 5: FMG calculations for flow in a branching duct, prescribed number of inner iterations

Grid	Solver	Number of outer iter. (total work units) run time		
		10 Proc.	20 Proc.	40 Proc.
6016 CV	GS	141 (10548) 398 s	151 (11296) 255 s	151 (11296) 189 s
24064 CV	GS	-	151 (11431) 857 s	161 (12188) 558 s
6016 CV	SIP	114 (13376) 409 s	114 (13316) 233 s	121 (14134) 173 s
24064 CV	SIP	-	131 (15485) 949 s	131 (15485) 552 s

Results are presented for grids with 6016 and 24064 CVs, corresponding to 2 and 3 levels of refinement of the grid shown in Fig. 6. All three solvers introduced before were tested. Calculations were performed for a prescribed number of inner iterations (2 for velocities and 10 for P') and for a prescribed convergence criterion for the inner iterations (reduction of absolute residual sum by a factor of 5). As in the previous example, calculations with a prescribed number of inner iterations are more efficient – both in terms of numerical and parallel efficiency – than when using convergence checks for inner iterations. The savings are: 20% on computing effort and 25% on run time for SIP, 27% on computing effort and 40% on run time for GS. Large savings on time for GS are mostly due to avoiding global communication in solver, since GS performs about twice as many inner iterations as SIP. The PCG solver converges in about the same number of outer and inner iterations as the SIP solver under the same conditions, but requires about twice as much of computing time. This is due partly to a larger number of computing operations per node, and partly to the necessity to perform additional global communication in each inner iteration (calculation of scalar products). The PCG solver is therefore found unsuitable for the present kind of solution methods.

Results for prescribed number of inner iterations are summarized in Table 5. The GS solver is numerically more efficient than SIP, although it requires more outer iterations. However, due to more iterations and less computation per iteration, the parallel efficiency is lower for GS. The total efficiency is almost the same for GS and SIP, whereas for variable number of inner iterations SIP is more efficient.

5 Conclusions

Results of test calculations with the parallelized version of an implicit multigrid finite volume code and the analysis presented in preceeding sections allow for the following conclusions to be made:

- The test calculations showed that the numerical efficiency of both the single grid and multigrid algorithm is close to 100% when the domain decomposition is used as the basis for parallelization on MIMD computers and data is exchanged between processors after each inner iteration. The parallel efficiency is worse for the multigrid (especially the FMG version) than for the single grid algorithm due to the frequent use of coarse grids with a large number of processors and more communication per outer iteration. In spite of that, the total computing times with the parallel multigrid algorithm are orders of magnitude shorter than with the single grid algorithm.

- It appears to be computationally more efficient, both in terms of parallel and numerical efficiency, to prescribe the number of inner iterations rather then checking convergence for each inner iteration. The possible increase in the number of outer iterations is more than compensated by the reduction of the number of inner iterations and avoiding of much of the global communication.

- If the comunication is very slow (like in a cluster of workstations), the exchange of data between processors after each outer iteration – which reduces substantially the numerical efficiency but increases the parallel efficiency – may turn out overall more efficient. This applies, however, only to small numbers of processors (up to 10).

- Efficiency of parallel computation can be greatly improved by overlaying communication and computation. Present solution algorithm allowes this to a large degree and the new generation parallel computers offer that possibility.

It is expected that the efficiencies will be higher for three-dimensional applications, since the number of floating point operations per CV and iteration is then much higher than in a two-dimensional case, while the effect of latency diminishes.

6 Acknowledgements

The Comission of the European Communities sponsored via "Parallel Computing Action" a part of the Meiko Computing Surface used in this study and the Deutsche Forschungsgemeinschaft provided the financial support through the "Sonderforschungsbereich 182". The authors thank for this support.

7 References

1. P. Bastian and G. Horton, "Parallelization of robust multi-grid methods: ILU factorization and frequency decomposition method", in W. Hackbusch and R. Rannacher (eds.), *Notes on Numerical Fluid Mechanics, Vol. 30*, Vieweg, Braunschweig, 1989, pp. 24-36.

2. I. Demirdžić and M. Perić, "Finite volume method for prediction of fluid flow in arbitrarily shaped domains with moving boundaries", *Int. J. Num. Methods in Fluids*, **10**, 771-790 (1990).

3. I. Demirdžić, Ž. Lilek and M. Perić, "Fluid flow and heat transfer test problems for non-orthogonal grids: bench-mark solutions", *Int. J. Num. Methods in Fluids*, in print (1992).

4. M. Hestens and E. Stiefel, "Methods of conjugate gradients for solving linear systems", *Nat. Bur. Standards J. Res.*, **49**, 409-436 (1952).

5. M. Hortmann, M. Perić and G. Scheurer, "Finite volume multigrid prediction of laminar natural convection: bench-mark solutions", *Int. j. numer. methods fluids*, **11**, 189-207 (1990).

6. S. V. Patankar and D. B. Spalding, "A calculation procedure for heat, mass and momentum transfer in three-dimensional parabolic flows", *Int. J. Heat Mass Transfer*, **15**, 1787-1806 (1972).

7. M. Perić, M. Schäfer and E. Schreck, "Computation of fluid flow with a parallel multi-grid solver", in K.G. Reinsch et al. (Eds.), *Proc. Int. Conference on "Parallel Computational Fluid Dynamics"*, Elsevier, Amsterdam, 1991.

8. E. Schreck and M. Perić, "Computation of fluid flow with a parallel multi-grid solver", Report *LSTM 327/N/91*, Lehrstuhl für Strömungsmechanik, University of Erlangen, Germany, 1991.

9. H.L. Stone, "Iterative solution of implicit approximations of multi-dimensional partional differential equations", *SIAM J. Numer. Anal.*, **5**, 530-558 (1968).

Direct Numerical Simulation of Turbulence on the Connection Machine

J. Blair Perot

Department of Mechanical Engineering, Stanford University, Stanford, CA 94305, USA

Abstract
Detailed performance measurements of the direct numerical simulation of turbulence on the Connection Machine 2 are presented. These are compared to similar simulations being performed on the Cray Y-MP. The current and future utility of the Connection Machine as a tool in the direct numerical simulation of turbulence is discussed.

1. INTRODUCTION

The aim of this paper is to evaluate the performance of the Connection Machine 2 (CM-2) in the context of the direct numerical simulation (DNS) of turbulence. The study of turbulence through the use of direct computer simulations has, from its inception in the early 1980's, accounted for a significant percentage of scientific supercomputer usage. The fundamental questions that can be answered about turbulence are closely related to the current level of supercomputer performance. In fact, turbulence has been identified as one of the Grand Challenge problems [1] that could benefit significantly from increased supercomputer performance and, in particular, from massively parallel computers.

However, the issue of whether massively parallel machines can actually fulfill their performance expectations when it comes to turbulence simulation is still an open question. It is now fairly apparent that unlike vector supercomputers, the performance of massively parallel computers is not closely related to their peak performance. Instead, massively parallel computers are inevitably dominated by communication overhead. Their performance varies wildly from application to application, depending on how well the communication patterns of the application match the communication patterns implied by the particular architecture. For instance, the architectures of the CM-2 and Intel Hypercube are optimized for vastly different communication patterns. As a result, stencil type operations work very well on the Connection Machine, and Fast Fourier Transforms (FFT's) with large intermediate data rearrangement work well on the Intel Hypercube, but not vice versa.

Direct numerical simulation of anything more complicated than isotropic decaying turbulence typically involves a number of very different communication patterns and data structures. Whether these all can be mapped fairly efficiently to the Connection Machine is an interesting question. Any single bottleneck could dramatically effect the performance of the simulation as a whole. Possible bottlenecks that will be investigated include the effect of implementing boundary conditions on a SIMD machine, and the impact of regular but long distance communication. So, despite the fact that turbulence simulation involves massively parallel data (on the order of 10^6 to 10^7 nodes), has no load balancing problems, and involves only regular communication, it is not a foregone conclusion that the CM-2 is the the supercomputer of choice. Only through the actual testing of an existing DNS code, and comparison to current vector supercomputer performance (as represented by the Cray Y-MP) can the potential of the CM-2 for turbulence simulations be evaluated.

2. NUMERICAL METHOD

The purpose of this section is not to present a new numerical method, although the method does differ in some fundamental ways from classical DNS methods. Instead, the purpose is to reveal the variety of solution algorithms, communication patterns and data structures that are used. An understanding of the basic numerical scheme will also help to put the various performance timings into the proper context.

2.1 Spatial Discretization

The spatial discretization of the incompressible Navier-Stokes equations is a primative variable, finite volume method on a staggered mesh. It is very much in the spirit of the discretization first introduced by Harlow and Welch [2]. The mesh is cartesian but not necessarily uniform, and in the results that follow one of the three directions will be non-uniform. A two dimensional representation of the spatial discretization is shown in Figure 1.

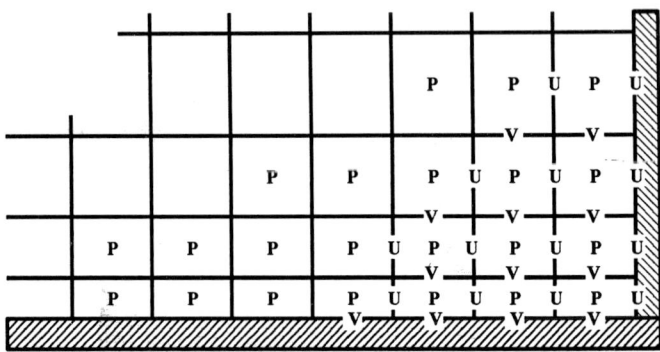

FIGURE 1. Locations for the discrete velocity and pressure variables on a 2-D staggered mesh.

The finite volume discretization is a departure from the traditional use of spectral methods in direct turbulence simulations. It is motivated by the desire to implement more complicated boundary conditions (and eventually more complicated geometries). Fortuitously, this also makes the method far more amenable to implementation on the CM-2 (as the timing results will show). To date, there is still no convincing evidence that the accuracy of spectral methods is far superior to that of second order methods, *when the grid spacing is at the limit of resolving the flow*. And in any case, this is a secondary issue in the context of this paper.

2.2 Temporal Discretization

The temporal discretization advances the nonlinear convective terms with a second order Adams-Bashforth method. An explicit method considerably simplifies the advancement of the non-linear terms, but also imposes a stability limit on the CFL number. The stability limit is not overly restrictive and corresponds roughly to the condition that the temporal accuracy match that of the spatial accuracy in a Taylor's hypothesis sense. The diffusive terms are advanced implicitly with the trapazoidal (Crank-Nicolson) method. An implicit method removes the very severe stability restrictions that would otherwise be imposed by this term, and computationally only requires a single matrix inversion, because the diffusive terms are linear in the velocity. Finally, the pressure is solved for by using a fractional step method [3]. Mathematically the time discretization can be written as,

$$\frac{\mathbf{v}^* - \mathbf{v}^n}{\Delta t} + \left(\frac{3}{2}(\mathbf{v}^n \cdot \nabla)\mathbf{v}^n - \frac{1}{2}(\mathbf{v}^{n-1} \cdot \nabla)\mathbf{v}^{n-1}\right) = \frac{1}{2Re}\nabla^2(\mathbf{v}^* + \mathbf{v}^n), \tag{1a}$$

$$\frac{\mathbf{v}^{n+1} - \mathbf{v}^*}{\Delta t} = -\nabla p^{n+1} \tag{1b}$$

where $\mathbf{v}^*$ is a temporary intermediate variable. The pressure is found from a Poisson equation obtained by taking the divergence of equation (1b). Ultimately the procedure breaks down into three fundamental parts,

$$\mathbf{r} = -\left(\frac{3}{2}(\mathbf{v}^n \cdot \nabla)\mathbf{v}^n - \frac{1}{2}(\mathbf{v}^{n-1} \cdot \nabla)\mathbf{v}^{n-1}\right) + \frac{1}{Re}\nabla^2(\mathbf{v}^n) \tag{2}$$

$$\left(1 - \frac{\Delta t}{2Re}\nabla^2\right)\frac{\mathbf{v}^* - \mathbf{v}^n}{\Delta t} = \mathbf{r} \tag{3}$$

$$\Delta t \nabla^2 p = \nabla \cdot \mathbf{v}^* \quad \text{and} \quad \mathbf{v}^{n+1} = \mathbf{v}^* - \Delta t \nabla p. \tag{4}$$

The first part (Eqn. 2) involves stencil type, nearest neighbor communication to evaluate the derivatives. All the communication is local and involves only the six neighbors of a 3-D cartesian volume. In contrast, the second part (Eqn. 3) involves a matrix inversion, and will ultimately involve long distance communication. The matrix inversion can be factored into a series of three tridiagonal inversions without any loss in the order of accuracy. The final part of the solution (Eqn. 4), the Poisson equation, is the most computationally intensive. For a simple domain it can be solved explicitly using a combination of discrete Fourier transforms, cosine transforms and tridiagonal matrix inversion. So, even though this is not a spectral method, an opportunity exists in this stage to evaluate the performance of FFT's on the CM-2.

2.3 Problem Configuration

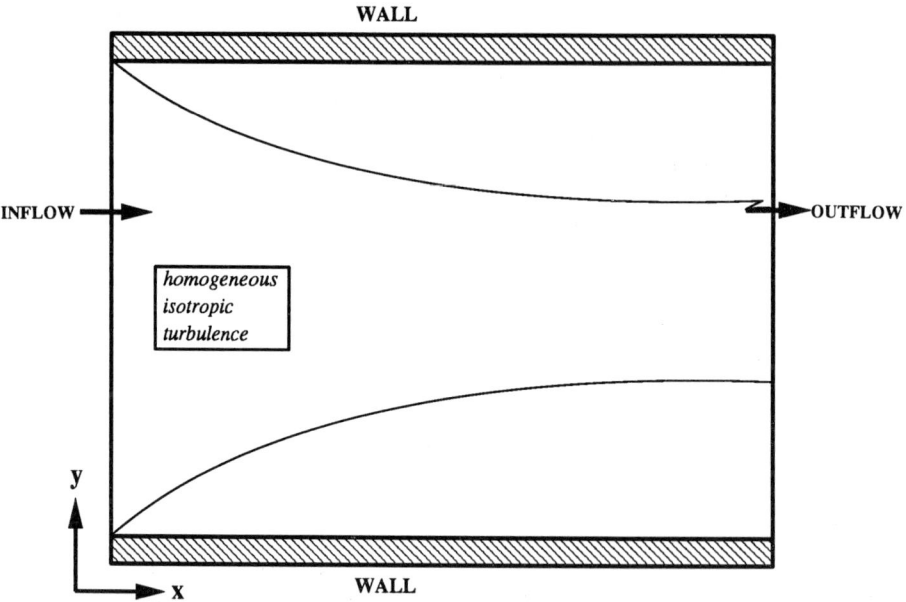

FIGURE 2. Problem configuration for the turbulent boundary layer with large free-stream turbulence.

The problem configuration used to test the computer performance is shown in Figure 2. It is a study of spatially decaying turbulence in the presence of a wall, or conversely, a boundary layer in the presence of high free-stream turbulence. The isotropic, homogeneous turbulence, that enters the domain at the left, interacts with the walls and decays as it is convected downstream. The domain is periodic in the spanwise (z) direction, and at the top and bottom faces of the domain no-slip boundary conditions are imposed. The inflow and outflow boundary conditions are topics unto themselves, and will not be discussed here or included in any of the timing results. The grid is stretched in the wall normal (y) direction in order to resolve the boundary layer. A fairly complicated configuration was chosen in view of the fact that future simulations will require at least this degree of versatility.

To confirm that the method is accurately resolving the turbulence a plot of the decay of turbulent kinetic energy and dissipation is shown in Figure 3. The slopes of the curves are nearly linear and very close to the accepted values of $(-1.2\ to\ -1.4)$ for the kinetic energy, and $(-2.2\ to\ -2.4)$ for the dissipation. This implies that both the large scales (responsible for the kinetic energy) and the small scales (responsible for the dissipation) are being adequately resolved.

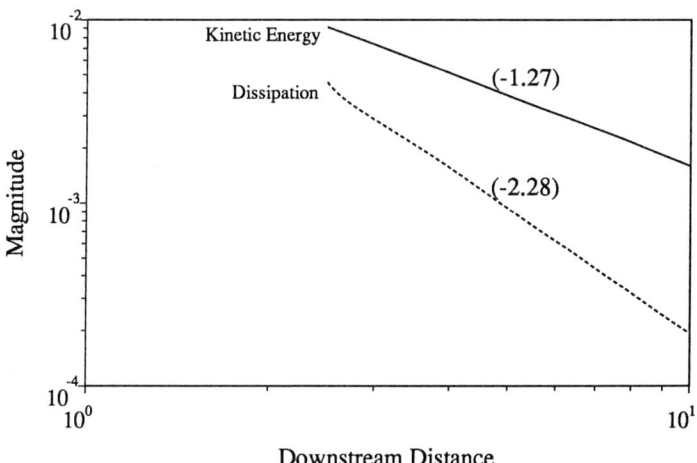

FIGURE 3. The decay of kinetic energy and dissipation as a function of downstream location.

3. GENERAL PERFORMANCE RESULTS

A good measure of performance for turbulence simulations is the normalized CPU-time, or CPU-time per time-step per computational node. In this way, simulations with vastly different numbers of grid points can be reasonably compared. Figure 4 shows a plot of the normalized CPU-time as a function of the the problem size (or number of nodes), for the Cray Y-MP and three different sized CM-2 configurations. For a 32^3 grid (2×10^4 nodes) the single processor Y-MP takes 5 μs per time-step per node, which corresponds to 145 Mflops. As the problem size increases, the performance of the Y-MP increases slightly due to the increased vector lengths. At problem sizes corresponding to 6×10^5 nodes it has reached 175 Mflops or almost 4 μs per time-step per node. However, as the problem size increases further there is an abrupt decrease in performance. This is because the Y-MP runs out of core memory and data must now be swapped in and out of memory from disk. The CM-2 does not have this problem, because its core memory is so much cheaper it can provide on the order of 100 times more memory to the user.

The outstanding feature of the CM-2 timings is the dramatic increase in performance with increasing problem size, almost an order of magnitude for the 32k CM-2 when the problem size is increased from 32^3 to 256^3. This is in contrast to the Cray Y-

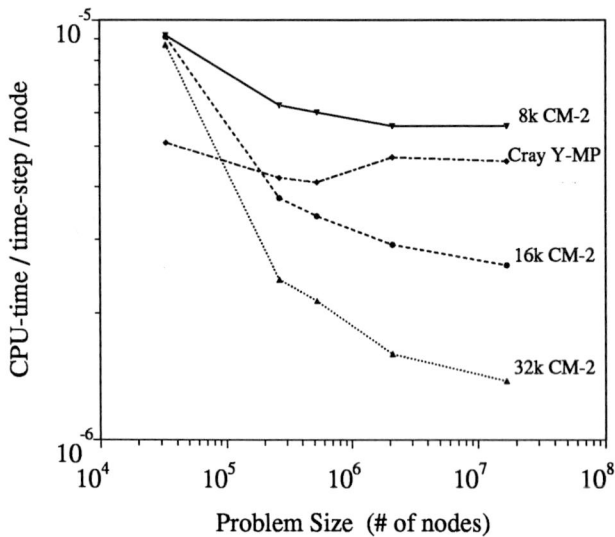

FIGURE 4. CPU-time per time-step per node as a function of the problem size for the Cray Y-MP and sections of the CM-2.

MP which is fairly independent of the grid size. The reason for this behavior is because the CM-2 performance is communication limited. Only with very large problems does the amount of computation begin to amortize the communication overhead. The effect of communication can also be seen in the fact that the Cray obtains, at its worst, 44% of its peak processor speed, but the CM-2 never obtains better than 22% of its peak.

In figure 5 the Mflops ratings of the Y-MP have been used to calculate a Y-MP equivalent Mflops rating for the CM-2. The dependence on problem size is still clear. For larger problems (greater than 128^3 nodes) the 16k CM-2 is roughly equivalent to one processor of the Y-MP. For smaller problem sizes, however, it is no longer clear that the CM-2 has great performance advantages over the Y-MP. This conclusion may be important to other types of turbulence simulation, such as large eddy simulation, which tend to use smaller grid sizes.

Finally, in figure 6 the normalized CPU-time is given as a function of the machine size. The slope (given in parentheses) of each line represents the speedup obtained for various problem sizes. Only the larger grid sizes manage to overwhelm the communication overhead and obtain reasonable speedups. With the smallest grid of 32^3 almost no speedup was obtained by adding more processors. Note that while the 128^3 simulation only takes about ten seconds per time-step, a single simulation takes on the order of a hundred time-steps, and reasonably converged turbulence statistics require on the order of a hundred simulations. So, the 128^3 problem actually requires about 30 CPU-hours of computer time.

It is important to mention some of the factors that may effect the CM-2 timings

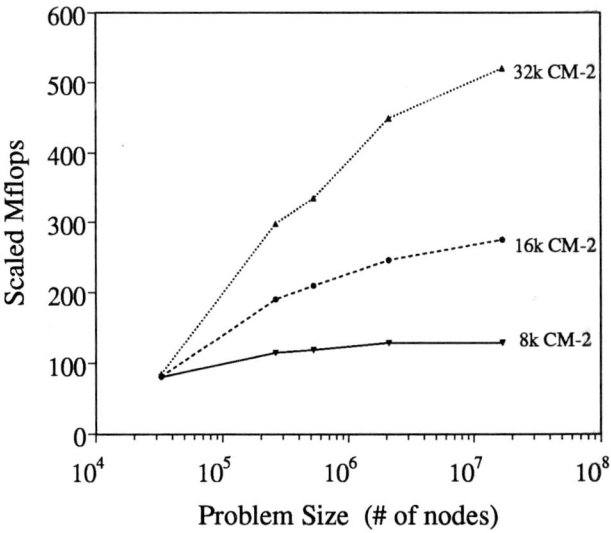

FIGURE 5. Cray Y-MP equivalent Mflops as a function of the problem size for various sections of the CM-2.

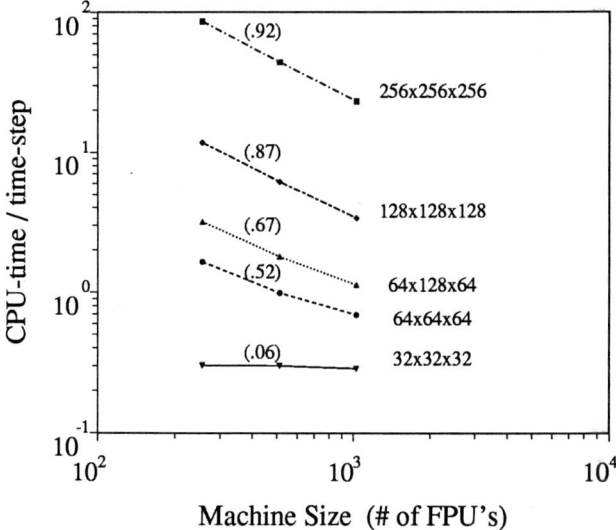

FIGURE 6. The speedup of the CM-2 for various problem sizes.

that are presented above. The number of grid points along each axis are a power of two. Many CM-2 subroutines run more efficiently with power of two axes. 32-bit floating point was used, as opposed to 64-bit on the Cray Y-MP, since 32-bit was deemed to be sufficiently accurate. The CM-2 actually has a 64-bit floating point unit, so 32-bit only saves about 25% on the computation time. Comparing 64-bit and 32-bit simulations is legitimate because the ultimate concern is the time it takes to arrive at a sufficiently accurate solution. Finally, the timings were performed on an unloaded Sun workstation front end. The loading and type of front end were found to effect performance by as much as 50%.

4. DETAILED TIMINGS

The code divides naturally into three parts each of which are fairly typical of DNS, (and computational fluid dynamics, in general). By analyzing each of these parts in detail, it will become clear where bottlenecks lie, and how well individual algorithms map to the CM-2.

4.1 First Stage

The first stage of the algorithm involves explicit stencil type operations to calculate the derivatives of the convective and diffusive terms. The operation is given by,

$$\mathbf{r} = -\left(\frac{3}{2}(\mathbf{v}^n \cdot \nabla)\mathbf{v}^n - \frac{1}{2}(\mathbf{v}^{n-1} \cdot \nabla)\mathbf{v}^{n-1}\right) + \frac{1}{Re}\nabla^2(\mathbf{v}^n). \tag{5}$$

This involves nearest neighbor communication which is very efficiently implemented on the Connection Machine provided NEWS communication grid. For a 128^3 grid only 25% of the total time was spent in this portion of the code (as opposed to 30% for the Y-MP), and a speedup of 96% was obtained (as opposed to 89% for the overall code) when the grid was doubled. This excellent performance is not due to a lack of communication in this portion of the code, fully 50% of the total time is spent in communication (within the CSHIFT function). It is due to the fact the CM-2 is well suited to cartesian nearest neighbor communication.

In addition, this portion of the code is where the majority of boundary conditions are implemented. Boundary conditions on a SIMD machine require that all interior processors lie idle while the boundary nodes do something special. This can impact the overall performance significantly. For this stage of the algorithm it was found that the four domain faces that require boundary conditions took on the order of 30% of the time. For this reason, algorithms that can naturally incorporate boundary conditions into the solution procedure are very desirable on the CM-2.

4.2 Second Stage

The second stage of the algorithm requires the solution of a Helmholtz equation,

$$\left(1 - \frac{\Delta t}{2Re}\nabla^2\right)\frac{\mathbf{v}^* - \mathbf{v}^n}{\Delta t} = \mathbf{r}. \tag{6}$$

There are a number of techniques for the solution of this matrix equation. On the Cray, the matrix is factored into a set of three tridiagonal matrices and then Guass elimination is used to solve the tridiagonals. On the CM-2 the same procedure can be used but a parallel tridiagonal algorithm such as cyclic reduction [4] must be substituted for serial Guass elimination. Or, on the CM-2, the matrix can be left unfactored and simply solved using an iterative technique such as conjugate gradients [5]. In either case, this portion of the code takes about 25% of the overall time, compared to about 30% for the Y-MP.

Cyclic reduction involves long distance communication. For the 128^3 grid, 85% of the cyclic reduction algorithm was spent in communication and a speedup of only 88% was obtained. In contrast, the conjugate gradient solution technique involves only nearest neighbor communication, and had a very good speedup. However, the conjugate gradient (CG) algorithm involves iteration. It was found that because the matrix is highly diagonally dominant, only 2-4 iterations are required, and the conjugate gradient algorithm is competitive with the more complicated matrix factorization and cyclic reduction technique. This is a useful fact for those cases, such as unstructured meshes, where the matrix can not be factored.

4.3 Third Stage

The final stage of the algorithm involves the solution of a Poisson equation for the pressure,

$$\Delta t \nabla^2 p = \nabla \cdot \mathbf{v}^*. \tag{7}$$

This is the most computationally intensive portion of the algorithm. For the simple geometry of this flow, this equation can be solved explicitly through the use of a Fourier transform in the z-direction, a cosine transform in the x-direction, and a tridiagonal matrix inversion in the y-direction [6]. Both the Fourier transform and cosine transform involve calls to the Connection Machine FFT subroutines.

FFT's involve regular (butterfly), but long distance communication. The FFT's have linear speedup and scale linearly in the number of data points but are very slow. Fully 50% of this portion of the code (and 25% of the total time) is spent in four FFT calls. A simple diagonally preconditioned conjugate gradient solution of the Poisson equation was found to be only three times to ten times slower than the transform method. This is because 15 iterations of the CG method can be accomplished in about the time it takes to do a single FFT. With better preconditioning, the CG method would be very competitive with the transform method.

5. ADDITIONAL ISSUES

In an evaluation of the CM-2 as a tool for turbulence research there are other issues besides performance. For instance, the issue of memory has already been mentioned. Being able to perform simulations entirely in core memory not only saves time swapping out to disk, but considerably simplifies the programming burden.

In terms of programming, the CM-2 is very similar to the Y-MP. The SIMD architecture allows a serial programing style, and similar programming environment. This is a great advantage of SIMD machines that should not be overlooked by the application programmer when evaluating the utility of parallel architectures. Both machines compile Fortran 90, and the CM-2 extensions to Fortran 90 are a very useful addition. Optimization on the Y-MP is much easier than on the CM-2, because the Connection Machine fortran compiler is still fairly unpredictable. But the debugging environment of the CM-2 tends to make up for this deficiency. And as we have seen, the lack of versatility of the CM-2 is made up for by the fact that it performs many brute force algorithms very efficiently.

Another useful measure besides performance is performance per dollar, or Mflops per Mdollar. A Cray Y-MP, such as the one used in this study costs approximately 25 million dollars (3.125 million per processor), and averages about 160 Mflops, or about 50 Mflops/Mdollar. The CM-2 costs about 5 million dollars. If a performance of 300 to 500 Cray equivalent Mflops is assumed (remembering, of course, that CM-2 performance is very problem dependent), then the CM-2 gets from 60 Mflops/Mdollar to 100 Mflops/Mdollar. These numbers are very approximate, and probably only accurate to within a factor of two. Nonetheless, they do indicate that the Connection Machine is already very competitive with vector architectures, not only in performance but in terms of cost, as well.

6. CONCLUSIONS

The results of this study indicate that the Connection Machine is a viable tool for the direct numerical simulation of turbulence. Its performance and performance per dollar can be very comparable to similar generation vector supercomputers. However, unlike vector supercomputers the performance of the CM-2 is a strong function of the problem size. Traditional performance measures need to be altered to account for these types of communication effects which are common to all massively parallel computers.

The architecture of the CM-2 limits its versatility. In short, it is a brute force machine, overwhelming problems with quantity rather than quality. As a result it tends to perform brute force algorithms, such as conjugate gradients or finite volume discretizations, just as efficiently as more cleverly constructed algorithms. This fact, is not entirely negative. It means that in those cases were tricks such as matrix factoring and variable transformation can not be applied (ie. general geometries), the CM-2 will not experience any loss in performance. Therefore, despite its lack of versatility the CM-2 may actually be more useful for complicated problems, because these problem are only amenable to brute force methods.

Direct numerical simulation of turbulence has always been closely tied to the performance of state of the art supercomputers. Many assumptions about turbulence simulation have been based on the vector nature of supercomputers. Many of these assumptions, such as the superiority of spectral methods, will be challenged as the availability of massively parallel computing begins to open computational horizons. It appears that the CM-2 and its successors will be key partners in these new explorations of turbulence simulation.

Acknowledgements

This work was funded by a grant from the National Science Foundation. All computer time on the CM-2 and Cray Y-MP was provided courtesy of the NAS division of NASA-Ames Research Center.

References

1. Committee on Physical, Mathematical, and Engineering Sciences, Grand Challenges: High Performance Computing and Communications, OSTP FCCSET report. National Science Foundation.

2. F. H. Harlow and J. E. Welsh, *Phys. Fluids* **8**, (1965) 2182-2189

3. J. B. Perot, An analysis of the fractional step method. Submitted to the *Jour. Comp. Physics*, (1992).

4. R. W. Hockney and C. R. Jesshope, Parallel Computers 2 Bristol and Philadelphia: Adam Hilger, (1988). 475-489.

5. W. H. Press, B. P. Flannery, S. A. Teukolsky, and W. T. Vetterling, Numerical Recipes, Cambridge University Press, (1986). 70-73.

6. B. L. Buzbee, G. H. Golub, and C. M. Nielson, On Direct Methods for Solving Poisson's Equation, *SIAM J. Numer. Anal.* **7-4**, (1970) 627-656.

PARALLEL CFD IN A NETWORKED ENVIRONMENT.

G. Prisco, D. Ferrer Pellicer, J. P. Huot, R. Molina and M. Roest.

European Space Agency, ESTEC, Keplerlaan 1, Noordwijk, The Netherlands.

ABSTRACT.
This note describes in detail a high performance/cost computing platform, assembled by linking commercial workstations with fast networks and efficient communication software, and used for detailed Computational Fluid Dynamics (CFD) analysis in aerospace design.
It is shown that this approach permits gaining experience in parallel processing without having to purchase expensive integrated multiprocessors, performing engineering computations at a speed per unit cost much better than that offered by mainframes, and optimizing the use of existing workstations, which may be underused or idle most of the time.
The hardware configuration is a cluster of IBM RS/6000 workstations connected by fast optical links. The software components (communication libraries, parallel programming environments and data visualizers) used for this work are discussed in detail.
The CFD solvers running on the cluster, which are briefly described, are state-of-the-art finite element codes able to account for real gas effects associated with hypersonic flows. They have been ported to the distributed memory platform by using a domain decomposition technique where the computational domain is partitioned into subdomains which are assigned to different hosts for processing, and message passing is used for updating field values at nodes on common boundaries.
The resulting parallel solvers have been optimized in terms both of load balancing and communication/computation ratio. The parallel decomposition strategies which have proved more efficient on the cluster are discussed in detail, and timing results are given.

1. INTRODUCTION.
Computer simulation is now seen as an approach to scientific problem solving comparable to theory and experiment. In the aerospace field, the last few years have seen a gradual shift from wind tunnel testing to CFD simulations. Though testing will remain a central component of aerodynamic strategies, computation often permits some results to be obtained at a much lower cost. Moreover detailed analysis, such as the separation of the contributions of different components to a given aerodynamic effect, are only accessible through computation.
On the other hand, computation of aerodynamic effects is not a trivial task. A state of the art CFD project is expensive in terms of both cost and time lapse. For example, a complete aerothermodynamic study of the Hermes spaceplane would require a CFD solution of the full Navier-Stokes equations coupled to a complex chemical model of the Earth's atmosphere, and to the Maxwell's equations when the temperature becomes sufficiently high for ionization of the air to take place. The

computation must be performed over a grid containing a few million points at least and requires hundreds of CPU hours on the most powerful supercomputers presently on the market. Note that CFD analysis must be iterated over a large number of points in the parameter space (Mach number, Reynolds number, angle of attack and others) in order to understand the aerothermal behaviour of a given design. This brief discussion shows that the performance of a computer system capable of running a CFD project must be very high.

Since VLSI technology alone does not seem able to advance computer performance much further, leading vendors are now committed to parallel processing. Parallel processing is a very broad term which covers a wide range of computer architectures, having in common the fact that the machine is able to do more than one thing at the same time. Some systems (MIMD) have the capability of executing different instruction streams (tasks) in parallel, and that of permitting communications between tasks. Note that, by suitably partitioning the resources of a single CPU, parallelism may also be obtained uniquely at the software level (multitasking). On the other hand, parallel software will actually run faster only on machines containing different CPUs each running one or several tasks.

Before beginning the not so trivial task of parallelizing an application code for a given multiprocessor system, it is worth remembering that the number of available processors N is only a higher limit for the speedup resulting from parallelization, and that the actual speedup may well be much lower. This is due to the fact that, in general, a set of tasks cooperating to the execution of a computational algorithm need to synchronize and exchange data. A quite common case is that of a task which has to wait until other tasks have completed the production of some data it needs in order to continue. It may also happen that the application programmer must implement complex handshaking sequences in order to ensure that two tasks do not wait for each other forever (deadlock) or corrupt data placed in shared memory by accessing them at the wrong time. Finally, one must remember that, if data are exchanged via a physical process (distributed memory), also the transmission of data takes time. All these factors can result in a relevant decrease in the parallel efficiency (actual speedup divided by N) of an application code running on a parallel system. In order to obtain the desired performance from parallel codes, it is important to design them in such a way as to keep all processors equally busy (load balancing) and to maximize the time spent in computation with respect to parallel overhead such as the time spent in task management or communications.

Parallel systems can be roughly divided into moderately parallel architectures (Alliant, Cray, IBM) and massively parallel architectures (DEC, Intel, Parsytec, Thinking Machines) This classification is based on the number of processing nodes (respectively low and high) and the processing power of the single nodes (respectively high and low). For optimal mapping, the level of parallelism of the hardware should match the number/complexity of tasks (granularity) resulting from a given parallel decomposition of a computational algorithm. There are both MIMD (Multiple Instructions Multiple Data) systems and SIMD (Single Instruction Multiple Data) systems where an instruction is broadcast to all processors which execute it on the data assigned to them. Vector processing may be viewed as a subset of SIMD. The distinction between MIMD and SIMD will prodably fade as more and more vendors adopt the MIMD approach.

Different nodes may be allowed to access the same physical memory address space (shared memory) or, alternatively, each processor may have its own private memory (distributed memory). A large and fast multiaccess memory for a shared memory multiprocessor system may well account for a significant fraction of the total cost. This is the main reason why distributed memory machines on the market offer more processing power per unit cost. In a shared memory system nodes may communicate via shared variables, whereas in a distributed memory system all necessary communications must be explicitly implemented via message passing, and the topology of the internode communication network (bus, crossbar, hypercube) largely determines the amount of traffic which can be handled within acceptable time penalties. The development of system software is easier for distributed memory system (in principle, it is sufficient to implement a small set of communication/parallelization primitives). On the other hand, for the application programmer it is easier to port existing codes to a shared memory environment.

2. SYSTEM REQUIREMENTS AND DESIGN.

Choosing a parallel environment for a given class of applications is an optimization exercise with many parameters (SIMD vs. MIMD, shared vs. distributed memory, power/number of processors). Without claiming it to be the best possible configuration, it is reasonable to state that a distributed memory MIMD system with powerful processing nodes, such as that described in this note, is a suitable platform for many aerospace computations. In fact, concerning the shared vs. distributed memory issue, some more effort in porting application codes (which will decrease with the availability of more programmer friendly environments) is a price worth paying for the greater performance per unit cost offered by distributed memory systems. Concerning the SIMD vs. MIMD issue, it is a fact that the asynchronous execution of different instruction streams possible on MIMD architectures permits a wider class of computational algorithms to be decomposed for multiprocessing in a more natural way. Finally, the requirement of powerful single nodes can be appreciated by noting that the granularity of parallel CFD solvers is low. The standard method for parallelizing a CFD code is the domain decomposition technique where the computational domain is partitioned into subdomains, each subdomain being assigned to a processing node, and message passing is used for interfacing different subdomains. It is clear that the system performs more efficiently if the time spent in computational tasks is much greater than the time spent in handling communications (volume to surface effect). If an excessive number of subdomains is used the processing nodes spend more and more of their time in exchanging messages which do not contribute directly to the desired results. This can be avoided if the processing power of the single node is sufficiently high.

A platform satisfying these requirement can be purchased from some vendors as an integrated system, or alternatively built by clustering existing machines (possibly manufactured by different vendors and running different operating systems) with suitable links. It should be noted that, even in corporate or academic environments without large number-crunching requirements, networked systems are replacing shared memory mainframes as main data processing resource. Users of a networked system can access resources on other nodes by using tools for file transfer and remote login. An application (client) can use the Remote Procedure Call (RPC) technique for linking to a tool (server) located elsewhere on the network. There are standards for network computing such as Network Computing System (NCS) and graphical user interfaces such as X11 and Motif. For users with large number-crunching requirements, the cluster alternative permits some familiarity with parallel processing to be gained without having to purchase an expensive multiprocessing platform. Consequently, it is well suited to users wishing to explore distributed computing technology but not ready to a major financial committment. Moreover, it is possible to use commercial workstations already installed at the user site for this purpose, thus building a solid foundation for future developments at virtually no extra cost while retaining the advantage of stand-alone systems running a familiar user interface. IBM and Hewlett-Packard are current market leaders in high performance UNIX workstations with mainframe-like computational capability (comparable to that of an IBM 3090 CPU with vector unit) for a much lower price (tens of Kdollars). For parallel applications, it is convenient to replace Ethernet or Token Ring Local Area Networks (LANs) with links of higher performance (hundreds as opposed to tens of Mbits/sec) based on optical fiber technology. Fiber Distributed Data Interface (FDDI) is becoming a standard in high performance LANs and FDDI products are offered by many vendors. Communication overhead can be also reduced by exchanging data over a private LAN not accessible from other users.

When the cluster alternative is chosen, a parallel programming environment must be purchased separately or developed in-house by using nested software layers. For the majority of commercial workstations, the lower level layers are Transmission Control Protocol/Internet Protocol (TCP/IP) and the socket libraries (a standard, programmer friendly interface to TCP/IP). These layers provide the basic communication facilities (sending and receiving data). On top of them, one may find a higher level of software tools (parallelization primitives) which make life easier for the application programmer. Typically, a parallel programming environment contains facilities such as semaphores and barriers for coordinating the execution of user tasks. Express (Parasoft), Linda (Scientific Computing Associates) and PVM (Oak Ridge National Laboratories) are among the available products.

Since CFD runs produce very large multi-dimensional data sets, visualization of data is a central issue in aerospace design. A computational resource based on commercial workstations is attractive in this respect, because it can support both computation and visualization. Research on the pattern-recognition capabilities of the eye-brain system has permitted selecting, among the endless possibilities offered by computer graphics, those more convenient for the visualization of large data set. Graphic tools such as colour-coding, shading, zooming, directed light sources, sweeping planes or volume rendering permit an understanding of output data difficult to reach otherwise.

The term post-processing is used for the visualization of static data. In post-processing, interactivity means being able to change such parameters as viewpoint, lighting or sweeping plane position and orientation. The hardware must be powerful enough to permit interactive use by ensuring a short response time to user inputs. Real-time tracking is a more powerful technique intended for dynamic data sets such as those generated by time-dependent flow solvers. The same technique may be used for monitoring the convergence of a computation to a steady state. In tracking, the time-dependent output of a process (client) is routed in real-time to an interactive data visualizer (server). Typically, no processing other than sending the data in a suitable format is done by the client, and consequently the visualization hardware must be powerful enough to display a frame before the next one is received. The term steering is used when the user is also allowed to change some computational parameters in real-time. A networked cluster permits running a data visualizer in parallel with a CFD solver, on a workstation (equipped with adequate visualization hardware and display unit) which may or may not also participate to the execution of the solver. If a commercial visualizer is used, it must have built-in or external facilities for retrieving data through the network interface.

3. HARDWARE COMPONENTS.

The IBM RISC System 6000 workstation family features desk-top, desk-side and rack-mounted computers. While computational power, memory and disk space vary across the family, all models run the same operating system (AIX 3) and are object-code compatible. All models feature a CPU composed by several individual RISC (Reduced Instruction Set Computer) processors, whose operations may overlap yielding more instruction executed per clock cycle (pipelining). 52 bit wide addressing permits a huge address space (theoretically up to 4 Terabytes). Maximum memory segment size is 4 megabytes.

RS 6000 models can suit the needs of different types of users. For example, the 320 is a cheap entry level machine, the 550 is a powerful machine for floating point computation with roughly the computational performance of an IBM 3090 mainframe, and the 730 is a very powerful graphic computer (120K Gouraud-shaded triangles/sec). All models can accommodate a moderate number of users (2 to 10) and allow remote users to run graphic applications in a X-Windows environment from personal X-stations.

Measured floating point computational performance (ref. 2) for a single workstation ranges from 38 Mflops (320) to 70 Mflops (550), same as for an IBM 3090 with vector unit. It may be seen that a typical mainframe installation can be outperformed by clustering a small number of workstations, at a cost reduced by one order of magnitude at least.

RS 6000 workstations have slots to insert adaptors for communications. Adaptors interface to an enhanced Micro Channel Architecture I/O bus. In principle, all communication hardware can be used for clustering workstations. In practice, only fast data links can provide the high performance required for NIC applications. On the other hand, it is useful to take advantage of cheaper and immediately available links (Token Ring, Ethernet) for preliminary developments.

High performance Token Ring and Ethernet adaptors are available for RS 6000 workstations. Performances are 10 Mbits/sec (Ethernet) and 16 Mbits/sec (Token Ring). Both cards can be installed in a single unit, which may act as a bridge between Token Ring and Ethernet LANs.

Fast optical links have become available for RS 6000 workstations in March 1991 with the IBM announcement of the Serial Optical Channel Converter (SOCC) adaptor. The unit plugs in the backplane of the machine and contains two links with a speed of 220 Mbits/sec (more than 20 times Ethernet speed and twice FDDI speed). Optical fiber cables with length up to 100 meters are available. It is possible to cluster workstations with SOCC links in a ring topology without additional hardware. A star configuration where each workstation has a fast data path to all other ones my be built by using the routing unit DX4290 of Network Systems Corporation (NSC). This is a high performance interface between IBM RS 6000 workstations and mainframes which supports 8 (optionally 16) SOCC links and has message routing capabilities. The maximum throughput of he unit is 800 Mbits/sec. This means that no more than 4 workstation should be transmitting on SOCC links at the same time, and that clusters containing more workstations should include more switching units.

The hardware components of the workstation cluster described in this note have been made available to ESTEC by IBM Holland under an ACIS (ACademic Information System) study contract. IBM Holland has also acted as interface with other IBM researchers and software developers for the early procurement of some of the software components described in the next section. In an ACIS contract, the study partner receives some hardware and software items which enable it to carry out a study project interesting for IBM. In exchange it undertakes a number of obligations, among which those of providing feedback to IBM, of acknowledging IBM contribution in technical papers or presentations related to the project, and of acting as a reference site on request of IBM.

The cluster configuration selected for this project consists of 3 workstations model 520 on a SOCC ring. Each workstation has 32 Mbytes of memory and 355 Mbytes of disk space. One workstation, acting as main user console and system controller, has some enhancements such as additional 355 Mbytes of storage, keyboard, mouse and display unit. The remaining two workstations, for which no disk access is foreseen, could eventually be replaced by cheaper diskless workstations. More details can be found in (ref. 1).

The cost of the hardware setup described, together with that of necessary system software such as AIX operating system, FORTRAN compiler and libraries,is approximately 150 Kdollars. Note that using discless workstations could significantly reduce the cost of the cluster leaving the performance unchanged.

4. SOFTWARE COMPONENTS.

The basic communication facilities in many commercial workstations, including the RS 6000 family, are the socket libraries which are provided with the system software. Sockets are the standard UNIX programming interface to TCP/IP protocols. A socket is a communication endpoint that can be accessed like a file by a process wishing to communicate across the network. Once a socket has been created and additional specifications such as local port address or remote destination address have been given, a process can write data to a socket or read data from a socket. Any parallel application for a networked environment could in principle be developed by using sockets. In practice, it is more convenient to use a higher level communication environment built on top of socket calls such as PVM (Oak Ridge labs), Linda (Scientific Computing Associates Inc.) or Express (Parasoft Corp.).

The Parallel Virtual Machine (PVM, see ref. 3) environment has been developed by the Mathematical Sciences Section of the Oak Ridge National Laboratories. It is a set of primitives for distributed computing on a heterogeneous network (task generation, communication, barriers, events, or emulated shared memory segments in future versions). PVM is available for most workstations and other computing platforms. An application under PVM can generate dynamically a set of tasks and define arbitrary communication and synchronization patterns. Since any task can exchange data with all others, the software designer is free to choose among a variety of computational models. Examples are tree models (communications are restricted to vertical and horizontal links of a tree defined by nested task generation) and crowd models (no structured communication pattern is superimposed). The parallel computational model more useful for domain decomposition in CFD is a regular crowd with multiple instances of the same component.

PVM is available free of charge through the Internet. The version currently released routes all data exchanged between user processes through UNIX daemons. Since this may be a bottleneck for applications requiring fast data links, researchers at the IBM European Center for Scientific and Engineering Computing (ECSEC) in Rome have developed a version which supports interrupt based point to point links between user processes. The basic message passing facilities are provided by the snd (send) and rcv (receive) calls. Both calls have a "message type" argument which can be used to identify messages. Typically, this can be chosen as equal to the process or processor identifier in order to differentiate messages coming from or being sent to different processes or processors. The snd call finds the message to be sent in a send buffer which must be explicitly managed by the programmer, which may use calls for clearing, filling or appending to the buffer. The rcv call is blocking, which means that once it has been issued the calling process is blocked until a message of the type specified has been received. The probe call may be used for checking whether a message has been received, thus enabling to keep the waiting process busy (if the computational algorithm being used permits doing other work while waiting for the new data). Synchronization patterns can be built also by using barriers (barrier call) and signals (ready and waituntil calls). The potential of PVM for effortless parallel application software development has been recently enhanced by the Heterogeneous Network Computing Environment (HeNCE, see ref. 3), a set of graphical CASE tools built on top of PVM which permits defining graphically task hyerarchies and coordination patterns.

A visualization system with built-in communication facilities is also a central component of the CFD environment. The 4-Dimensional Scientific Data Visualizer (4D-SDV) is a tracking/steering tool developed at IBM ECSEC. Its components are an interactive 3-dimensional data visualizer running on all RS/6000 workstations and a communication library which uses TCP/IP links for routing data from a client process to the visualizer. The communication library is written in C and built on top of socket calls. The visualizer is written in FORTRAN and built on top of graPHIGS. New frames are processed for display as soon as they are received. If the total (transmission and image generation) time needed for producing a new frame is not greater than the computational time step, the visualizer keeps displaying the current computational output and, if the process is fast enough, gives the impression of animated motion. User interaction with 4D-SDV takes place in windows containing dials and buttons which permit selecting among a number of display modes (isosurfaces, sweeping planes) and performing interactively all the usual image editing functions. There is also a steering window for changing in real-time the values of pre-selected parameters or stopping/restarting the client process. User input files contain such specification as type of fields (scalar, vector, spin) and type of grid (regular, deformed).

5. MESH GENERATION AND DECOMPOSITION.

A powerful mesh generator is a central component of the CFD analysis. The requirement of having a mesh embedding as smoothly as possible a complex and generally non convex three-dimensional surface, and that of producing it within a reasonable time, put strong gemands on the computational methods used.

In the popular multiblock technique, the computational domain is split into blocks which are meshed separately. The multiblock technique permits tuning the mesh generation process for each block and optimize it for the topology and geometry of the block. Our team uses the commercial CAD package CAEDS (IBM) for meshing the surface of each block. The three-dimensional mesh is then produced by a generator developed in-house which uses the input surface meshes as boundary conditions, thus ensuring the continuity of the mesh across block boundaries. The three-dimensional meshing algorithm used is based upon a modified Kennon spring analogy. The mesh is regarded as a system of springs connected at the nodes, whose potential energy function is to be minimized. The original formulation of this technique (Kennon 85) can produce overlapping for non convex domains. In our implementation overlapping is detected by computing the volume of each element. A negative volume indicates presence of overlapping. A stiffness parameter defined for each node (in the original formulation the stiffness parameter was associated with springs) is adjusted according to the results of the overlapping detection, and the mesh is improved iteratively until no further overlapping is found. This technique is routinely used for mesh generation around the European spaceplane Hermes, and results show that high quality meshes can be created within a reasonable time (ref. 4).

In order to run a parallel CFD solver where each processor handles one or more of a set of subdomains, one can either generate the subdomains independently or generate a global mesh and then partition it into subdomains. The first strategy comes for free when the mesh has been created by using a multiblock technique similar to that described above. In this case, one has simply to identify blocks with subdomains.

On the other hand, this simple approach is not necessarily the one which ensures the fastest execution of the parallel solver. The domain should in fact be decomposed with load balancing and minimization of communications with respect to computation in mind. Load balancing means that the times taken by the processors to reach the next synchronization point in the computation should be as similar as possible. In fact, as discussed in the next section, when a process has completed one iteration of the computational loop it remains idle until all other processes have completed the same iteration. The time taken by the solver is consequently a function of the time taken by the slowest process (the one which runs on the slowest processor or the one which has more work to do), regardless of the speed of the other processes. It should be noted that the time taken to complete an iteration of the loop includes, besides the time taken for computation, the time necessary for sending data to and receiving data from other processors. When all processor have the same computational power and all communication channels have the same capacity, load balancing can be ensured by simply assigning the same number of nodes and the same number of interface nodes to each processor. Otherwise, the different processing speeds and communication bandwidths must be accounted for in some way. Besides load balancing, the efficiency of a parallel application requires also minimizing the fraction of time spent in communications. The computational domain should consequently be decomposed in such a way as to minimize the total amount of data exchanged per iteration.

The decomposition resulting from a multiblock mesh generator is only meant to satisfy geometrical and topological requirements which have nothing to do with the requirements mentioned above, which consequently are not necessarily met. Due to this circumstance, it has been preferred to treat the mesh as a single entity and decompose it afterwards by taking the relevant set of requirements into account.

The method used for mesh decomposition is a two-stage process. First, a reasonable but rough decomposition is found by using a fast algorithm such as one way bisection or coordinate bisection (ref. 6). Then, the decomposition is improved iteratively by using simulated annealing (ref. 5). Simulated annealing is a non-deterministic optimization method which permits finding a near-optimal solution even for Non Polinomial (NP) complete problems. NP-complete problems, to which class mesh decomposition belongs, require a processing time which is not bound by any polinomial function of the problem size. This means that these problems become rapidly untreatable with the increase of their size, and this fact has prompted a wide interest in simulated annealing. A well known application of the technique is found in VLSI design where the placement of components on a chip is optimized in such a way as to minimize the mean length of hardwired connections.

A typical application of the simulated annealing technique to an optimization problem begins with an initial choice, which can be tought of as a starting point in phase space of the problem. An energy function is defined on the phase space in such a way as to assume lower and lower values for more and more optimal solutions. An initial temperature T is chosen and the representative point in the phase space initiates a random walk. A move to a lower energy state is always accepted, whereas a move to a higher energy state is accepted with probability:

$$P = \exp \frac{-\delta}{T}$$

where δ is the positive difference in energy. Uphill moves are conditionally accepted in order to permit backing out of a local energy minimum which can be far from the global minimum sought. The temperature is then lowered according to some rule (annealing strategy) until a sufficiently optimal solution has been found. Lowering the temperature means decreasing the acceptance probability of a typical uphill move. Terms such as energy and temperature are used in order to underline the analogy of this process to the physical process of slowly cooling a physical system toward a stable state. The simulated annealing technique can be shown to converge eventually to a globally optimal solution (van Laarhoven 88), but the computational cost necessary can be proibitively high for NP-complete problems if the starting point is not chosen properly. On the other hand if one can use a less computationally expensive method for finding a reasonable solution, simulated annealing can be used to improve it iteratively. For applying simulated annealing to a specific optimization problem, one has to choose an energy function and an annealing strategy (ref. 7). In both cases a trial and error approach may be necessary.

The energy function for the domain decomposition problem has been chosen as:

$$E = \alpha \sum_{i,j} (\frac{n_i}{p_i} - \frac{n_j}{p_j})^2 + \beta \sum_{i<j} (\sum_k \frac{c_{ik}}{s_{ik}} - \sum_k \frac{c_{jk}}{s_{jk}})^2 + \gamma \sum_{i<j} \frac{c_{ij}}{s_{ij}}$$

where n_i is the number of nodes assigned to processor i, p_i is proportional to its processing power, c_{ij} is proportional to the amount of data which must be sent from processor i to processor j and s_{ij} is proportional to the speed of the data transfer between processors i and j. The parameters α, β, γ can be used to tune the weight and normalization of the three terms. The first term in E corresponds to load balancing, the second to communication balancing and the third to minimizing of the total communications.

The random walk is generated by exchanging randomly chosen surface elements between subdomains. The annealing strategy chosen is to keep the temperature constant for a fixed number of iterations N, then decreasing it by a constant factor r. The termination condition used is to stop the computation when the energy has never changed at the current temperature or, more simply, to use a preselected number of temperature changes. More details can be found in (ref. 9).

6. PARALLEL CFD SOLVERS.

Once installed the hardware and software components described in the previous sections, parallelizing CFD codes initially developed for sequential machines is a rather straightforward task. It amounts in fact to inserting, between two successive iterations of the solver (computation module), a coordination module which handles the communications with other instances of the solver executing on the other processors and the merging of computed data and received data appropriate to the computational method used. The two modules together form the task which is assigned to all processors. PVM calls are used for starting tasks and handling communications.

The computation module is a sequential version of the solver. Consequently, the code developed for single processors does not have to be changed at all in passing to a distributed system. This modular approach is made possible by the MIMD nature of the networked cluster and the processing power of the single nodes, each able to handle a significant fraction of the computational domain. On the contrary, the solver should be extensively modified and eventually re-written for a SIMD system or a massively parallel array of less powerful processors.

Some advanced CFD solvers (ref. 8) have been modified as outlined above and installed on the cluster. These solvers are not academic codes developed uniquely for research purposes, but rather fully functional codes meant for production runs in aerospace design. As such, they are routinely employed for the computational analysis of hypersonic space transportation systems such as the Hermes spaceplane. Besides being able to use complex three-dimensional geometric models as described in the previous section, our CFD solvers are able to compute viscous, compressible and chemically reacting flows.

The development a set of advanced CFD solvers at ESTEC, started in the late eighties, has followed an evolutionary pattern similar to that found in other technology centers and areas of computational science. The first prototypes have been developed on an IBM 3090 manframe. When the mainframe has been equipped with a vector unit (1988) the codes have been extensively optimized for the vector hardware. The vectorization process has paid off in terms of an average speedup of three. After their appearance on the market, RISC workstations with mainframe-like processing power (IBM, Hewlett-Packard, Sylicon Graphics) have been increasingly used in place of the mainframe for both development and operation of computationally intensive applications. The development of parallel solvers on distributed systems made by clustering workstation is a logically following evolutionary step which can either coexist with or prepare the way for the development of solvers on more tightly coupled platforms. It should be noted that parallelizing CFD solvers has required less effort and at the same time permitted a larger reduction of time lapse than the vectorization.

After each iteration of the computation module, the coordination module begins by sending data to all processors handling a subdomain which interfaces to the current one along common boundaries. For each interface, field values at all node points are sent. Similar data are then received from the same set of processors. A barrier call is inserted in order to ensure that all necessary data have been collected before continuing. At this point, new field values at all interface points are defined by summing up the contributions of all subdomains. It should be noted that the communication patterns used by the coordination module are defined by the cross-reference lists of interfacing subdomains and interfacing nodes built by the domain decomposer described in the previous section.

The 4D-SDV visualizer is used routinely for tracking/steering. It has been found that the possibility of interacting in real-time with the solver and receiving immediate visual feedback is of great help in reducing the turnaround time of the trial and error loop necessary for setting up properly the input parameters of a computation, and that it provides very useful hints for the development of the solver.

7. RESULTS AND CONCLUSIONS.

The simulated annealing based tool for domain decomposition has been successfully tested on several complex two-dimensional and three-dimensional meshes. Real-time visualization of its output has been provided by coupling it to the 4D-SDV data visualizer. The steering capabilities of 4D-SDV can be used for an interactive choice of the exit point and an interactive development of the annealing strategy. The first videotaped sequence presented at the Conference shows a far from optimal initial decomposition changing smoothly into a much better one. The work for linking the domain decomposition tool to the CFD solvers is almost complete and results will be available soon.

Due to unforeseen delays in the delivery of the SOCC hardware, which has eventually been installed and successfully tested in the first week of May, timing results for CFD solvers are available only for an Ethernet cluster. More detailed timing results for the SOCC cluster will be available soon. The test run discussed is a reasonably realistic computation of a Mach 10 flow around the front part of the Hermes spaceplane (containing the cabin) at 30 degrees angle of attack. The multiblock mesh used contained a total of 120000 node points in four blocks. Since it would not have been possible to keep good load balancing with four blocks and three processors, a 730 workstation has been added to the cluster.

Part of the time-dependent output, which has been successfully compared to that of a sequential version of the solver, is shown in the second videotaped sequence presented at the Conference.

The computation has been carried on for 4000 iterations and has required 45000 seconds, which is to be compared with the corresponding time lapse of 56000 seconds on an IBM 3090 mainframe with vector unit. The time required for a single iteration on the cluster is 11.3 seconds with approximately two seconds spent in exchanging data for a total of 1 Mbyte, and the rest spent in computation.

These results look quite encouraging when one takes into account the facts that the four machines were placed on a shared rather than private LAN with one workstation (the 730) physically removed from the others, that optical links were not used, and that the load was not optimally balanced.

ACKNOWLEDGMENTS.

The valuable contributions of J. Lewis and G. Webb (European Space Agency), R. Houweling, K. van Unen, J. van Zijl and B. Vos (IBM Holland), M. Bernaschi, R. di Antonio and P. Santangelo (IBM Italy) are gratefully acknowledged.

REFERENCES.

1. G. Prisco, D. Ferrer, R. Molina, J. P. Huot and M. Roest, Parallel Computing on a Networked Risc Cluster, ESTEC WP 1645, ESTEC, Noordwijk (1992).

2. J. Dongarra, Performance of Various Computers Using Standard Linear Equations Software, available on the Internet from netlib@ornl.gov.

3. A. Geist et al., PVM and HeNCE documentation and user guide, available on the Internet from netlib@ornl.gov.

4. D. Ferrer, R. Molina and J. P. Huot, "A Mesh Generator for Computational Fluid Dynamics Based on Modified Potentiel Energy Minimization", Proc. of the 3rd Int. Conf. on Numerical Grid Generation, A. S. Arcilla, J. Haeuser, P. R. Eiseman and J. F. Thompson, eds., Barcelona, North-Holland Elsevier (1991).

5. S. Kirkpatrick, C. D. Gelatt and M. P. Vecchi, "Optimization by Simulated Annealing", Science 220, 671 (1983).

6. H. D. Simon, "Partitioning of Unstructured Problems for Parallel Processing, Computing Systems in Engineering 2, 2/3, 135 (1991).

7. M. E. Johnson, ed., "Simulated Annealing and Optimization", Am. Journ. of Mathematics and Management Sc. 8, 3/4 (1988).

8. J. P. Huot, R. Molina and D. Ferrer, Fast Finite Element Solvers for Reentry flows, in Proc. of the 1st European Symp. on Aerothermodynamics for Space Vehicles, ESA SP-318, Noordwijk (1991).

9. M. Roest, Domain Decomposition Using Simulated Annealing, in preparation (1992).

LARGE EDDY SIMULATIONS OF TURBULENCE ON A MASSIVELY PARALLEL COMPUTER

J.H. Robichaux[a], S.P. Vanka[b], and D.K. Tafti[c]

[a]Graduate Student in the Department of Mechanical and Industrial Engineering at the University of Illinois at Urbana-Champaign.

[b]Associate Professor in the Department of Mechanical and Industrial Engineering at the University of Illinois at Urbana-Champaign.

[c]Senior Research Engineer at the National Center for Supercomputing Applications in Urbana-Champaign.

Abstract
In this study the implementation of a finite volume code for large eddy simulations (LES) of turbulence on the Connection Machine Model 2 (CM-2) is discussed. This code is applied to simulate fully developed channel flow with one inhomogeneous and two periodic directions. Implementation and parallelization of the explicit and implicit portions of the algorithm will be discussed. Performance comparisons between a similar code run on the CRAY-2 indicate that for this application the CM-2 is competitive with the CRAY-2.

1. INTRODUCTION

Large eddy simulations (LES) of turbulent flow are a cost effective method for numerical simulation of turbulent fluid flows (Voke & Collins [1]). In this method the Navier-Stokes equations are filtered to remove the small scales of turbulent fluctuations. The unresolved or small scales that have been filtered out are then represented by a model that relates their effects to the resolved or large scales. The merits of this approach are that by filtering out the smallest scales of motion the computational task of exactly integrating the Navier-Stokes equations, as in direct numerical simulations (DNS), is significantly reduced. However, large eddy simulations still require several hours of CPU time on the CRAY class machines to obtain reasonable flow statistics. The new massively parallel computers, such as the Connection Machine - 2, offer a possible solution to reduce the integration time and enable large eddy simulations to be used as an engineering tool.

The focus of this paper is the implementation on the CM-2 of a finite volume LES code for the simulation of incompressible turbulent flow (Tafti & Vanka [2], Robichaux et al. [3]). The code is applied to simulate the fully developed flow between two rigid walls in which a mean pressure gradient is applied to drive the flow in the x (streamwise) direction. Walls are placed at the boundaries in the y (cross stream) direction. The flow domain is assumed to be periodic in the x and z (spanwise) directions. A schematic of the channel geometry is shown in Figure 1.

A general description of the CM-2 computing system is first presented. The time integration of the governing equations is performed by a mixed explicit/implicit fractional step method (Moin & Kim [4]). This requires the solution of several tridiagonal matrix equations and a Poisson equation. The implementation of these algorithms is discussed. Timings of the code run on the CM-2 and a similar code run on the CRAY-2 are compared. Finally, some numerical results are presented for the simulation of channel flow with two walls and a channel with a shear free surface.

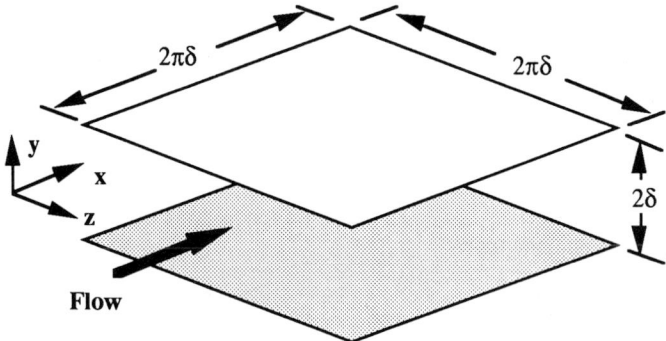

Figure 1. Channel Geometry

2. THE CONNECTION MACHINE - 2

The Connection Machine Model 2 (CM-2), manufactured by Thinking Machines Corporation, is a massively parallel computing system based on the data parallel computing model (Thinking Machines [5]). In the data parallel model each data element is associated with one individual processing element. All processing elements execute the same instruction simultaneously. Different processing elements can only share data by interprocessor communication.

The main architectural elements of the CM-2 are the front end computer and the parallel processing unit. The front end computer performs computations that may best be done serially, and sends instructions to the parallel processing unit. The parallel processing unit (ppu) performs all the data parallel operations. A ppu may contain 16K, 32K, or 64K (K = 2^{10}) bit serial processors. Sets of sixteen processors are placed on a single chip, and are fully interconnected. The chips are connected through a hypercube communication network. Floating point computations are performed by a floating point processor which is shared by two chips (32 processors).

Programmers are permitted to allocate parallel arrays larger than the number of physical processors. If the number of data elements is larger than the number of physical processors, virtual processors are created. The virtual processor facility enables a physical processor to simulate more than one processing element. The ratio of virtual processors to physical processors is called the Virtual Processor (VP) Ratio.

Data in parallel arrays may be allocated to processors in order to take advantage of the *router* or NEWS grid communication mechanism available on the CM-2. The *router* is a general communication mechanism, whereas the NEWS grid is a structured communication mechanism which allows communication along the axis of rectangular grids. NEWS addressing is a more structured form of addressing which allocates the processing elements into a grid such that neighboring elements in an array are actually nearest neighbors on the physical grid. For this application the NEWS grid is the preferred allocation mode.

Software can be written for the CM-2 using parallel extensions of Fortran, C, and Lisp. Thinking Machines also provides subroutines in the Connection Machine Scientific Software Library (CMSSL) for computationally intensive algorithms, such as fast Fourier transforms, matrix inversions, etc.

3. GOVERNING EQUATIONS

For incompressible flows the non-dimensionalized Navier-Stokes equations can be written in tensor notation as

$$\frac{\partial u_i}{\partial x_i} = 0 \tag{1a}$$

$$\frac{\partial u_i}{\partial t} + \frac{\partial}{\partial x_j}(u_i u_j) = -\frac{\partial p}{\partial x_i} + \frac{1}{Re_\tau}\frac{\partial^2 u_i}{\partial x_j \partial x_j} + \delta_{i1} \tag{1b}$$

where the index i=1, 2, 3 refers to the x (streamwise), y (crosstream), and z (spanwise) directions, respectively. For the present channel geometry (Figure 1), the spatial coordinates have been non-dimensionalized by the channel half width δ, while the velocities u_i have been scaled with the wall shear velocity u_τ. Appropriately, the time coordinate is non-dimensionalized with δ/u_τ. The Reynolds number Re_τ is based on the friction velocity u_τ and the channel half width δ. Further, the mean driving pressure gradient in the streamwise (x) direction is separated from the total pressure (mean + fluctuating) which then assumes a non-dimensional value of unity in the x momentum equation.

In the present study, the finite volume form of the governing equations of motion are filtered according the the procedure described by Schumann [6]. After applying the filtering procedure to the Navier-Stokes equations and using the Smagorinsky subgrid scale model [7] the filtered finite volume form of the continuity and momentum equations for the resolved scales can be written as

$$\partial_j {}^j\overline{u_j} = 0 \tag{2a}$$

$$\frac{\partial}{\partial t}{}^v\overline{u_i} + \partial_j({}^i\overline{u_i}\,{}^j\overline{u_j}) = -\partial_i({}^i\overline{P}) + \partial_j\left[\left(\frac{1}{Re_\tau}+v_T\right){}^j\overline{\frac{\partial u_i}{\partial x_j}}\right] + \partial_j\left[v_T\,{}^j\overline{\frac{\partial u_j}{\partial x_i}}\right] + \delta_{i1} \tag{2b}$$

Here P is the sum of the hydrodynamic pressure and the subgrid kinetic energy. The presuperscript v denotes a volume average over the finite volume cell, and the presuperscript i and j denote surface averages of the cell, the normals of which are in the i and j direction, and ∂_i and ∂_j are the second order finite difference operators for the first derivative in the i and j direction. The sub-grid viscosity v_T is given by

$$v_T = l^2 \sqrt{2\, S_{ij}\, S_{ij}} \tag{3}$$

where S_{ij} is the resolved strain rate tensor and l is the characteristic length scale of the subgrid scales. For this problem the length scale is prescribed to be

$$l = C_s \left[1 - \exp(-y^{+3}/A^{+3})\right]^{1/2} (\Delta x\, \Delta y\, \Delta z)^{1/3} \tag{4}$$

where C_s is the Smagorinsky constant, and $A^+ = 25$.

4. NUMERICAL ALGORITHM

4.1 Time Integration

In the present method time integration from time step n to n+1 is accomplished by treating the convective, subgrid scale terms, and diffusion terms in the periodic directions explicitly using a second order Adams-Bashforth scheme. The diffusion term in the y (cross stream) direction is treated implicitly using a Crank-Nicolson formulation. A fractional step method (Kim & Moin [4]) is used to avoid the coupled solution of velocity and pressure. The three steps of the fractional step method begin by solving for an intermediate velocity field ($\tilde{u}_i$) for each component from the equation

$$\frac{\overline{\tilde{u}_i} - \overline{u_i^n}}{\Delta t} = \frac{3}{2}H_i^n - \frac{1}{2}H_i^{n-1} + \frac{1}{2}\partial_2\left[\left(\frac{1}{Re_\tau} + \nu_T\right)\overline{\frac{\partial \tilde{u}_i}{\partial x_2}}^2\right] \quad (5)$$

The H_i^n terms represent all of the terms that are treated explicitly at time step n

$$H_i^n = -\partial_j \overline{u_i}^j \overline{u_j}^j + \partial_1\left[\left(\frac{1}{Re_\tau} + \nu_T\right)\overline{\frac{\partial u_i}{\partial x_1}}^1\right] + \frac{1}{2}\partial_2\left[\left(\frac{1}{Re_\tau} + \nu_T\right)\overline{\frac{\partial u_i}{\partial x_2}}^2\right]$$

$$+ \partial_3\left[\left(\frac{1}{Re_\tau} + \nu_T\right)\overline{\frac{\partial u_i}{\partial x_3}}^3\right] + \partial_j\left[\nu_T \overline{\frac{\partial u_j}{\partial x_i}}^j\right] + \delta_{i1} \quad (6)$$

After the intermediate velocities are determined, a Poisson equation for a pseudo pressure given by

$$\nabla^2 \phi^{n+1} = \frac{1}{\Delta t}\frac{\partial \overline{\tilde{u}_i}}{\partial x_i} \quad (7)$$

is solved. The velocity components at time step n+1 are subsequently updated using

$$\overline{u_i^{n+1}} = \overline{\tilde{u}_i} - \Delta t \frac{\partial \phi^{n+1}}{\partial x_i} \quad (8)$$

4.2 Poisson Equation Solver

Solution of the Poisson equation for the pressure (Equation (7)) is the most computationally intensive portion of the present algorithm and thus requires particular attention. The uniform grid spacing and the periodic boundary conditions in the two directions make direct Fourier methods (Hockney [8]) applicable to this problem. To illustrate the method, consider the three-dimensional discretization of the Poisson equation using a second order seven point stencil.

$$\frac{u_{ix-1,iy,iz} - 2 \cdot u_{ix,iy,iz} + u_{ix+1,iy,iz}}{\Delta x^2} + a_{iy} \cdot u_{ix,iy-1,iz} + b_{iy} \cdot u_{ix,iy,iz} + c_{iy} \cdot u_{ix,iy+1,iz}$$

$$+ \frac{u_{ix,iy,iz-1} - 2 \cdot u_{ix,iy,iz} + u_{ix,iy,iz+1}}{\Delta z^2} = f_{ix,iy,iz} \tag{9}$$

The indices ix, iy, iz correspond to the x, y, and z directions respectively. The variables a_{iy}, b_{iy}, c_{iy} form the appropriate discretization operator in the y direction for the non uniform spacing. Δx and Δz are the grid spacings in the x and z directions. The index ix ranges from 0 to nx-1, iy ranges from 1 to ny, and iz ranges from 0 to nz-1, where nx, ny, and nz are the number of finite volume cells in the x,y, and z directions respectively.

We define a two dimensional discrete Fourier transform of the variable u for each plane where iy = constant by

$$u_{ix,iy,iz} = \sum_{kx=0}^{nx-1} \sum_{kz=0}^{nz-1} \hat{u}_{kx,iy,kz} \exp(-\alpha \cdot ix \cdot kx) \exp(-\beta \cdot iz \cdot kz) \tag{10}$$

where $\alpha = \frac{2 \cdot \pi \cdot i}{nx}$, $\beta = \frac{2 \cdot \pi \cdot i}{nz}$, and $i = \sqrt{-1}$

$\hat{u}_{kx,iy,kz}$ is a complex Fourier coefficient. Taking the Fourier transform of equation (9) and simplifying we get an equation for each (kx,kz) Fourier mode of the form

$$a_{iy} \cdot \hat{u}_{kx,iy-1,kz} + \hat{b}_{kx,iy,kz} \cdot \hat{u}_{kx,iy,kz} + c_{iy} \cdot \hat{u}_{kx,iy+1,kz} = \hat{f}_{kx,iy,kz} \tag{11}$$

$$\hat{b}_{kx,iy,kz} = b_{iy} - \frac{4}{\Delta x^2} \sin^2\left(\frac{\pi \cdot k x}{nx}\right) - \frac{4}{\Delta z^2} \sin^2\left(\frac{\pi \cdot k z}{nz}\right) \tag{12}$$

The above procedure has decoupled the problem into nx·nz independent tridiagonal equations for the Fourier coefficients. The final solution is obtained from the inverse Fourier transform of $\hat{u}$.

4.3 Implementation

The algorithm described in the preceding sections has several properties which make it inherently parallel. If each finite volume cell is associated with a single processing element, computation of the explicit terms in equation (6) and updating the velocity field in equation (8) are inherently parallel and require only nearest neighbor communication. Solving for the intermediate velocity field from equation (5) requires independent tridiagonal matrix inversions (TDMI) in the y direction for each x-z location which can also be done in parallel.

The Poisson solver also contains a large degree of parallelization. The 2-D Fourier transforms can be performed independently for each plane iy = constant. Furthermore, the tridiagonal matrix inversions in equation (11) can be performed independently for each (kx,kz) mode.

The tridiagonal matrix inversions, and Fourier transforms are very computationally and communication intensive and require special attention for implementation on parallel computers. The tridiagonal solutions may be parallelized using parallel cyclic reduction (Jespersen & Levit [9]) or odd-even cyclic reduction (Ortega [10]). One dimensional fast

Fourier transform (FFT) is inherently parallel (Jameson et al. [11]) and the degree of parallelism is increased for multi-dimensional transforms.

The data locality of the second order finite volume method and the inherent parallelism in the numerical kernels (TDMI and FFTs) make the current numerical procedure ideally suited for implementation on the CM-2.

4.4 Performance

The entire code is written in CM Fortran (version 0.8). Connection Machine Scientific Software Library (CMSSL) routines are used to perform the discrete Fourier transforms, and tridiagonal matrix inversions. The CMSSL routines are found to perform very well especially when compared to similar routines written in CM Fortran. Table 1 gives speeds in seconds per time step for the current LES code run with different VP ratios for two different grids along with a comparison with a similar code run on a CRAY-2. The present implementation shows that the use of the CM-2 (32 K processors) decreases the CPU time by a factor of 2.1 over a single processor CRAY-2. It can also be observed from Table 1 that the execution speed depends strongly on the VP ratio. If the 64^3 problem is run on a full 64K processor CM-2 with a VP ratio of 4, a speed up of 3.23 over a CRAY-2 can be expected.

Table 1
Timings of LES code

VP Ratio	32 x 64 x 32 Grid			64 x 64 x 64 Grid		
	# Processors	sec./step	Cray-2 time / CM time	# Processors	sec./step	Cray-2 time / CM time
4	16 K	0.62	1.45	64 K*	0.62	3.23
8	8 K	0.95	0.95	32 K	0.95	2.10
16	--	--	--	16 K	1.60	1.25

* Scaled up to 64K

Since the completion of this study Thinking Machines has introduced the CM-200, and the CM-5 which contain many improvements over the CM-2 used in this study. It is expected that if the code is run on these machines much larger speed ups can be obtained.

5. NUMERICAL RESULTS

The LES code was applied to simulate turbulent flow in a wall bounded channel The simulation was run at a shear Reynolds number (Re_τ) of 180. This corresponds to a Reynolds number based on half width (Re_δ) of approximately 3300. No slip boundary conditions were applied on the walls. The simulation was conducted for a grid size of 64x64x64. Results are compared to a 4 million node direct numerical simulation conducted by Kim et al. [12] using spectral methods. Plots of the root mean squared (rms) fluctuating velocities in the x and y direction (u" and v") are shown in Figure 2. It can be observed that comparisons with the direct numerical simulation are very good. Other studies (Tafti & Vanka [2], Robichaux et al. [3]) show that the LES code simulates the dynamics of the bursts and sweeps in the wall region very well.

The program was then applied to solve for the flow in an open channel. In this case one of the surfaces is a wall and the other is a shear free gas-liquid interface. This flow is of particular engineering interest because of the large heat and mass transfer rates that occur near gas-liquid interfaces. The enhanced heat and mass transfer rates are thought to be due to the ejection of turbulent fluid from the bottom wall which impinges on the interface and then flows back into the rest of the channel (Rashidi et al. [13]).

In order to simulate this flow no slip boundary conditions are applied at the wall, while on the shear free surface the following conditions are applied

$$\frac{\partial \bar{u}}{\partial y} = \frac{\partial \bar{w}}{\partial y} = \bar{v} = 0 \qquad (13)$$

For simplicity, a fixed zero penetration boundary at the free-surface is assumed which did not allow the surface to move or waves to form. While this is not the proper boundary condition at a shear free surface, because it neglects surface tension effects, it is easy to implement and can capture many qualitative features of the flow.

The simulation was performed at a Reynolds number Re_h based on the full channel height h (h = 2δ) of 6600. As in the previous simulation the grid size used was 64x64x64. In Figure 3 the rms fluctuating velocities u" and v" are ploted and compared with experimental measurements of Rashidi et al. [13] at Reynolds numbers of 5000, and 7500. The variation of the Reynolds stress <-u"v"> is shown in Figure 4.

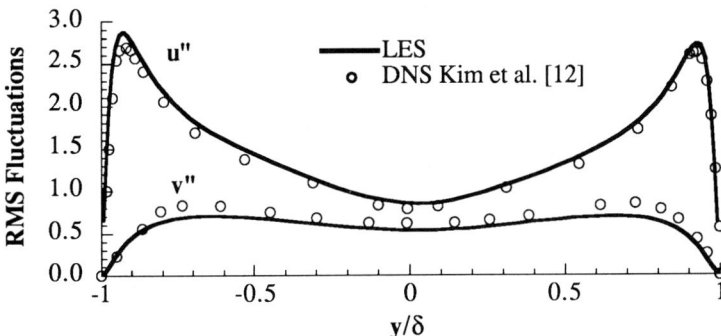

Figure 2. RMS Velocity Fluctuations in Channel Flow

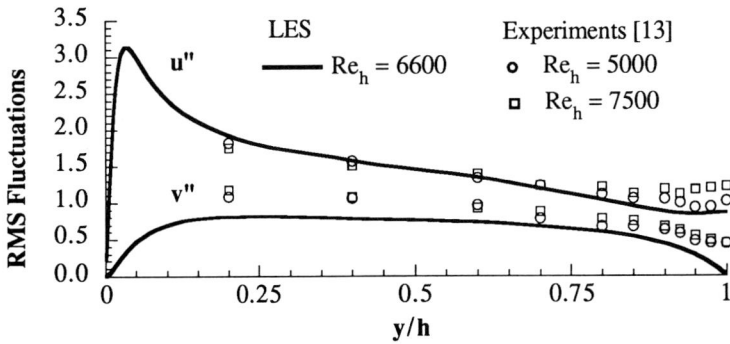

Figure 3. RMS Fluctuations for Channel with Free Surface

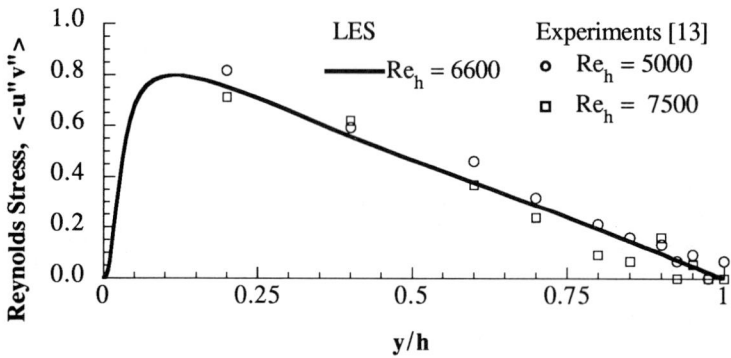

Figure 4. Reynolds Stress for Channel Flow with Free Surface

The rms velocity fluctuations show trends consistent with the experimental measurements. The Reynolds stress varies linearly outside of the viscous wall region as expected. There are some differences near the free surface which can be attributed to the approximation of the free surface boundary condition. However, the qualitative agreement is quite good.

6. CONCLUSIONS

The CM-2 has been found to be effective in reducing the computational task of performing large eddy simulations of turbulence. At the time this study was conducted a speed up of 2.1 could be obtained on a 32K processor CM-2 compared to a single processor CRAY-2. As improvements are made in massively parallel computing, the cost of performing turbulent simulations will decrease thus enabling engineers and scientists to perform more detailed studies of turbulent flows.

Acknowledgements

This study was sponsored by an Office of Naval Research Graduate Fellowship. Access to the Connection Machine-2 was provided by the National Center for Supercomputing Applications at the University of Illinois at Urbana-Champaign.

REFERENCES

1 P. R. Voke, and M. W. Collins, Large-Eddy Simulation: Retrospect and Prospect, *PhysicoChemical Hydrodynamics*, vol. 4, no. 2, pp. 119-161, 1983.
2 D.K. Tafti, and S. P. Vanka, Large Eddy Simulation of Channel Flow using Finite-Difference Techniques, Report No. CFD 90-01, Department of Mechanical and Industrial Engineering, University of Illinois at Urbana-Champaign, 1990.
3 J.H. Robichaux, D.K. Tafti, and S.P. Vanka, Large Eddy Simulation of Turbulence on the CM-2, *Numerical Heat Transfer Part B*, vol. 21, pp. 367-388, 1992.
4 J. Kim, and P. Moin, Application of a Fractional-Step Method to Incompressible Navier-Stokes Equations, *Journal of. Computational Physics*, vol. 59, pp. 308-323, 1985.
5 Thinking Machines Corporation, *Connection Machine, Model CM-2 Technical Summary*, Version 5.1, May 1989.

6. U. Schumann, Subgrid Scale Model for Finite Difference Simulations of Turbulent Flows in Plane Channels and Annuli, *J. Comp. Phys.*, vol. 18, pp. 376-404, 1975.
7. J. Smagorinksy., General Circulation Experiments with the Primitive Equations. I. The Basic Experiment, *Monthly Weather Review*, vol. 91, pp. 99-164, 1963.
8. R.W. Hockney, A Fast Direct Solution of Poisson's Equation using Fourier Analysis, *Journal of the Association for Computing Machinery*, vol.8, pp. 95-113, 1965.
9. D. C. Jesperson, and C. Levit, A Computational Fluid Dynamics Algorithm on a Massively Parallel Computer, AIAA Paper No. 89-1936-CP, 1989.
10. J.M. Ortega, *Introduction to Parallel and Vector Solution of Linear Systems*, Plenum Press, New York, 1988.
11. L.H. Jameson, P.T. Muller, and H.J. Siegl, FFT Algorithms for SIMD Parallel Processing Systems, *Journal of Parallel and Distribulted Computing*, vol. 3, pp. 48-71, 1981.
12. J. Kim, P. Moin, and R. Moser, Turbulence Statistics in Fully Developed Channel Flow at Low Reynolds Number, *J. Fluid Mech.*, vol. 177, pp. 133-166, 1987.
13. M. Rashidi, and S. Banerjee, Turbulent Structure in Free Surface Channel Flows, *Physics of Fluids*, vol. 31, no. 9, pp. 2491-2503, 1988.

Parallel Computational Fluid Dynamics on Unstructured Meshes using Algebraic Multigrid.

Guy Robinson.

Applied Science and Mathematical Modelling Department, AEA Technology, Dounreay, Caithness, KW14 7TZ.

Abstract

This paper reports on the implementation of a *fluid dynamics code* ASTEC by *geometric/topological decomposition* on a *distributed memory Multiple Instruction Multiple Data* architecture. The decomposition of the *unstructured mesh* onto the processor network is described. An *algebraic multigrid* solver which allows the accurate solution of large problems and offers improved scaling compared to the conventional techniques is described.

1. INTRODUCTION.

The *ASTEC* code performs calculations of fluid flow and scalar transport in and around complex geometries in three dimensions [1]. The code uses the k,e model for turbulent flows, and has additional models for porous media, compressible/incompressible flows, combustion, radiation transfer, and free surface flows. It employs a finite volume discretisation on an entirely unstructured finite element mesh. *ASTEC* has been applied to a wide range of flows, in complex geometries such as reactor systems, car engine blocks and blood flows in arteries. Figure 1 shows the accurate mesh representation of complex three dimensional geometries.

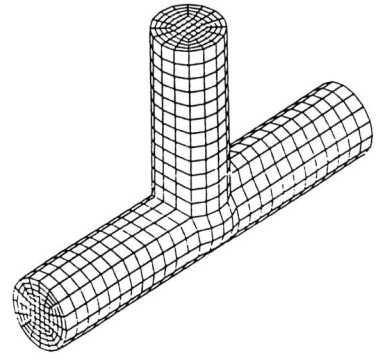 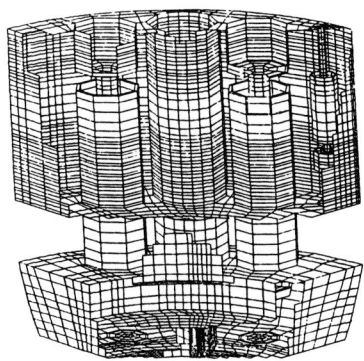

FIGURE 1. Example ASTEC Meshes Showing Unstructured Mesh and Geometrical Representation of Problems.

All these problems require large amounts of processing effort, especially transients, which can only be obtained from high performance computing devices. This paper describes how the combination of Multiple Instruction Multiple Data (MIMD) architectures with advanced multigrid solvers is enabling ever larger and more complex flow problems to be tackled economically in terms of cpu time, real time and cost.

2. METHODS.

The general aim of any parallel code is to ensure that all the processors are performing useful work at all stages of the computation. The techniques are determined by the hardware and programming models available. Here a distributed memory Multiple Instruction Multiple Data (MIMD) machine supporting the Communicating Sequential Process (CSP) paradigm is considered[2]. The system used in the development and application of the *ASTEC* code is a *Meiko Computing Surface* [3] based around the Inmos transputer (*t800*), which can either be used directly as a compute element, or as a communication server for a high performance processor such as the Intel *i860*. The *i860* processor offers ~10-20 times the performance of a *t800* processor.

2.1 Geometric/Topological Decomposition.

For mesh based computation the best technique is usually a geometric or topological decomposition [4], where each processor runs the same code, working on different physical regions of the problem. Each region is responsible for a set of nodes and elements known as the core data. In order for each region to update its core data correctly it requires data from neighbouring regions, the halo data. Messages containing data from the core are sent to those regions where it is required in the halo data.

Thus a general method is required which can distribute any *ASTEC* mesh over a range of processor numbers. When considering the decomposition of data it is important that each region is of roughly equal computational complexity at *all* stages of the calculation, to ensure efficient use of the resources, and that the data to be exchanged between regions must be within the communication bandwidth of the available hardware.

The technique adopted to divide the datasets is Weighted Recursive Layer Decomposition (WRLD). Here a seed layer is selected, and alternating layers for nodes and elements are grown outwards until all of the mesh is allocated. A typical two dimensional example of the layers is shown in figure 2. Notice how a single layer may not be continuous.

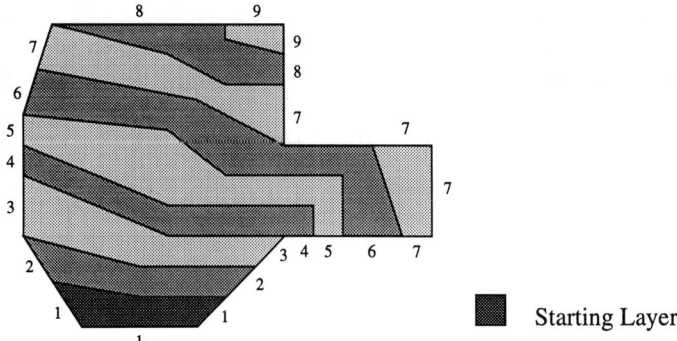

FIGURE 2. Simple Layering of 2 Dimensional Plain Figure.

The work load for each layer/region is estimated by counting up the number of data items which require processing within it, e.g. the number of nodes, elements etc. On the basis of such estimates a layer is chosen at which to divide the region into two sub-regions, the size of each being weighted so that upon further sub-division the required number of regions is achieved. The same layered growth scheme is used to divide each sub-region until the necessary number of regions is achieved. This process can divide any mesh into an arbitrary number of regions, each of roughly equal size and topological complexity with a small data exchange set.

3. PERFORMANCE.

On a single processor the *ASTEC* code scales as $n^{1.7}$, where n is the number of nodes in the problem. This scaling is due to the nonlinear nature of the coupled equations being solved and the nature of the linear solvers. The graph in figure 3 shows the performance for different sized problems being run over a range of processor types and numbers.

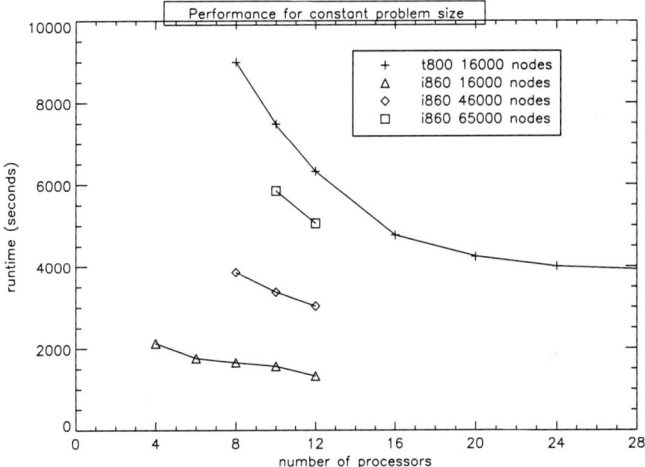

FIGURE 3. Scaling for Constant Size Problem.

Here the runtime for each problem decreases as the number of processors is increased. For the 16000 node problem on *t800* processors the runtime roughly halves as the number of processors is doubled. However for large numbers of processors this speedup is less than expected. On *i860* processors little or no speedup is achieved as the number of processors is increased, only on larger problems does the speedup return as shown in figure 4. Note that in figure 4 the number of nodes per processor is kept constant, i.e. the number of processors is increased linearly with the problem size. Since the *ASTEC* code scales as $n^{1.7}$ on a single processor, here the code should scale as $n^{0.7}$.

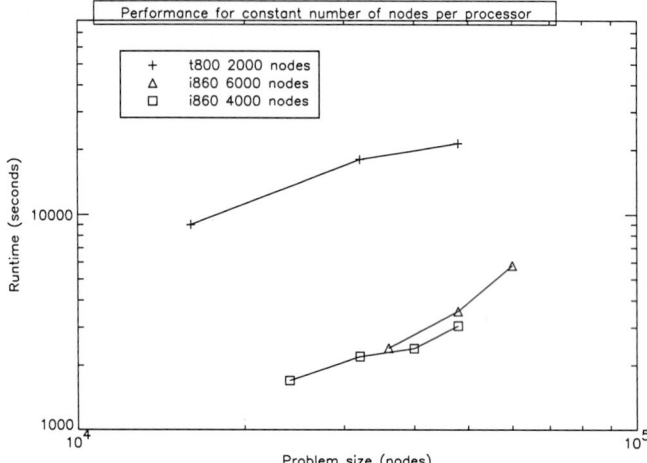

FIGURE 4. Scaling for Constant Number of Nodes per Processor.

3.1 Comments on Performance Characteristics.

Each of the lines in figures 3 and 4 show a slightly rough behavior due to inexact loadbalance. This causes the performance to degenerate for high processor numbers, where the amount of work per processor varies considerably and it is hard to achieve loadbalancing when there is only a small amount of mesh per processor.

Interlacing is a simple technique whereby data is passed to other processors whilst the processors are performing useful computation. In figure 3 the amount of work each processor has to perform between communication events decreases as the number of processors is increased. In the 16000 node problem on *t800* processors there is sufficient work to hide the majority of messages until the problem is distributed over more than twenty processors. Hence there is no significant gain in performance by employing more than twenty processors to this problem. Scaling is only possible when there is sufficient work on each processor to hide all communication events.

Also data is often required immediately so that the next stage of the computation can proceed, and the time taken to send and receive messages is important to the performance of the algorithm. This is known as the latency of the message and depends on the number of processors and the processor type involved. Latency is also important when there is a relatively small amount of work per processor, or a large number of messages to be exchanged. If the time taken to exchange the message is significantly greater than the computational effort required the system is communication bound.

The majority of communication events in the *ASTEC* code are exchanges of halo data between neighbouring regions. It is sensible for best use of the scalable communication architecture to place neighbouring regions on adjacent or nearest neighbouring processors, often resulting in a topology for the processor network which mirrors the physical topology of the mesh.

3.2 Halo Data, Computation or Communication?

Geometric/topological decomposition using halos to communicate information is a simple and convenient model, however it has several drawbacks. Obviously the halo data represents duplication of storage on each processor. Also in many areas of the code the halo data is considered explicitly, rather than indirectly as the boundary dataset, this is duplication of effort. If the halo represents a sizeable fraction of the region assigned to the processor this duplication will have a significant impact on the overall efficiency. If there is available communication bandwidth this duplication of effort can be avoided.

3.3 Summary of Performance Gains.

With geometric/topological decomposition of CFD problems onto MIMD architectures, impressive speedups have been achieved for large problems using an appropriate number of processors. A system of 12 *i860* processors with 16 megabytes of memory per processor can perform computations on meshes of upto 80 000 nodes at speeds ~3 times that of a *Cray 2* processor. Higher speedups could be achieved for larger problems by adding processors to the system.

4. MULTIGRID.

The fundamental problem with conventional solvers, such as Gauss Seidel, is that only local data is used in updating the values at each point, this point being drawn into a local equilibrium. However in many real physical systems, including fluid dynamics, long range affects must be considered and long range errors must be corrected in the iterative process.

4.1 Overview.

The general aim of the multigrid algorithm is to improve the communication of long range effects by using a hierarchy of grids to reduce the long range error on the finest grid. This is achieved by mapping both the equation and errors from fine to coarse grids, solving for the correction using a conventional solver and mapping this back to the fine grid.

The only input requirement for algebraic multigrid [5,6,7] is the equation set for the fine grid; equation sets for the coarse grid hierarchy are generated from this without recourse to geometric factors. Previous schemes had required regular meshes or complex interpolation schemes [8]. However unstructured meshes or problems where there is no regular geometric structure require a solver capable of generating the grids, or rather the equations that would be associated with the grids without explicit reference to geometry. Whilst the following procedure refers to nodes and grids for clarity, it must be emphasised that it is an entirely algebraic operation on the equations.

Algebraic multigrid allows the grid hierarchy to be more flexible and to accurately represent the problem at all stages of the computation, coarsening automatically occurring in a way which depends on the physics of the problem. Thus in the particular application being investigated here, CFD, the coarsening depends on the dynamics of the fluid flow, and is therefore also dynamic, changing to properly reflect the problem from iteration to iteration and from time cycle to time cycle in transient calculations.

4.1.1 Coarsening Algorithm and Data Mapping.

A simple method has been chosen. Coarse grid nodes are created by combining selected fine grid nodes into a single node. The equation for the coarse grid node is obtained simply by adding the terms in the contributing fine grid equations. The equation is then translated, any reference to a fine grid node being replaced by the coarse grid node to which it is mapped. It is necessary to combine nodes which are strongly connected, i.e. node a is strongly connected to node b if the coefficient of node b in the equation for a is large. It is convenient to represent the matrices graphically, as a network of interacting nodes, figure 5. Each node has a weight given by the diagonal matrix entry and coupling strength to neighbours given by the off diagonal entries. Strong coupling may be signified by close proximity, otherwise the disposition of the nodes is arbitrary. On the left the fine mesh is shown, with arrows indicating the preferred combinations that result in the mesh coarsening shown on the right.

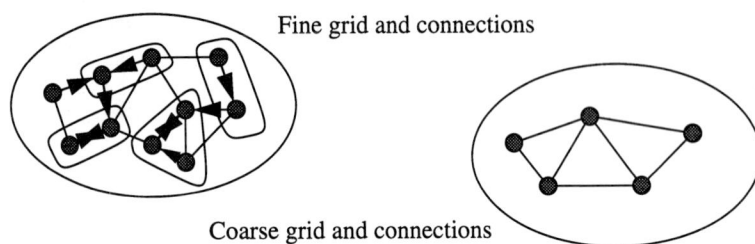

FIGURE 5. Simple Fine Mesh and Associated Coarse Grid.

The reduction in the number of nodes between successive grids is arbitrary, but here the algorithm approximately halves the number of nodes and stops once the coarsest grid consisting of a single node has been reached. This algorithm is simple, cheap and if the original fine mesh equation is conservative then the coarse grid equations are also conservative. This also provides a method for mapping data between the grids.

4.1.2 Smoother and Cycle Shape.

A Gauss Seidel smoother reduces the error on each grid and removes the short wavelength errors for that grid. The simplest cycle structure is known as a V cycle. Here the residuals or remaining errors are passed down from the fine mesh to the coarsest grid. Then the smoother is applied to the coarse grid and the corrections mapped back to the fine mesh applying the smoother on each grid. The cycle shape chosen depends on the wavelength of errors for which the problem has most work to do. Each grid in the cycle resolves errors of the wavelength associated with that grid. If errors of a particular wavelength dominate the problem then the solver cycle should concentrate on that wavelength. In nonlinear problems where the solution feeds back into the equations, errors at one wavelength may generate errors at another. The V cycle is suitable for solving problems with a range of wavelengths which do not interact. The W cycle can be used when long wavelength errors dominate or interact with short wavelength errors and the F cycle can be used when the correction of short wavelength errors can generate longer wavelength errors.

4.2 Parallel Algorithms.

Two algorithms have been developed; a complete scheme following the original algorithm, and a second restricted version which performs better in the parallel environment.

4.2.1 Grid Coarsening in the Complete Scheme.

Each region of the problem coarsens its set of core nodes, coarse grid nodes are composed of core and halo nodes from the fine grid depending on the connection strengths.

This requires the communication of matrix rows for the halo nodes and the determination of a new halo at each stage of the grid coarsening. Each processor reduces its region of the grid to roughly the same number of nodes, so that each level of the grid hierarchy is load balanced across the processors. The process results in a single identical node on each processor which represents the base correction to the entire problem. Note only a fully connected mesh will be reduced to a single node, if the mesh is divided into several distinct regions the process will stop when no further combination of nodes can occur. Figure 6 shows how the division of the problem into several regions affects the grid coarsening.

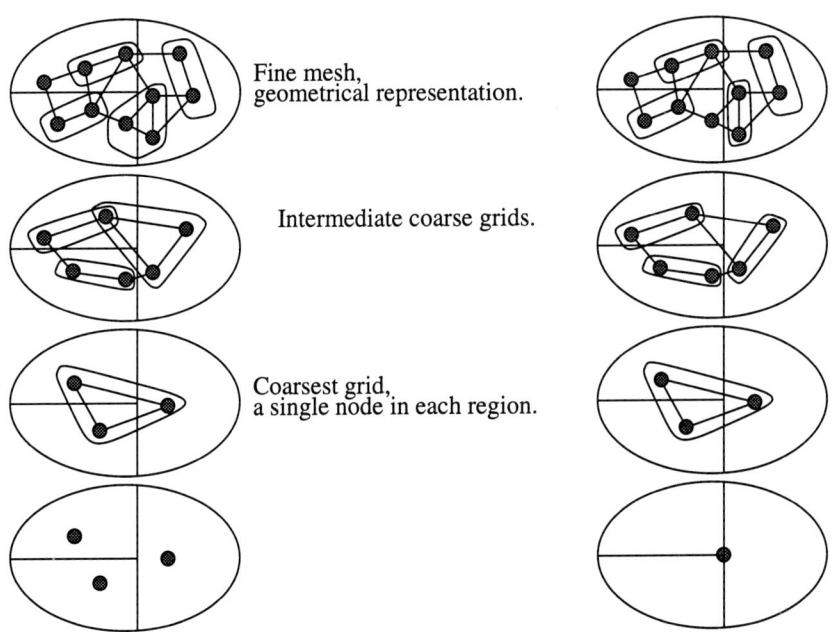

a) complete scheme
single identical node in each region.

b) restricted scheme
single node composite node

FIGURE 6. Grid Hierarchy Derived From a Simple Mesh.

4.2.2 Grid Coarsening in the Restricted Scheme.

Each region of the problem coarsens its set of core nodes; nodes are only allowed to combine with nodes in the same region. However when determining which node it would prefer it also considers the halo nodes, this avoids nodes with weak connections being combined because of the decomposition. This leads to a final grid with a single node on each processor, or overall a mesh consisting of p nodes, where p is the number of processors. If all these nodes are interconnected they may be combined into a single supernode.

4.3 Solution Scheme.

Note that two types of information transfer will be referred to; transfer between grids which will be referred to as data mapping, and transfer between processors which will be referred to as communication. Communication ensures that halo data is correct at a given level. The mapping of data between grids may use both core and halo data on the source grid but results in only the core data on the destination grid being correct. Hence for both algorithms it is necessary to communicate after a mapping to ensure that the halo data on the destination grid is correct.

The necessary communications in the complete scheme are summarised as follows;

- Before any calculation, of new residuals or of new solution, a communication of the current solutions must have occurred since the last update to ensure the correct halo data is used in calculating the new data.
- Before any fine to coarse mapping, communication is required because halo nodes of the fine grid contribute to the core nodes on the coarse grid.
- Before a coarse to fine mapping no communication is necessary because only core data from the source grid is required.

Many of these demands overlap, figure 7 shows the communications necessary in an F cycle.

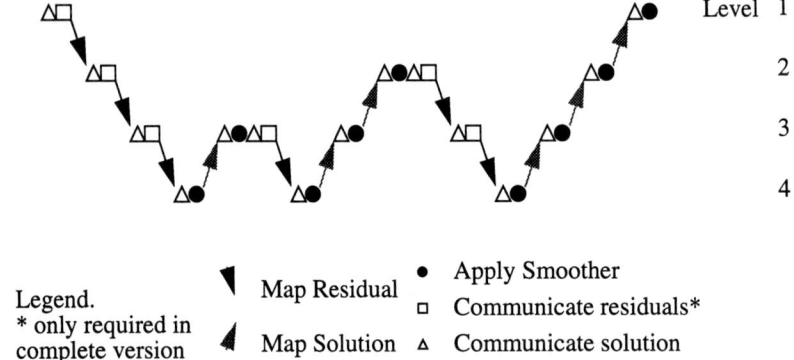

FIGURE 7. Processes and Communication Inside F Cycle.

The reduced scheme requires the same communication events as the complete scheme, except no communication is required before any fine to coarse mapping because only core nodes of the fine grid contribute to the core nodes on the coarse grid.

5. PERFORMANCE WITH LINEAR SCALAR EQUATIONS.

The graph below, figure 8, shows the absolute sum of residuals against solver sweeps for a simple test problem. The top curve shows the residual for the Gauss Seidel (GS) solver applied to the fine grid only. Clearly this solver performs very poorly.

The dotted lines show the performance of multigrid solver without the coarsest grid. As the number of processors applied to the problem increases the reduction in residual with each solver cycle decreases due to long wavelength errors not being corrected below the coarsest grid. This is improved when the supernode approach is adopted, shown by the dashed lines. The performance degrades as the number of nodes/regions/processors combined to form this supernode increases due to intermediate wavelengths being obscured during the final coarsening process.

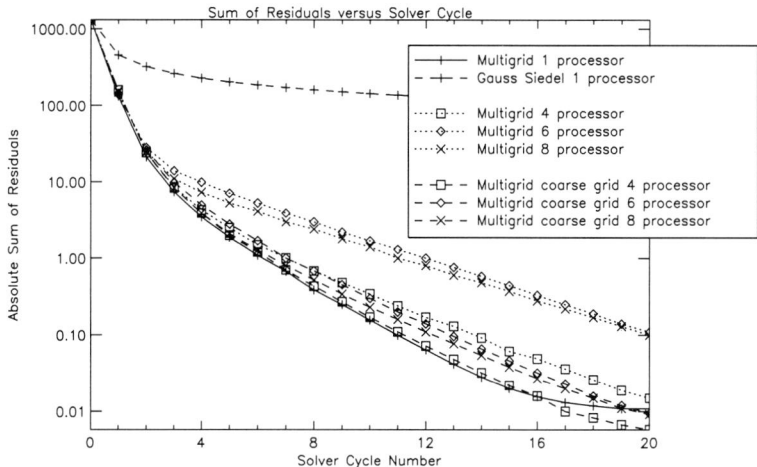

FIGURE 8. Graph Showing sum of Residuals Versus Solver Cycle.

6. PERFORMANCE WITH COUPLED EQUATIONS.

In the SIMPLE scheme [9] much effort is spent dealing with the velocity-pressure coupling, i.e. enforcing continuity via derived pressure and velocity corrections. Often the accuracy of the pressure correction has been the crucial factor in the performance of the SIMPLE scheme. Figure 9 shows how the Multigrid solver offers both superior numerical performance and stability compared to the Preconditioned Conjugate Gradient solver.

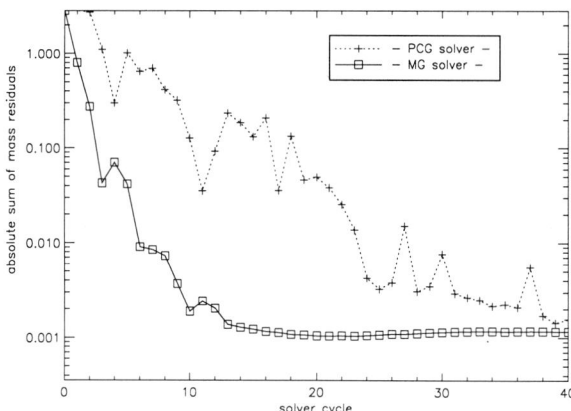

FIGURE 9. Performance of Preconditioned Conjugate Gradient compared with Multigrid.

7. DISCUSSION.

In multigrid algorithms the varied grid sizes and variety of data to be communicated cover the entire spectrum of parallel problems; from coarse grained with little communication to fine grained where communication dominates. As the grids become smaller the communication to calculation ratio increases leading to a communication bound process. Indeed the coarsest grid of the approximate/reduced method is a fine grained problem. This is somewhat balanced by spending relatively less time in this section of code overall. The relative proportion of the time spent in this region depends on the cycle shape.

Figure 10 shows relative fraction of communication and processing at each level of multigrid solver cycle for the restricted scheme. The top bar is the computation with the communication underneath. The computation is divided into two parts, one of which can be interlaced with the communication. The cost of computation is roughly the number of nodes in each mesh with some dependence on the complexity of the matrix. The communication costs remain approximately constant on each grid since the size of the message is small.

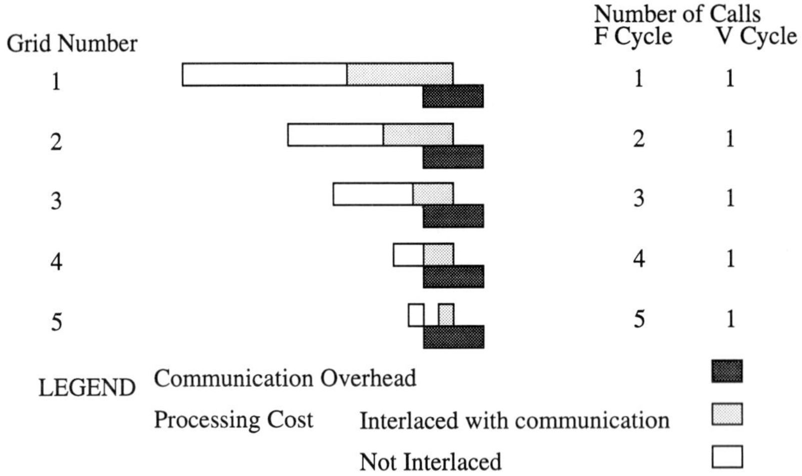

FIGURE 10. Relative Proportion of Computation to Calculation on Different Grids in Hierarchy.

More complex cycles, such as the F cycle, spend more time around the coarse grid compared to the V cycle, increasing the relative proportion that is communication bound. It may be possible to tailor the cycle shape to optimise the overall performance. This will depend on the exact nature of the matrix generated by the problem and on which wavelength errors are important. For the complete scheme the amount of communication at each level is increased greatly. Also there is considerable additional communication when determining the gird hierarchy and communication tables.

The proposed restricted algorithm requires only nearest neighbour communication and this pattern of communication is fixed at all levels. This will scale well in a message passing architecture with both problem size and processor number. This is not the case for the complete scheme, where the inclusion of halo terms in each region when generating the coarse grids leads to a complex communication pattern. This pattern requires global interconnection, every region must exchange data with every other region in the problem. This rapidly absorbs the available bandwidth even in scalable architectures.

Smoothing on a particular grid is most effective in reducing errors of a particular wavelength characteristic of that grid. If any grid is ignored, errors of that wavelength are not efficiently reduced. Both the serial and the complete parallel schemes reduce the problems to a single node, i.e. they cover the entire range of wavelength possible from the fine grid wavelength to infinity. However the restricted algorithm stops at a coarse grid with the number of nodes approximately equal to the number of processors. Smoothing on this grid will not be efficient at removing errors of a greater wavelength. If longer wavelength errors are to be removed effectively, it is necessary to continue the grid hierarchy further. This is achieved by creating a single node which represents the coarse grid in the serial and complete parallel cases. However this sudden jump between grids will not consider errors of intermediate wavelength and additional smoothing sweeps on each grid or V cycles may be necessary overall. A direct solution of the matrix for the coarsest grid by Gaussian Elimination or some other method removes all longer wavelengths.

8. SUMMARY.

The multigrid solver described here will perform best when there are many nodes and many grids per processor. In the reduced scheme relatively few grids are missing from the hierarchy and the correction that these would make is approximated by the supernode. If many grids are present only the coarsest will be communications bound, the runtimes of the solver cycle being dominated by the finest grids.

Conversely, if comparatively few nodes are allocated to each processor there will be few grids in the hierarchy and many grids will be missed out from the hierarchy. The creation of a single node will be unable to correct for the wide wavelength range of errors being ignored, leading to poor numerical performance. Also the few grids will each contain a small number of nodes and most of the solver cycle will be communications dominated.

With the current computation to communication ratios of processing elements and the available topologies it is difficult to implement the complete scheme efficiently. However future technology may permit this scheme to be developed.

9. CONCLUSION.

This paper has shown that MIMD architectures offer significant gains in performance for large and complex CFD calculations on unstructured meshes.

It has also shown that algebraic multigrid solution schemes, that give close to optimum scaling of solution time with problem size, can also run on MIMD architectures and give similar performance gains over single processor calculations.

Multigrid solvers on MIMD architectures offer a practical route to achieving the computing performance required for CFD calculation on the very large unstructured meshes necessary for resolving complex flow fields in complex engineering geometries.

10. ACKNOWLEDGEMENT.

The author wishes to express his thanks to Dr. B. R. MacGregor and Dr. R. Webster of AEA Technology for their help and advice in this work and the preparation of this paper.

An extended version of this paper is available from the author.

11. REFERENCES.

[1]   LONSDALE, R.D. and WEBSTER, R. Application of finite volume methods for modelling three dimensional flows on an unstructured mesh. 6th International conference on numerical methods in laminar and turbulent flows, Swansea, (1989).
[2] HOARE, C.A.R. Communicating sequential processes. Prentice Hall, (1985).
[3] Meiko Ltd., Literature on Meiko Computing Surface.
[4]   ROBINSON, G. and LONSDALE R.D. Fluid dynamics in parallel using an unstructured mesh. First international conference on parallel processing for computational mechanics. Southampton, (1990).
[5]   BRANDT, A. Multilevel adaptive computations, AIAA paper 79-1455 (1979).
[6]   RUGE, J, and STRUBEN, K. Algebraic multigrid, Arbeitspapiere 2010, (1986).
[7]   LONSDALE, R.D. An algebraic multigrid scheme for solving the Navier-Stokes equations on unstructured meshes. 7th international conference on numerical methods in laminar and turbulent flow, University of Stanford, California USA, (1991).
[8]   OSAMA, E.G. and HOPKINS, T. Generally configurable multigrid implementation for transputer networks, Occam User Group 11, Edinburgh, Scotland, (1989).
[9]   PATANKAR, S.V. Numerical heat transfer and fluid flow. Hemisphere, (1980).

Parallel Computational Fluid Dynamics '92
R.B. Pelz, A. Ecer and J. Häuser (Editors)
© 1993 Elsevier Science Publishers B.V. All rights reserved.

PARALLEL SIMULATION OF INCOMPRESSIBLE VISCOUS FLOWS BY HIGHEST ORDER FINITE DIFFERENCE SCHEME

C. Shu[a] and B. E. Richards[b]

[a]Department of Mechanical and Production Engineering, National University of Singapore, 10 Kent Ridge Crescent, Singapore 0511

[b]Department of Aerospace Engineering, University of Glasgow, Glasgow G12 8QQ, U.K.

ABSTRACT

This paper introduces a global method of the highest order finite difference (HFD) scheme for the discretization of any order spatial derivative. The weighting coefficients in this scheme can be determined by a simple algebraic formulation or by a recurrence relationship. For the parallel computation, the multi-domain HFD scheme was developed, and successfully applied to solve incompressible Navier-Stokes (N-S) equations. For the test problem of a driven cavity flow, three methods for dealing with the interface between subdomains (i.e. patched with enforcing continuity to the function and its normal derivative; patched with using a Lagrange interpolation scheme; and overlapped) were studied comparatively. In addition, an idea to develop a general code which can be run on any array of processors without modification to the program was presented.

1. INTRODUCTION

The numerical simulation of incompressible viscous flows can usually be done by low order finite difference and finite element methods. These low order methods often require the use of a large number of grid points to obtain the accurate numerical solutions. As a result, a large computational effort and virtual storage are needed. To save the computational effort and the virtual storage, the global method of the highest order finite difference (HFD) scheme may provide a promising way. The key problem in the HFD scheme is how to determine the weighting coefficients for the discretization of any order derivative.

In seeking an efficient method for the solution of a partial differential equation, Bellman et al [1] recently presented an attractive method of differential quadrature (DQ). DQ is also a type of global method, which discretizes the derivative using the same form as the HFD scheme. The key to DQ is the determination of weighting coefficients for any order derivative discretization. Bellman et al suggested solving an algebraic equation system to determine the weighting coefficients for the first order derivative. But unfortunately, when the order of the system is large, the matrix of the system is ill-conditioned. Thus it is very difficult to obtain the weighting coefficients using this approach. To overcome the drawbacks of DQ, the current authors [2] have developed the technique of generalized differential quadrature (GDQ), where the weighting coefficients of the first order derivative are determined by a simple algebraic formulation, and the weighting coefficients of the second and higher order derivatives are determined by a recurrence relationship. It has been proved that the algebraic equation system for the weighting coefficients derived from the HFD scheme is equivalent to that derived from GDQ. Thus the weighting coefficients for discretizing any order derivative in the HFD scheme can be given from the results of GDQ.

For the parallel simulation on a distributed parallel computer, a multi-domain HFD scheme was developed, and applied for the driven cavity flow simulation. This paper will demonstrate the application of a multi-domain HFD scheme for the comparative studies of the interface treatment on the transputer-based distributed Meiko Computing Surface.

2. NUMERICAL METHOD

2.1 Highest Order Finite Difference (HFD) Scheme

The one-dimensional case is considered for demonstration. It is supposed that there are N grid points in the whole domain. The discretization of the first order derivative of function f(x,t) with respect to x at x_i can be given by the HFD scheme as a linear sum of all the functional values at N grid points, i.e.

$$f_x(x_i,t) = \sum_{j=1}^{N} a_{ij} \cdot f(x_j,t), \quad \text{for } i = 1, 2, \cdots, N, \tag{1}$$

where $f_x(x_i,t)$ indicates the first order derivative of f(x,t) with respect to x at x_i, a_{ij} the weighting coefficients. In the same fashion as used in the design of the low order finite difference schemes, a_{ij} can be determined by the following Taylor series expansion

$$f(x_j,t) = f(x_i,t) + f^{(1)}(x_i,t) \cdot (x_j - x_i) + \cdots + f^{(k)}(x_i,t) \cdot (x_j - x_i)^k / k! + \cdots +$$
$$f^{(N-1)}(x_i,t) \cdot (x_j - x_i)^{N-1} / (N-1)! + R_N \tag{2}$$

where $f^{(k)}(x_i,t)$ is the kth order derivative of f(x,t) with respect to x at x_i, R_N is the truncated error, which can be written as

$$R_N = f^{(N)}(\xi_j,t) \cdot (x_j - x_i)^N / N! \ , \quad \xi_j \in [x_i, x_j] \tag{3}$$

Substituting eq. (2) into eq. (1) and keeping the (N-1)th order accuracy leads to the following algebraic equations

$$\begin{cases} \sum_{j=1}^{N} a_{ij} = 0 \\ \sum_{j=1}^{N} a_{ij} \cdot (x_j - x_i) = 1 \\ \sum_{j=1}^{N} a_{ij} \cdot (x_j - x_i)^k = 0, \ k = 2, 3, \cdots, N-1 \end{cases} \quad (4)$$

Equation set (4) is an equation system for the determination of the weighting coefficients of the first order derivative in the HFD scheme. Similarly in the domain $[x_1, x_N]$, the mth order derivative of function f(x,t) with respect to x at x_i can be discretized by the HFD scheme as

$$f_x^{(m)}(x_i, t) = \sum_{j=1}^{N} w_{ij}^{(m)} \cdot f(x_j, t), \quad \text{for } i = 1, 2, \cdots, N; \ m = 2, 3, \cdots, N-1, \quad (5)$$

where $f_x^{(m)}(x_i, t)$ indicates the mth order derivative of f(x,t) with respect to x at x_i, $w_{ij}^{(m)}$ the weighting coefficients. Substituting eq. (2) into eq. (5), and keeping the $(N-m)$th order accuracy, we obtain

$$\begin{cases} \sum_{j=1}^{N} w_{ij}^{(m)} = 0 \\ \sum_{j=1}^{N} w_{ij}^{(m)} \cdot (x_j - x_i)^m = m! \\ \sum_{j=1}^{N} w_{ij}^{(m)} \cdot (x_j - x_i)^k = 0, k = 1, 2, \cdots, N-1, \ k \ne m \end{cases} \quad (6)$$

Equation set (6) is an algebraic equation system for determining the weighting coefficients of the second and higher order derivatives in the HFD scheme. In the following, we will show that all these weighting coefficients can be determined by a simple algebraic formulation or by a recurrence relationship.

2.2 Weighting Coefficients of the First Order Derivative

For the efficient solution of a smooth problem, Bellman et al [1] introduced a technique of differential quadrature (DQ), which uses the same form as eq. (1) to discretize the first order derivative. They determine the weighting coefficients a_{ij} by letting eq. (1) be exact for test functions $g(x) = x^k$, $k = 0, 1, \cdots, N-1$, which leads to a set of algebraic equations as follows

$$\sum_{j=1}^{N} a_{ij} \cdot x_j^k = k \cdot x_i^{k-1}, \quad \text{for } i = 1, 2, \cdots, N; \ k = 0, 1, \cdots, N-1. \quad (7)$$

This equation system has a unique solution because its matrix is of Vandermonde form. We will prove that the algebraic equation system (7) given from DQ is equivalent to the equation system (4) given from the HFD scheme. It is obvious that the first two equations of equation sets (7) and (4) are the same. Now, assuming that the first p+1 equations of the two systems are the same, that is

$$\sum_{j=1}^{N} a_{ij} \cdot (x_j - x_i)^k = \sum_{j=1}^{N} a_{ij} \cdot x_j^k - k \cdot x_i^{k-1} = 0, \quad \text{for } k = 2, 3, \cdots, p; \ i = 1, 2, \cdots, N, \quad (8)$$

then using the binary formulation

$$(a-b)^p = a^p - c_p^1 \cdot a^{p-1} b + \cdots + (-1)^k c_p^k a^{p-k} b^k + \cdots + (-1)^p \cdot b^p, \quad (9)$$

the $(p+2)$th equation of equation set (4) can be written as

$$\sum_{j=1}^{N} a_{ij} \cdot (x_j - x_i)^{p+1} = \sum_{j=1}^{N} a_{ij} \cdot x_j^{p+1} - c_{p+1}^1 \cdot x_i \cdot \left[\sum_{j=1}^{N} a_{ij} \cdot (x_j^p - \frac{1}{2} \cdot c_p^1 \cdot x_j^{p-1} \cdot x_i + \cdots + \frac{(-1)^p \cdot x_i^p}{p+1}) \right]$$

$$= \sum_{j=1}^{N} a_{ij} \cdot x_j^{p+1} - (p+1) \cdot x_i^p \quad (10)$$

Equation (10) demonstrates that the $(p+2)$th equation of the two systems are exactly the same. Since p is an arbitrary integer only if $p \leq N-2$, it has been proved that the two systems (4) and (7) are the same. But although the weighting coefficients a_{ij} can be determined by the equation system (4) or by the equation system (7), the solution of either (4) or (7) is not easy to be obtained for a large N. We will use the results of GDQ to calculate them.

In accordance with the Weierstrass polynomial approximation theorem, a continuous function in a closed domain can be approximated by an infinite polynomial accurately. In practice, a truncated finite polynomial may be used. Following this approach, it is supposed that any smooth function in the domain can be approximated by a $(N-1)$th order polynomial. And it is easy to show that the polynomial of degree less than or equal to N-1 constitutes an N-dimensional linear vector space V_N. From the concept of a linear vector space, there exist N base polynomials. Here if $r_k(x)$, $k = 1, 2, \cdots, N$, are the base polynomials, any polynomial in V_N can be expressed uniquely as a linear combination of $r_k(x)$, $k = 1, 2, \cdots, N$. And if all the base polynomials satisfy a linear constrained relationship such as eq. (1), so does any polynomial in the space. In the linear vector space, there may exist several sets of base polynomials. It is found that, if the base polynomial $r_k(x)$ is chosen to be x^{k-1}, the same results given by Bellman et al can be achieved. For generality, GDQ chooses the base polynomial $r_k(x)$ to be the Lagrange interpolated polynomial

$$r_k(x) = \frac{M(x)}{(x - x_k) \cdot M^{(1)}(x_k)} \quad (11)$$

where $M(x) = (x - x_1) \cdot (x - x_2) \cdots (x - x_N)$, $\quad M^{(1)}(x_k) = \prod_{j=1, j \neq k}^{N} (x_k - x_j)$

For simplicity, we set

$$M(x) = N(x, x_k) \cdot (x - x_k), \ k = 1, 2, \cdots, N \quad (12)$$

with $N(x_i, x_j) = M^{(1)}(x_i) \cdot \delta_{ij}$, where δ_{ij} is the Kronecker operator. Thus we have

$$M^{(m)}(x) = N^{(m)}(x, x_k) \cdot (x - x_k) + m \cdot N^{(m-1)}(x, x_k), \ \text{for } m = 1, 2, \cdots, N-1; \ k = 1, 2, \cdots, N \quad (13)$$

where $M^{(m)}(x)$, $N^{(m)}(x,x_k)$ indicate the mth order derivative of $M(x)$ and $N(x,x_k)$. Substituting eq. (11) into eq. (1) and using eq. (13), we obtain

$$a_{ij} = \frac{M^{(1)}(x_i)}{(x_i - x_j) \cdot M^{(1)}(x_j)}, \text{ for } j \neq i, \qquad a_{ii} = \frac{M^{(2)}(x_i)}{2M^{(1)}(x_i)} \qquad (14)$$

Equation (14) is a simple formulation for computing a_{ij}. If x_i is given, it is easy to compute $M^{(1)}(x_i)$, thus a_{ij} for $i \neq j$. The calculation of a_{ii} is based on the computation of the second order derivative $M^{(2)}(x_i)$ which is not easy to obtain. However, from the equation system (4), a_{ii} can be obtained easily from the following formulation

$$\sum_{j=1}^{N} a_{ij} = 0 \qquad (15)$$

2.3 Weighting Coefficients of the Second and Higher Order Derivatives

For the second and higher order derivatives, eq. (5) can be applied. To deduce a recurrence relationship for the weighting coefficients, the following linear constrained relationship is also applied

$$f_x^{(m-1)}(x_i, t) = \sum_{j=1}^{N} w_{ij}^{(m-1)} \cdot f(x_j, t), \text{ for } i = 1, 2, \cdots, N; \; m = 2, 3, \cdots, N-1. \qquad (16)$$

Substituting eq. (11) into eqs. (5), (16), and using eqs. (13), (14), a recurrence formulation is obtained as follows

$$w_{ij}^{(m)} = m \cdot \left(a_{ij} \cdot w_{ii}^{(m-1)} - \frac{w_{ij}^{(m-1)}}{x_i - x_j} \right), \; j \neq i \qquad (17)$$

Again, from the equation system (6), $w_{ii}^{(m)}$ can be obtained by

$$\sum_{j=1}^{N} w_{ij}^{(m)} = 0 \qquad (18)$$

2.4 Multi-Domain HFD Parallelism

Multi-domain HFD parallelism is an extension of the concept of geometric decomposition. It is supposed that the physical domain of a problem is represented by Ω, and the boundary by Γ. The multi-domain HFD scheme decomposes the domain Ω into several subdomains Ω_i, $i = 1, 2, \cdots, K$, where K is the number of subdomains. In each subdomain, a local mesh is generated with stretching near the boundary and the local HFD scheme is applied, in the same fashion as the application of the scheme in a single domain. The solutions for interior grid points are independent for each subdomain. Globally, the information exchange between subdomains is required. This can be done across the interface of subdomains.

In the multi-domain HFD parallelism, each subdomain is allocated to a slave processor. With the boundary conditions at the solid boundary and the interface, the evolution of the fluid in the interior of each subdomain is determined entirely by data which is present in the processor's local memory. Global communication between processors is required for exchanging the data to give the new functional values at the interface as the iteration progresses. The formulation to determine those values at the interface is related to the topology of the interface. 3 basic cases of interface treatment will be studied in this paper.

2.4.1 Interface Type I : Patched with Enforcing Continuity Condition

As shown in Fig. 1, Γ_{ij} is the interface between the subdomain Ω_i and Ω_j. The patching condition is enforced at the interface Γ_{ij} so that both the function and its first order derivative normal to Γ_{ij} are continuous along the normal direction of the interface, i.e.

$$f(x_N^i) = f(x_1^j) \; ; \quad f_n(x_N^i) = f_n(x_1^j) \quad on \; \Gamma_{ij} \qquad (19)$$

where $f(x_N^i)$, $f(x_1^j)$ represent the values of the function f at the interface of the i subdomain and the j subdomain, and $f_n(x_N^i)$, $f_n(x_1^j)$ the values of the first order derivative of f with respect to n at the interface. For the cases selected for study, each subdomain is rectangular. Then the normal direction to the interface is parallel to one coordinate axis in the local coordinate system. For simplicity, this coordinate axis can be assumed as the x axis, and along this direction, there are N grid points in the i subdomain and M grid points in the j subdomain. The weighting coefficients of the first order derivative along the x direction are written as a_{mn}^i in the i subdomain and a_{mn}^j in the j subdomain. After discretizing the derivative by HFD, (19) can be combined to give

$$\bar{f} = \frac{\sum_{k=1}^{N-1} a_{Nk}^i \cdot f(x_k^i) - \sum_{k=2}^{M} a_{1k}^j \cdot f(x_k^j)}{a_{11}^j - a_{NN}^i} \qquad (20)$$

where $\bar{f}$ is the value of the function f at the interface Γ_{ij}, and $f(x_k^i)$, $f(x_k^j)$ represent the values of the function f at x_k^i in the i subdomain and x_k^j in the j subdomain.

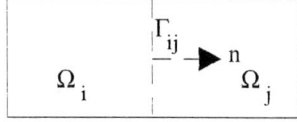

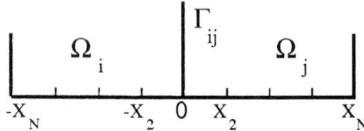

Fig. 1 Topology of a Patched Interface   Fig. 2 Global Coordinates for Ω_i and Ω_j

2.4.2 Interface Type II : Patched with Using Lagrange Interpolation

The topology of the interface type II is the same as the above case. But the functional values at the interface are obtained using a high order Lagrange interpolated polynomial. For simplicity, it is assumed that subdomains Ω_i and Ω_j have the same structure of grid, and globally, the coordinates in both subdomains are symmetric to the interface Γ_{ij}. Thus, we can establish a frame whose origin is on the interface, as shown in Fig. 2. Using the Lagrange interpolated polynomial, the functional value at the interface can be obtained by the extrapolation from both sides of the interface, which is given as

$$\bar{f} = \sum_{k=2}^{L} L_k^r \cdot f(x_k^j) + \sum_{k=2}^{L} L_k^l \cdot f(x_{N-k+1}^i) \tag{21}$$

where (L-1) is the number of functional values in a subdomain being used for interpolation,

$$L_i^r = L_i^l = \frac{1}{2} \cdot \prod_{j=2, j \neq i}^{L} \frac{x_j^2}{x_j^2 - x_i^2} , \quad \text{with } x_k = x_k^j \tag{22}$$

2.4.3 Interface Type III : Overlapped

The overlapped topology of the interface, as shown in Fig. 3, is an easy way to implement parallel computation using the domain decomposition approach. In Fig. 3, subdomain ABCD is overlapped with subdomain EFGH (shaded area). It is noted that the right boundary of subdomain Ω_i, BD, is in the interior of subdomain Ω_j, and the left boundary of subdomain Ω_j, EG, is in the interior of subdomain Ω_i. Thus, if the solution in the interior of the subdomains is known at a time step, then the functional values along the lines of BD and EG are known. These values are then transferred into neighboring subdomains as new boundary conditions to get the solution in the interior of subdomains at the next time step. This process continues until converged solutions in all subdomains are obtained. For the overlapped topology, the functional values at the interface are given from the solution of the governing equations, but the functional values within the overlapped region may not be unique. They can be determined from the solution of a subdomain, and can also be given from the solution of another neighboring subdomain.

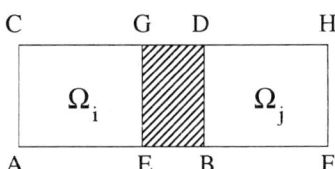
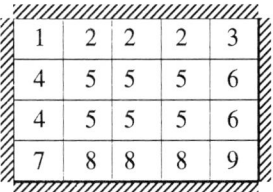

Fig. 3 Topology of an Overlapped Interface   **Fig. 4 Nine Basic Cases of a 2D Problem**

3. PARALLEL SIMULATION

3.1 Parallel Architecture

A transputer-based Meiko Computing Surface, which is a MIMD concurrent processor, is used to tackle the parallel simulation. The Meiko Computing Surface is a modular, reconfigurable transputer array, which provides system software to control the computing surface in single or multiple mode, using MEIKOS, a UNIX-like system. In the system, there are compilers for Occam, C, Fortran and Pascal. For the conventional languages, communication can be made by an Occam harness or by using the CS Tools. Using CS Tools is easier for a new user, but may reduce the efficiency since the configuration may not be optimal to some physical problems. In the following application, the program in each transputer is written in Fortran, and run from an Occam harness, which control the placement of, and the communication between transputers.

3.2 Development of A General Computer Code

A parallel program, which can be run on any array of processors without any modification to the program, is very attractive in engineering. To develop this, we firstly consider the special features of parallel computation. These new considerations include: (1) how the data is to be distributed in the memory; (2) how computations are distributed among the processors; (3) inter-processor communications; and (4) inter-processor connections. If a multi-domain technique and a distributed parallel computer are used, the first two items are easy to implement since each processor is assigned to be responsible for the computation of a subdomain and the data used in the computation is stored in the local memory of that processor. The problem lies in determining how the master processor divides the whole computational field into required subdomains, each with its local properties which include the way that the local slave processors ascertain the location of the physical boundary and the interface in a subdomain. It is very important to know the position of the interface in the solution of a differential equation. This is because, on the one hand, the interior solutions are greatly affected by the boundary conditions, not only in their values and types, but also in their positions. On the other hand, since the data communication is required across the interface, the inter-processor communications and connections are related to the position of interfaces. In other words, depending on the position of the interfaces in a subdomain, the local slave processor needs to know where the communications and connections are required to the neighboring subdomains. In terms of the topology of subdomains, there are several basic cases for **2D** and **3D** problems. Thus we can generate a code which can automatically distribute the data and computation to each slave processor, and produce a configuration for the communications if all the basic cases are included. To demonstrate this, we consider a rectangular computational field for simplicity. For this case, if it is assumed that there are at least 2 subdomains in both x and y directions, then all topologies of subdomains can be described within a set of 9 basic cases, as shown in Fig. 4. If the code involves all these 9 basic cases, it can then be run on any array of processors (again only if there are at least 2 subdomains in both x and y directions).

4 COMPARATIVE STUDIES OF THE DRIVEN CAVITY FLOWS

For the application of the multi-domain HFD scheme to simulate incompressible viscous flows in parallel, the vorticity-velocity formulation was taken as the governing equations. In each subdomain, after all the spatial derivatives are discretized by the local HFD scheme, the resultant set of ordinary differential equations for vorticity are then solved by the 4th order Runge-Kutta scheme, and the set of algebraic equations for velocity are solved by LU decomposition.

For the test case, we first study the influence of the order of interpolated polynomials and the number of grid points being overlapped, on the accuracy of results and the operational time. The comparison of the accuracy and the operational time needed between the three types of interfaces is also studied. In all the following cases, the program is produced using the general scheme proposed in section 3.2, which can be run on any array of transputers without modification to the program. All the codes have been successfully run on the array of 2X2, 3X3, 4X4, 5X5, 6X6 slave processors representing different kinds of domain decomposition. Some selected results are shown as follows. The numerical results of Ghia et al [3] will also be included for comparison.

4.1 Different Order of Lagrange Interpolated Polynomials

Using interface type II, it was found that, as L increases from 2 to 3, the accuracy of results is improved, and more operational time is required; and as L increases above 3, the accuracy of results keeps nearly the same, and a little more operational time is needed. Thus in balance, to reduce the operational time and obtain accurate results, the use of L=3 is recommended. Fig. 5 displays the velocities through the geometric centre of the cavity, where the array of 2X2 slave transputers, and a local mesh size of 13X13 for Re=100, 17X17 for Re=400 in each subdomain, were used. Fig. 8 shows the non-dimensional operational time, where the reference time T_{ref} is the operational time taken by the case of L=2.

4.2 Different Number of Grid Points Overlapped

Using interface type III, it was found that, as NO increases from 2 to 3, where NO is the number of grid points overlapped, the numerical results are nearly the same, but the operational time is greatly reduced; and when NO increases above 3, both the accuracy of results, especially in the overlapped region, and the operational time, are reduced. The reason for reduction of the accuracy in the overlapped region is that solutions in this region may not be unique since they can be obtained in a subdomain and in other neighbouring subdomains. Although the physical positions in the overlapped region are the same, solutions derived from different subdomains may not be consistent with each other. This is particularly true when NO becomes relatively large. Hence, to obtain accurate results with less operational time, the use of NO=3 is recommended. Fig. 6 shows the velocities through the geometric centre of the cavity, where the array of 2X2 slave transputers and the local mesh size of 13X13 for Re=100, 17X17 for Re=400 in each subdomain, were used. Fig. 9 shows the non-dimensional operational time, where the reference time T_{ref} is the operational time taken by the case of NO=2.

4.3 Comparison of the Interface Treatment

The performance of the three types of interface treatment introduced in section 2.4 and optimized as described in section 4.1 and 4.2, has been studied. Numerical experiment showed that the interface III gives the most accurate results and needs the least operational time, and the use of interface II presents more accurate results and requires less operational time than that of interface I. For the reason of this behavior, it is analyzed that the solutions in the interior of each subdomain can be affected by the boundary conditions and by the solutions at the interface, thus any error introduced at the interface may spread into the interior solutions. Hence, although the interior solutions are obtained by HFD with high order accuracy, the low order solutions at the interface may produce low order solutions in the whole computational field. Since the use of interfaces II and III gives solutions at the interface with high order accuracy, and the use of interface I gives solutions at the interface with accuracy of order one, the interfaces II and III provide more accurate results than interface I. On the other hand, although solutions at the interface are obtained by high order polynomials for interface II, they may not satisfy the governing equations. This is not the case for interface III. As a result, the use of interface III gives more accurate results than interface II. From the numerical experiment, it was found that the higher order accuracy of solutions at the interface may require fewer time steps, and therefore less operational time, to steady state resolution. Fig. 7 gives the velocities through the geometric centre of the cavity, where the array of 2X2 slave transputers and the local mesh size of 13X13 for Re=100, 17X17 for Re=400 in each subdomain, were used. Fig. 10 shows the non-dimensional operational time, where the reference time T_{ref} is the operational time taken by using interface I.

The first author was supported by Glasgow University and ORS Scholarships.

5. REFERENCES

1. R. Bellman, B.G. Kashef and J. Casti, J. Comput. Phys. 10 (1972) 40
2. C. Shu, Generalized Differential-Integral Quadrature and Application to the Simulation of Incompressible Viscous Flows Including Parallel Computation, PhD Thesis, University of Glasgow, U.K., July 1991.
3. U. Ghia, K.N. Ghia, and C.T. Shin, J. Comput. Phys. 48 (1982) 387

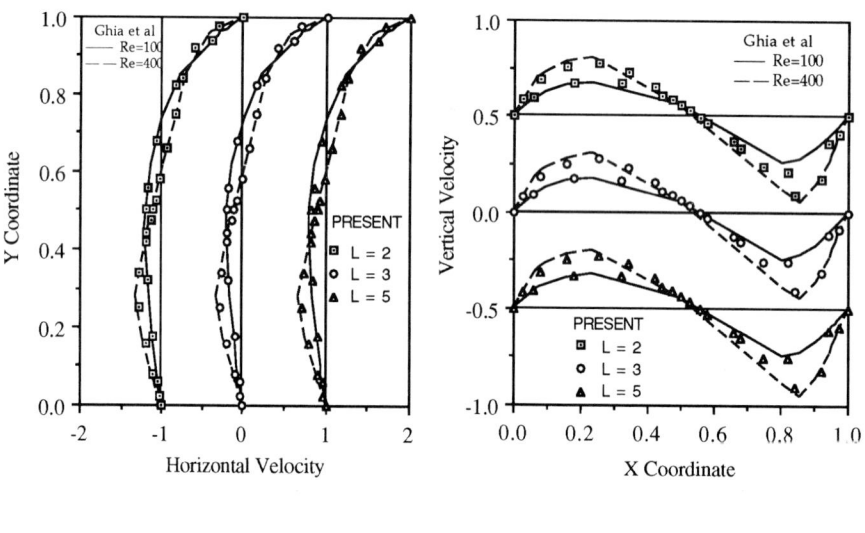

(a) Horizontal Velocity Profile (b) Vertical Velocity Profile

Fig. 5 Comparison of Velocities Through the Geometrical Center of a Cavity for Interface II

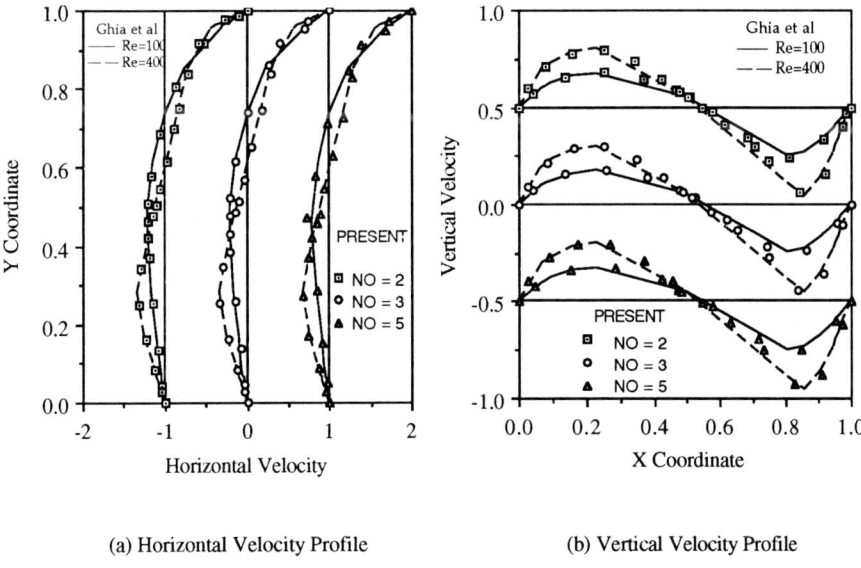

(a) Horizontal Velocity Profile

(b) Vertical Velocity Profile

Fig. 6 Comparison of Velocities Through the Geometrical Center of a Cavity for Interface III

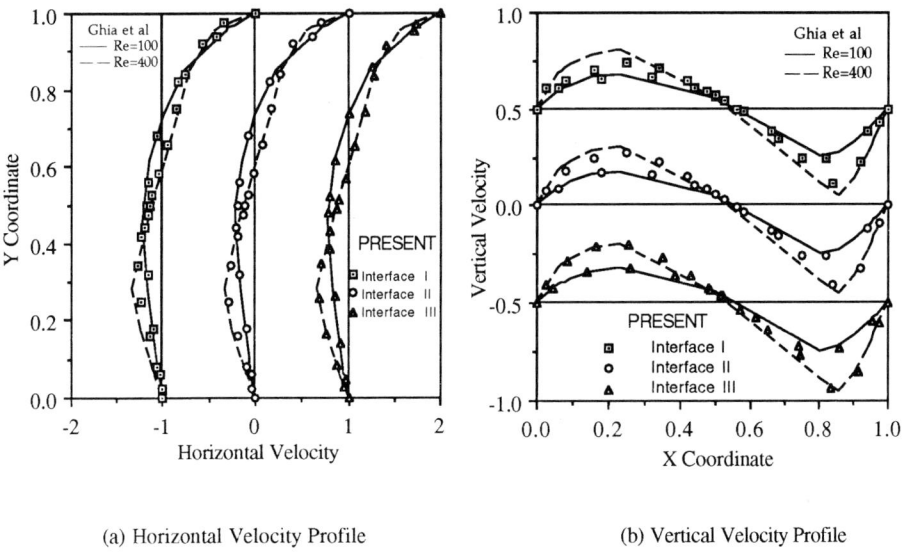

(a) Horizontal Velocity Profile

(b) Vertical Velocity Profile

Fig. 7 Comparison of Velocities Through the Geometrical Center of a Cavity for Three Interfaces

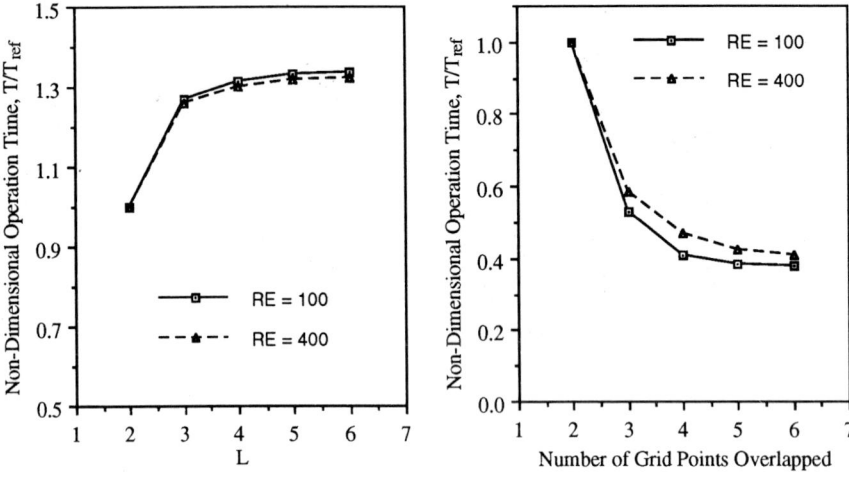

Fig. 8 Comparison of the Non-Dimensional Operational Time for Interface II

Fig. 9 Comparison of the Non-Dimensional Operational Time for Interface III

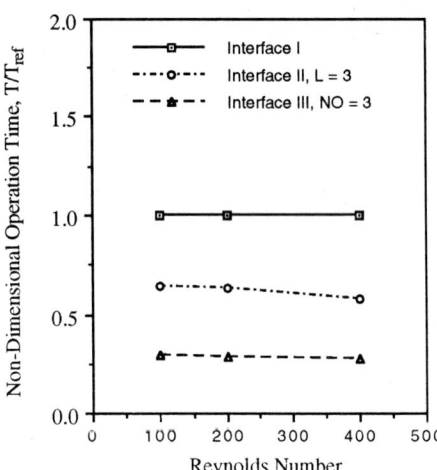

Fig. 10 Comparison of the Non-Dimensional Operational time for Three Interfaces

Parallel Computational Fluid Dynamics '92
R.B. Pelz, A. Ecer and J. Häuser (Editors)
© 1993 Elsevier Science Publishers B.V. All rights reserved.

Efficient Parallelisation of Implicit and Explicit Solvers on a MIMD computer

D.M. Smith and S.P. Fiddes

Department of Aerospace Engineering, University of Bristol, Bristol BS8 1TR, U.K.

Abstract

The phenomenon of speeddown is demonstrated for the finite volume solution of Laplace's equation, to illustrate the importance, and some of the difficulties, of predicting parallel performance. We present some parallel results which we hope will allow the development of some models of parallel performance in a future paper. Specifically, we have compared a number of parallel implicit solvers, including conjugate gradient and multigrid, and have parallelised an explicit Euler solver using both domain and demand driven decomposition.

1 Introduction

The overall objective of this work is to gain experience of parallel processing for computational fluid dynamics (CFD) problems, with a view to assessing its future potential. For example, will massively parallel computers make the direct numerical simulation of the turbulent flow around complete aircraft a feasible proposition? And if so:

- how many processors will be needed?
- how fast at calculating and communicating must they be?
- what will be the best solution technique? and
- how user-friendly may its parallel implementation be without compromising efficiency too much?

In theory, it is possible to predict the execution time of any parallel system given certain processor parameters, such as calculation speed and communication rate, and hence to answer the above questions. Unfortunately, the relation between parallel speed and the number of processors is not a simple one. For example, it is possible for a system to run slower on a large number of processors than it does on a small number of processors. This phenomenon, known as *speeddown* (Hockney & Jesshope [8]), has been demonstrated by Chalmers [3] and Chalmers et al. [4], and is shown by some results presented below. Furthermore, on computers which take advantage of memory caching, it is possible, as shown below, to obtain *superlinear* speedups (ie. speedups greater than the number of processors). Thus, mathematical models of parallel speed are likely to be quite complicated, and will require much validation before we may have confidence in them. In this paper, we present some parallel results for a number of different CFD solution techniques, which we hope will contribute towards the development of mathematical models of parallel performance in a future paper. Two different types of processor have been used, and we compare two different environments for the FORTRAN implementation of parallel codes.

In section 2 we present some results which illustrate some of the difficulties of predicting parallel performance. In section 3 we present a comparison of some implicit CFD solvers, including preconditioned conjugate gradient and multigrid, and in section 4 we present some results for an explicit CFD method, using both data driven and demand driven decompositions.

2 Predicting Parallel Performance: Some Difficulties

Most computer codes contain an inherently sequential part, such as file access, and a part which may be parallelised. In this paper, we will concentrate only on the latter. It is well known that the parallel execution time for parallelisable code on a MIMD computer consists of two parts: a calculation time; and a communication plus synchronisation overhead. Thus, if T_p is the single processor execution time of the parallelisable code, and $T_o(n)$ is the communication and synchronisation overhead, then the execution time on n processors, $T(n)$, is given by (see, for example, Flatt & Kennedy [5])

$$T(n) = T_p/n + T_o(n) \tag{1}$$

Generally, T_p depends mainly on the number of floating point operations required by a particular solution technique, and is quite straightforward to model if the floating point speed of the processor is known. However, if the size of the problem is near to the limit of memory available, then the array access time may become significant, presumably because memory caching is not as effective as it is for smaller problems. This means that it is possible to achieve *superlinear* speedups, because when the problem is spread over more processors, memory caching is more effective and hence array access is quicker. This is demonstrated below for the finite volume solution of Laplace's equation.

The communication and synchronisation overhead $T_o(n)$ is much more difficult to predict (Annaratone et al. [1]), and depends not only on the parallel hardware, but also on the parallel environment in which the code is implemented.

There are two basic types of parallel environment on MIMD machines:

1. physically connected environments, and

2. logically connected environments

In a physically connected environment, shown in figure 1, messages must be explicitly assigned to processor links by the programmer. This is efficient, but can be difficult to use, especially when processors must communicate with ones to which they are not directly connected. For example, if processor 1 in figure 1 needs to send a message to processor 3, then processor 2 must be told to pass it on.

In a logically connected environment, shown in figure 2, *all* processors are logically connected so that the programmer only needs to specify the message destination, and does not have to worry about the message route (which is taken care of by some additional software). This obviously makes programming much easier, but only at the expense of efficiency: firstly, there is an additional startup time, not present in physically connected environments, for the router software to decide on the message route; and secondly, there is no guarantee that the message will take the most direct route. For example, if processor 1 in figure 2 needs to send a message to processor 2, it is possible for the message to go via processors 3 & 4, especially if the link between 1 & 2 happens to be temporarily busy.

To illustrate the influence of environment on communication speed, we have timed the sending and receiving back of a message between two processors which are directly connected. The physically connected environment was implemented on T800 transputers, and the logically connected environment was implemented on i860 processors with T800's to handle their communication. Thus, in both cases, communication is performed on T800's. Figure 3 shows the time to send and receive messages of different lengths for the two environments. It can be seen that the logically connected environment is significantly slower, especially for small message sizes, where the additional startup time is important. This may have serious implications for massive parallelisation because, the more processors that are used for a fixed problem size, the smaller will be the messages.

The relationship between time and message size in figure 3 is very nearly linear for both physically and logically connected environments, so that the communiciation time between adjacent processors may be predicted simply in terms of a startup time and a transfer rate, as used by Nicol & Willard [13] and Gropp & Smith [7]. When processors are not physically adjacent, predicting communication time will be more difficult. Scherson & Corbett [17] assumed that the routing time is a function only of the distance in hops between the processors, but this ignores the synchronisation overhead which will occur if an intermediate processor is not ready to pass on the message.

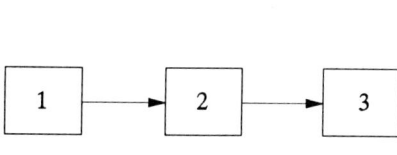
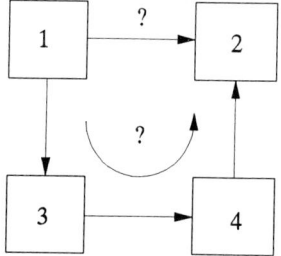

Figure 1: Physically connected parallel environment

Figure 2: Logically connected parallel environment

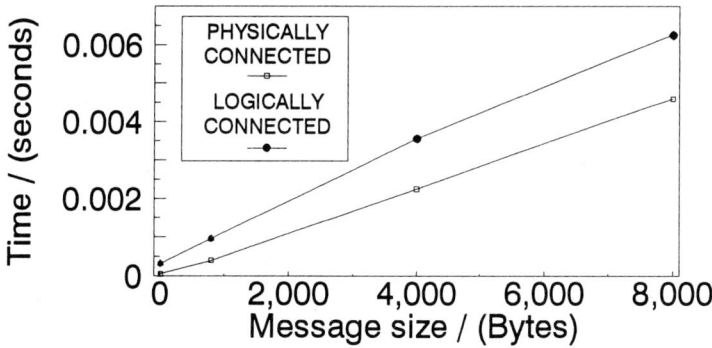

Figure 3: Interprocessor communication in physically and logically connected environments

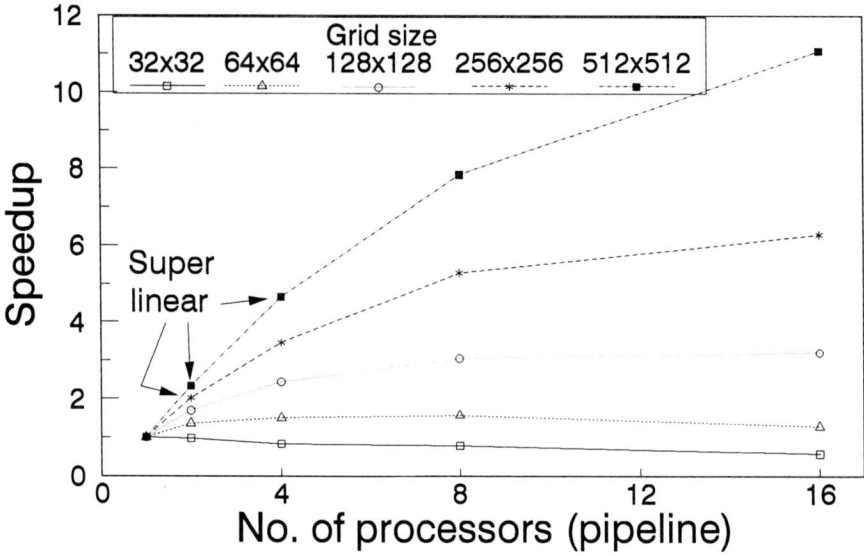

Figure 4: Finite volume solution of Laplace's equation; logically connected environment

Many CFD codes are based on finite volume or finite difference methods, and only require near-neighbour communication (except for convergence checking). In theory, the communication plus synchronisation overhead should follow a linear relationship with the message size, as shown in figure 3. In practice, however, this is not always the case, as we will demonstrate here. We have solved the two-dimensional Laplace equation using a finite volume method and a red/black Gauss-Seidel solver. Domain decomposition onto a pipeline of 2, 4, 8 & 16 processors was used, and the code was implemented in a logically connected environment. Each processor had the same amount of work to do (ie. each dealt with the same number of grid points), and the code was run for 100 iterations, so that there was no convergence checking. On 4 or more processors, each processor must exchange information with only two others, the two on either side of it in the pipeline (except the end processors, which only have to exchange with one other). Thus, in theory, the communication plus synchronisation overhead should *not* increase on 4 or more processors, for a fixed grid size. This means that the speedup should, at worst, approach a constant value, and not decrease. Figure 4 shows the speedup we obtained for a variety of grid sizes, and it can be seen that, for the 64 × 64 grid, there is a speedup on up to 8 processors, but a speeddown on 16 processors. This can only happen if the communication plus synchronisation overhead is, in fact, increasing with the number of processors. It is not obvious why this happens, although we believe it is caused by the routing software used by the logically connected environment. Thus, the prediction of communication and synchronisation overhead is not straightforward, and may require detailed knowledge of the routing software. It is, however, clearly important to be able to predict the parallel execution time so that speeddown is avoided: there is no point in using more processors if they result in the code running slower!

We have demonstrated some of the difficulties of predicting parallel performance, as well as its importance. Mathematical models of parallel speed must be callibrated and validated using real parallel results. In the next two sections we obtain some parallel results for both implicit and explicit CFD techniques.

3 Implicit Methods

In implicit CFD methods, the governing differential equations are discretised in such a way that values at a new time level depend on neighbouring values at the same time level. This technique is well known, and is described in detail by, for example, Patankar [15] and Roache [16]. The end result is the need to solve a large set of simultaneous equations, which may be written in matrix form as:

$$A\Phi = B \tag{2}$$

where A is a sparse (usually pentadiagonal for two dimensional situations) coefficient matrix, Φ is the solution vector, and B is the source vector.

A number of methods are available for solving equation 2, but it is not obvious which will be the fastest on a parallel computer. Thus, in this paper, we also compare the parallel performance of some widely used solvers.

As a test problem, we have taken the pressure equation from the well known two-dimensional lid driven cavity, at Reynolds number of 1000. The coefficient matrix A and source vector B in equation 2 were written out at convergence by a code which solves the full Navier-Stokes equations, and then read in as input to the parallel solver codes, which solved the resulting equation starting from a guess of $\Phi = 0$. The hybrid differencing scheme of Spalding [19] was used, which means that the value at any grid point depends only on values at its nearest neighbours, ie. that the coefficient matrix A is pentadiagonal.

Although we only solve a single scalar equation, much information may be deduced about the likely performance of the solvers in real CFD codes, which must solve a coupled set of equations for momentum and pressure. Indeed, many CFD codes are based on the well known SIMPLE algorithm (described by Patankar [15]) in which the momentum and pressure equations are each solved separately, in the form of equation 2, and their coupling is handled by repeating the solution using the latest available values of the other variables in a series of 'outer iterations'. It is well known that the the solution of the pressure equation is often by far the most time consuming part of this type of algorithm (see, for example, Van Doormal & Raithby [21]), and that is one reason why we chose it as the test problem.

3.1 Parallel Implementation of Solvers

The solvers we have investigated are well known and it is not necessary to describe them in detail here. Instead, we shall concentrate on aspects relating to their parallel implementation. We have used a physically connected environment.

3.1.1 Jacobi

Jacobi iteration, described by, for example, Jennings [10], is an ideal candidate for parallelisation because there are no data dependencies, ie. the solution at the new iteration depends only on values at the old iteration. The method may be parallelised by a simple domain decomposition, in which the computational grid of $N_i N_j$ cells is decomposed onto a processor grid of $p_i p_j$ processors, such that the first processor is allocated the points in the range ($i = 1$ to N_i/p_i, $j = 1$ to N_j/p_j), the next processor is allocated points ($i = N_i/p_i + 1$ to $2N_i/p_i$, $j = 1$ to N_j/p_j), and so on. As mentioned above, the differencing scheme we have used is such that the grid point value depends on its nearest neighbours. But when the grid is decomposed onto a number of processors, some of the neighbours for some of the grid points will be dealt with by a different processor to the one dealing with the actual grid point. This means that it is necessary for each processor to store a halo of information, and for this halo to be updated before each iteration. Updating the halo requires communication between the processors.

3.1.2 Gauss-Seidel

In the Gauss-Seidel method, use is made of the latest available information at points already visited. This does not parallelise using the normal lexicographic ordering because there is a data dependency: processors allocated parts of the grid with high i & j indices must wait for information from those dealing with lower i & j indices. It is well known, however, that the data dependency may be avoided by the so called red/black ordering (Young [22], Ortega & Voigt [14]), in which the grid points are coloured red or black in a chequered pattern, like a chessboard, and updated in two stages: first, all the red points are updated, using information from the black points at the old iteration; then the black points are updated using the latest information at the red points. Each step may be parallelised by domain decomposition, as for Jacobi iteration.

Red/black Gauss-Seidel iteration may also be *over-relaxed*, by multiplying the change between iterations by the over-relaxation factor ω (usually in the range 1 to 2), in order to increase convergence rate [10].

3.1.3 ADI

The alternating direction implicit (ADI) technique is based on the direct solution along lines of constant i or j, using the latest available information in neighbouring lines. It is a two stage method in that the new solution is obtained by first solving along lines of constant i and then along lines of constant j. Its parallel implementation is likely to be very inefficient because global, rather than just near-neighbour, communication is required for the processors to handle the alternating lines. Furthermore, Kershaw [11] has shown that ADI is slower than conjugate gradients on a single processor. For these reasons, we did not implement ADI in parallel.

3.1.4 Conjugate Gradient

The basic conjugate gradient algorithm has no data dependencies, and is straightforward to parallelise by domain decomposition. However, the rate of convergence is usually greatly improved by preconditioning, and this is more difficult to parallelise. One of the most successful preconditioners is the incomplete Choleski factorisation (ICCG) of Meijerink & van der Vorst [12], in which the preconditioning matrix is obtained by an incomplete LU decomposition of the coefficient matrix (where L & U are lower and upper triangular matrices). The inversion of L & U does not parallelise easily because of data dependency. One way around this, proposed by van der Vorst [20], is to approximate the inverse of L (& U) as a truncated Neumann Series. For example, if L is scaled such that its diagonal entries are 1, then

$$L^{-1} = I - L + L^2 - L^3 + \cdots \qquad (3)$$

By truncating the Neumann series after the L^2 term, L^{-1} may be computed in parallel using domain decomposition and communicating a halo of two cells between processors. Of course, the parallelisable algorithm is only an approximation, and its rate of convergence is slower than the original algorithm. For the test problem considered here, on a 32×32 grid the parallelisable algorithm required 3 times the computing time of the standard algorithm, to achieve convergence (defined as a residual level of 10^{-4}). However, as will be seen below, it parallelises very efficiently.

3.1.5 Multigrid

We have used red/black Gauss-Seidel, as described above, as the multigrid smoother. All the grid transfers have no data dependencies, and so are straightforward to parallelise by domain decomposition. However, multigrid was by far the most difficult solver to implement, using a physically connected environment, because the coarsest grids can be coarser than the processor grid, so that communication is necessary between processors which are not directly connected, and processors in between must be told to receive and pass on messages. We have used fixed V cycles (Brandt [2]), and a coarsest grid of 2×2 cells. One iteration per visit was performed on the finest grid, two on the next coarser grid, and 3 on all other grids.

3.2 Parallel Results

The parallel execution times to achieve convergence (a residual level of 10^{-4}) are presented in figures 5 and 6 for 64×64 and 128×128 grids respectively. It can be seen that on the 64×64 grid, red/black Gauss-Seidel with no over-relaxation and Jacobi are always about an order of magnitude slower than the other solvers, and so were not considered on the 128×128 grid. As expected, on a single processor multigrid is the fastest solver, followed by parallelisable conjugate gradient and then red/black Gauss-Seidel with optimum over-relaxation. However, on the 64×64 grid, both conjugate gradient and over-relaxed red/black Gauss-Seidel overtake multigrid on the 8×8 processor grid. This is because multigrid does not parallelise as

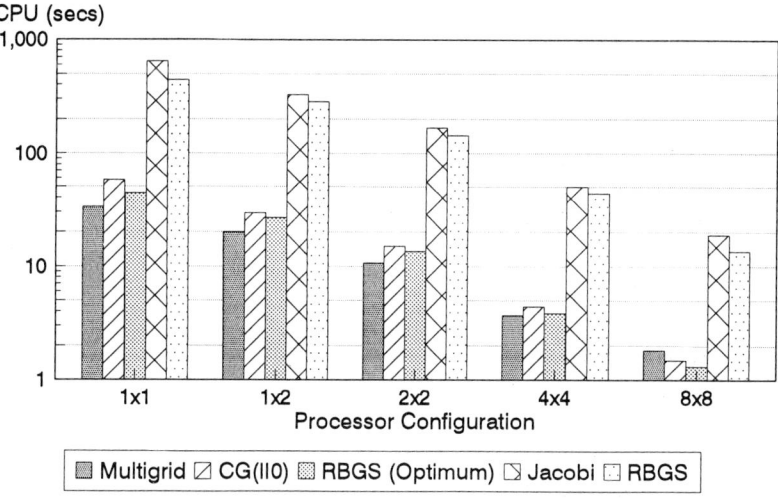

Figure 5: Parallel execution times for 64 × 64 grid

efficiently, because much of the work is done on coarse grids, on which the communication to calculation ratio is higher and hence the parallel efficiency lower. However, on the finer 128 × 128 grid, figure 6, multigrid remains the fastest solver even on 64 processors, although its parallel efficiency, shown in figure 7, is less than 60% on 64 processors, compared to over 80% for the other two solvers. It is very likely that, even on this fine grid, conjugate gradient and over-relaxed Gauss-Seidel would overtake multigrid on more than 64 processors.

3.3 Conclusions

The results presented here illustrate the importance of being able to predict parallel performance. For example, there is no point in using a multigrid solver, which is difficult to implement (especially in parallel), if a simpler technique will run faster for the grid size we are interested in and on the number of processors we have available. Conversely, it does not follow that the most efficient parallel techniques will always be the fastest. For example, on the 128 × 128 grid, multigrid was the least efficient solver tested on 64 processors, but was also the fastest.

4 Explicit Methods

4.1 Domain Decomposition

4.1.1 Implementation

Explicit CFD methods are generally straightforward to parallelise by domain decomposition, because values at the new timestep depend only on values at the old timestep, so there is no data dependency. We have parallelised an existing production code, based on the method of Jameson et al [9] which uses 4th order Runga-Kutta time stepping, and 2nd & 4th order artificial dissipation. It is necessary to exchange a halo of two layers of cells at the beginning of each timestep because of the 4th order dissipation, and one layer of cells after each Runga-Kutta step. Communication is also required to calculate and broadcast the circulation in order to enforce the far field boundary conditions.

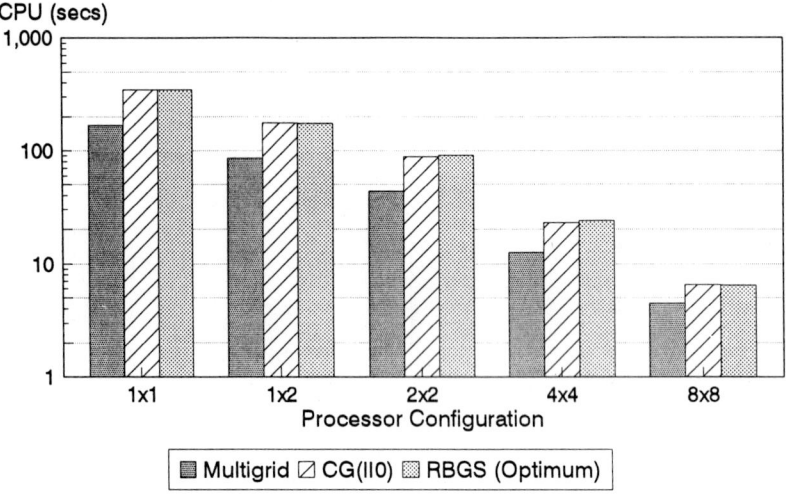

Figure 6: Parallel execution times for 128 × 128 grid

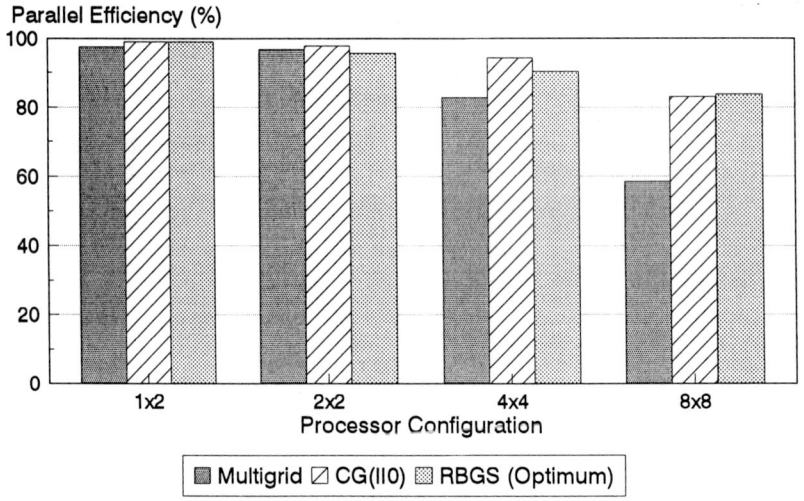

Figure 7: Parallel efficiency for 128 × 128 grid

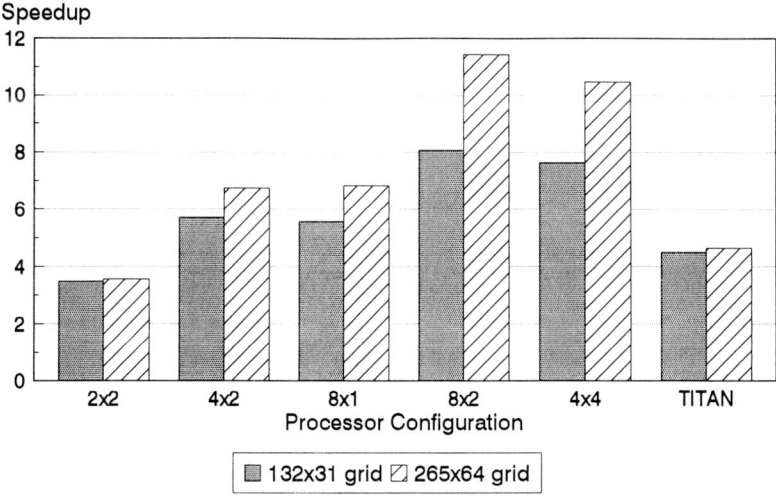

Figure 8: Speedup for domain decomposition of Euler solver

We have used a logically connected parallel environment on up to 16 i860 processors. The code was run for 1000 timesteps.

4.1.2 Results

The speedup for various processor configurations is shown in figure 8 for a 132×31 and 265×64 grid. For the coarser grid, a speedup of 3.5 was achieved on 4 processors, corresponding to a respectable parallel efficiency of 87%, but the speedup dropped to only 8 on 16 processors. However, for the finer grid a speedup of 11.4 on 16 processors was achieved, corresponding to a parallel efficiency of over 71%. Also shown in figure 8 is the 'speedup' relative to a single i860 processor obtained on an Ardent Titan 3000 minisupercomputer running in vectorised mode. The Titan is over 4 times faster than a single i860, but 2.5 times slower than 16 i860's for the fine grid problem.

4.2 Demand Decomposition

4.2.1 Implementation

It is well known that explicit methods are restricted in the timestep which they may take before becoming unstable. This means that it can take a very long time to reach a steady state solution, especially as finer grids are used. In order to try to increase the rate of convergence to steady state (ie. to decrease the amount of computational work), we have implemented a method similar to that used by Southwell [18] and Fox [6], in which only regions of high residual are updated each timestep. Thus, the domain is divided into many more blocks than processors, but each processor only updates those blocks with a residual larger than some factor, F ($0 < F \leq 1$), times the maximum residual of all its blocks. Of course, the residuals in blocks adjacent to an updated block must also be updated, but this only requires work roughly equivalent to one Runga-Kutta timestep, whereas updating a block involves 4 Runga-Kutta timesteps. In order that a processor dealing with low residual areas cannot dictate the number of blocks being updated, the maximum number of blocks updated by each processor is limited by the one dealing with the block with the overall highest residual.

This technique is not suitable for parallelisation by simple domain (or data) decomposition, because it is likely that the areas of highest residuals will be close together (around a shock wave, for example) and

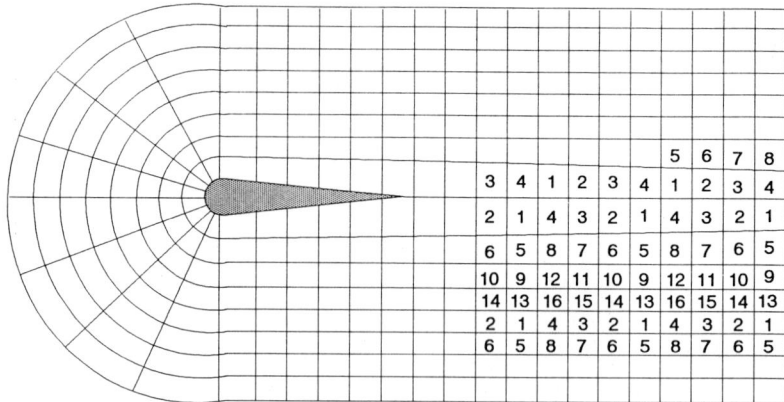

Figure 9: Scattered Decomposition on 4 × 4 Processor Grid. The numbers are the processors dealing with each block of cells.

may all be allocated to just one or two processors. This would lead to a serious load imbalance, because these processors would be doing most of the work. Instead, we have employed a scattered decomposition, shown in figure 9, in which processors are allocated blocks from all over the domain. In this way, every processor should have a number of blocks from the high residual areas. Care was taken to ensure that a processor only exchanges information with its nearest neighbours. For example, blocks adjacent to a block allocated to processor 9 in figure 9 are always allocated to processors 5, 10, 12 & 13, which are directly connected to processor 9. The decomposition was arranged such that this holds true even across the wake.

4.2.2 Overheads

The implementation of this technique involves some overheads not present in the standard Euler solver of the previous section.

- Firstly, the length of the message communicated each timestep (or Runga-Kutta step) now depends on the number of blocks updated by each processor. Thus, in the parallel environment we have used, it was necessary to send an extra message containing the size of the boundary information message to follow. This would not have been necessary if we had used Occam, which allows the size of the message to be part of the message itself.

- Secondly, as mentioned above, the residuals must be recalculated in blocks adjacent to an updated block.

- As explained above, the maximum number of blocks updated is limited by the processor dealing with the highest residual. The implementation of this requires the gathering of the maximum residuals from each processor, and then the broadcasting of the maximum number of blocks to update.

- Finally, for ease of implementation into the existing production code, we stored all of the blocks for each processor in one large array, and employed a subroutine to transfer data into the standard arrays used by the code. In some of the results presented below we have timed this subroutine, which would not be necessary were the code written from scratch, and subtracted it from the overall time.

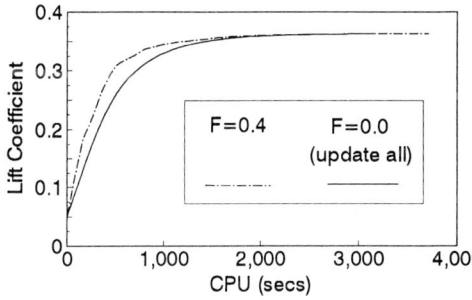
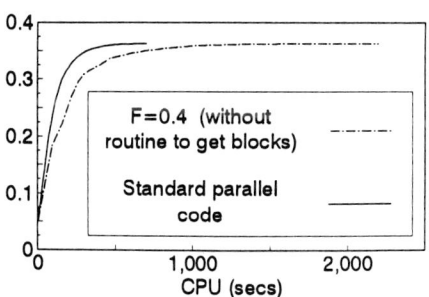

Figure 10: Comparison of updating all blocks ($F = 0$) with updating highest residual blocks ($F = 0.4$)

Figure 11: Comparison of updating highest residual blocks ($F = 0.4$) with standard parallel code

4.2.3 Results

As mentioned above, a block is updated if its residual is larger than F times the maximum residual of all the blocks held on a particular processor. In figure 10 we have plotted the lift coefficient against CPU time (on 4 × 4 processors, 132/$times$31 grid decomposed into 32 × 8 blocks) for $F = 0$ (ie. updating all blocks), and $F = 0.4$. It can be seen that the lift coefficient reaches its steady state value more quickly for the $F = 0.4$ case, indicating that there is an advantage in only updating where the residual is high.

However, it can be seen in figure 11 that the $F = 0.4$ case is slower than the standard parallel code of the previous section. In this comparison we have eliminated the time to transfer data into the standard arrays used in the code, as explained above, and assume that inefficiency of the $F = 0.4$ case is mainly due to the need to communicate extra messages, as discussed above.

5 Conclusions

We have compared the parallel performance of a number of implicit CFD solvers. For the test problem, multigrid was the fastest on up to 64 processors for a fine grid size of 128 × 128, but was overtaken on 64 processors by both conjugate gradient and red/black Gauss-Seidel (with optimal over-relaxation) for the coarser grid size of 64 × 64. This is because much of the multigrid work is done on coarse grids, on which the communication to calculation ratio is higher, and hence the parallel efficiency is lower.

We have demonstrated the parallelisation of a production Euler solver, achieving a speedup of over 11 on 16 processors for a 265 × 64 grid. A technique for increasing the convergence to steady state by concentrating only on high residual areas was not successful in parallel because of some additional communication.

We have also demonstrated that it is possible to slow down a parallel calculation by using too many processors. This illustrates the need for more research into parallel processing, and the need for mathematical models capable of predicting parallel performance on massively parallel computers. The results from such models should influence the design of future processors and parallel environments.

6 Acknowledgements

This work was funded by Rolls-Royce plc and SERC.

References

[1] M. Annaratone, C. Pommerell & R. Rühl, "Interprocessor Communication Speed and Performance in Distributed-memory Parallel Processors", proc. 16th Annual Int. Symp. on Computer Architecture, Computer Architecture News, vol. 17, no. 3, 315-324, June 1989

[2] A. Brandt, "Multi-level adaptive solutions to boundary-value problems", Math. Comp., 31, 333-390, 1977

[3] A.G. Chalmers, "A Minimum Path System for Parallel Processing", PhD Thesis, Bristol University, 1991

[4] A.G. Chalmers, S.P. Fiddes & D.J. Paddon, "Parallel Panel Methods", In 13th Occam User Group Conference, York, 1991

[5] H.P. Flatt & K. Kennedy, "Performance of Parallel Processors", Parallel Computing, 12, 1-20, 1989

[6] L. Fox, "A Short Account of Relaxation Methods", Quarterly Journal of Mechanics and Applied Mathematics, 1, 253-280, 1948

[7] W.D. Gropp & E.B. Smith, "Computational Fluid Dynamics on Parallel Processors", Computers & Fluids, 18, 289-304, 1990

[8] R.W. Hockney & C.R. Jesshope, *Parallel Computers 2: Programing and Algorithms*, Adam Hilger, 1988

[9] A. Jameson, W. Schmidt & E. Turkel, "Numerical Solutions of the Euler Equations by Finite Volume Methods using Runga-Kutta Time Stepping Schemes", AIAA-Paper, 81-1259, 1981

[10] A. Jennings, *Matrix Computation for Engineers and Scientists*, J. Wiley & Sons, 1977

[11] D.S. Kershaw, "The Incomplete Cholesky Conjugate Gradient Method for the Iterative Solution of Systems of Linear Equations", J. Comp. Phys., 26, 43-65, 1978

[12] J.A. Meijerink & H.A. van der Vorst, "An iterative solution method for linear systems of which the coefficient matrix is a symmetric M-matrix", Math. Comp., 31, 148-162, 1977

[13] D.M. Nicol & F.H. Willard, "Problem Size, Parallel Architecture, and Optimal Speedup", Journal of Parallel and Distributed Computing, 5, 404-420, 1988

[14] J.M. Ortega & R.G. Voigt, "Solution of Partial Differential Equations on Vector and Parallel Computers", SIAM Rev., 27, 149-240, 1985

[15] S.V. Patankar, *Numerical Heat Transfer and Fluid Flow*, Hemisphere, Washington D. C., 1980

[16] P.J. Roache, *Computational Fluid Dynamics*, Hermosa, Albuquerque, New Mexico, 1972

[17] I.D. Scherson & P.F. Corbett, "Communications Overhead and the Expected Speedup of Multidimensional Mesh-Connected Parallel Processors", Journal of Parallel and Distributed Computing, 11, 86-96, 1991

[18] R.V. Southwell, *Relaxation Methods in Theoretical Physics*, Clarendon Press, Oxford, 1946

[19] D.B. Spalding, "A novel finite difference formulation for differential expressions involving both first and second derivatives", Int. J. Num. Meth. Eng., 4, 551-559, 1972

[20] H.A. van der Vorst, "The performance of FORTRAN implementations for preconditioned conjugate gradients on vector computers", Parallel Computing, 3, 49-58, 1986

[21] J.P. Van Doormal & G.D. Raithby, "Enhancements of the SIMPLE method for predicting incompressible fluid flows", Num. Heat Transfer, 7, 147-163, 1984

[22] D.M. Young, *Iterative Solutions of Large Linear Systems*, Academic Press, New York, 1971

The Uniform Boundary Algorithm for Supersonic and Hypersonic Flows

Y.S. Weber,[a] J.W. Weber,[a] J.D. Anderson,[a] E.S. Oran,[b] and C. Li[b]

[a]Department of Aerospace Engineering, University of Maryland, College Park, MD 20742, USA

[b]Laboratory for Computational Physics and Fluid Dynamics, Naval Research Laboratory, Washington, DC, 20375, USA

Abstract

The efficient determination of nonperiodic boundary conditions for the computation of fluid dynamics problems is required for the effective use of massively parallel SIMD computers such as the Thinking Machines CM2. It has been found that some methods for treating the boundary conditions can increase the computation time of two-dimensional solutions on structured, cartesian meshes by more than 50%. The penalties associated with computing nonperiodic boundary conditions have been reduced by algorithms explicitly developed to address such problems. In the present work, two implementations of the Uniform Boundary Algorithm are incorporated with the Flux-Corrected Transport method in CM FORTRAN to simulate high-speed compressible flows with nonperiodic boundaries. In particular, variations of the Uniform Boundary Algorithm, where the boundary conditions are coupled and uncoupled from the fluid algorithm, are implemented. As a coupled approach, the boundary conditions are a constituent part of the integration method, and in the uncoupled approach the boundary conditions are calculated prior to the integration of the governing equations. The performance of the coupled and uncoupled versions of the Uniform Boundary Algorithm is assessed from the speed, memory requirements, and flexibility of each method. It has been found that for the Flux-Corrected Transport scheme, the fully coupled Uniform Boundary Algorithm increases the CPU time by only 10%, and is the most desirable approach for the tradeoffs considered.

1. INTRODUCTION

In recent years, the problems solved by researchers in computational fluid dynamics (CFD) have become increasingly complex. Solutions involving complex geometries are routinely produced, and in the high-speed regime calculations involving nonequilibrium chemically reacting gases are the current state of the art. As a result, the amount of computational time needed to obtain these solutions has increased dramatically, and currently a large percentage of the nation's supercomputing power is devoted to CFD. Thus, researchers have started looking toward massively parallel processing, with its promise of reduction in computing time, as an alternative to conventional supercomputers. The Connection Machine CM2 is a massively parallel computer that consists of thousands of individual scalar processors connected by a hypercube communications network into a single-instruction, multiple-data (SIMD) architecture. This means that all of the processors perform the same operations on varying data as the instruction stream is broadcast by a conventional serial computer which acts as a front end host. The most efficient mode of operation for this type of machine occurs when the data needed to complete an operation is stored locally in the processor memory. However, when data from other processors is needed, interprocessor communications must be employed. On the CM2, there are generally two types of communications, the router and the NEWS network. NEWS stands for north, east, west, and south and simulates a multidimensional cartesian grid. It is especially fast for communications between nearest neighbors.

Many of the fluid dynamic algorithms used in CFD are explicit and are applied to structured Cartesian meshes. These are especially well suited for efficient implementation in a parallel, SIMD environment. However, one of the most challenging problems for the programmer is the efficient determination of the boundary conditions. At the boundaries of the computational domain, special consideration must be given to satisfy physical or geometrical constraints. On the CM2, periodic boundary conditions are built into the hardware by the wrap-around communications network and they require no additional time or programming to implement. Thus the hardware ensures periodic boundary conditions along all specified dimensions of the problem. Although some interesting and important physical questions can be answered by solving problems in multiply periodic geometries, most physically realistic engineering problems require more complex boundary conditions.

There are several existing methods for handling nonperiodic boundary conditions on a SIMD machine [1]. The most obvious approach is the traditional serial approach in which the boundary cells are selected and computations for these cells are performed separately from those where the CFD algorithm is applied. This is the basic idea behind "local" boundary methodologies. Although this approach is the most obvious to implement, it can be the most inefficient, especially when the calculation of the boundary conditions is a measurable part of the total CPU time. When the number of boundary cells is less than the number of physical processors, inefficient load balancing will result, where many of the processors remain idle while only a small subset are active. Due to these shortcomings, an alternative approach to local boundary algorithms, which has proven more efficient for computing boundary conditions, was developed by Myczkowski [2] and Li [3].

Myczkowski et al. solved the acoustic wave equation in connection with seismic modeling [2]. They optimized a finite-difference algorithm in CM FORTRAN that was rated at approximately 5.6 Gigaflops including I/O operations. As part of the optimization, they developed a method for efficiently calculating boundary conditions. Their reasoning was based on the fact that both the interior and the boundary nodes obey the same physical laws, but the boundary nodes must satisfy additional conditions because of geometrical considerations. Thus, they expanded the SIMD program to incorporate all the boundary condition expressions into a finite difference stencil in such a way that both the boundary and integrated cells were simultaneously calculated at all points on the grid. The actual value assigned to a given node depended on the coefficients of the stencil which varied across the computational domain.

Simultaneously Li et al. [3] developed a similar methodology for applying nonperiodic boundary conditions in C*, an alternate programming language supported by the CM2. Li labeled this approach the Uniform Boundary Algorithm (UBA). They combined the Flux-Corrected Transport (FCT) flow solver with the UBA to solve compressible flow problems. Because NEWS communication can be costly, they concentrated on combining expressions for the boundary conditions directly into the FCT scheme, so as to not increase the number of communications needed. As a result, Li noted only small increases in computation time over periodic problems which were due to additional floating point operations. Additionally, since the coefficients of the stencil varied across the domain, they were able to solve flows around complex internal geometries without incurring additional computational cost.

In the present work, a local boundary method and the UBA are combined with the FCT algorithm in CM FORTRAN to calculate high-speed flows. The local boundary method represents an uncoupled approach where array subscripts are used to calculate boundary conditions for selected cells apart from the integrated domain. Variations of the UBA where the boundary conditions are coupled and uncoupled are also implemented. When the UBA is fully coupled, the boundary conditions are an integral part of the calculation of cell interface quantities and flux differences required by the FCT algorithm. As an uncoupled boundary algorithm, the boundary conditions are applied directly to the integrated fluid quantities and are independent of the integration routine. In all approaches, the boundary algorithm is designed for calculation of boundary conditions typical of supersonic and hypersonic flow. The performance of the coupled and uncoupled boundary algorithms is assessed from the speed, memory requirements, and program flexibility determined for each method. Timings are referenced to the FCT algorithm with multiply periodic boundary conditions.

2. BOUNDARY ALGORITHMS

2.1. Uncoupled Methods

2.1.1. Local Boundary Algorithms

Local boundary algorithms are the most obvious method for calculating boundary conditions since they represent the traditional approaches used in serial or vector codes. While they are the most obvious, local boundary methods ignore the SIMD architecture of the CM2 since separate instructions are issued to selected boundary cells. Because only one instruction stream can be issued to the CM processors at a time, the calculation of the boundary conditions does not occur in parallel with the integration routine. For flow problems where the boundary conditions are applied to a small percentage of the computational domain, considerable inefficiencies in load balancing will result since the majority of virtual processors are inactive during the boundary condition assignments.

The local boundary method is an uncoupled approach where array subscripts are used to calculate boundary conditions for the indicated cells apart from the integration scheme. Consider the following CM FORTRAN statement which calculates the boundary condition along the i=0 axis:

$$\rho(0,:) = fbc * \rho(1,:) + rhofbc + pbc * \rho(0,:) \tag{1}$$

By a suitable choice of values for the coefficients fbc, $rhofbc$, and pbc, the applied boundary condition can be a zeroth-order extrapolation, a reflecting wall, a specified inflow, or a periodic condition. If the parameters fbc, $rhofbc$, and pbc are vectors, then different types of boundary conditions can be implemented along the $i = 0$ axis. In this example, the boundary elements are selected by array indexing. Although only nearest-neighbor communication is required, it is an expensive process because the communication first requires address calculations. For exterior boundary conditions, only four such statements are required, but with complex interior boundaries the number of such statements rapidly increases thereby increasing the computation time.

2.1.2. The Uniform Boundary Algorithm

Some of the inefficiencies in the local boundary method can be avoided if the boundary elements are selected a priori with precomputed masks. In the UBA, array multiplier masks replace address calculations with floating point operations. Since this approach is independent of the CFD algorithm, and is applied separately from the integration, this version of the UBA can also be described as uncoupled. In an expression similar to that used for the local boundary algorithm, the UBA calculates the boundary conditions for a parallel array, ρ^n, as follows:

$$\rho_i^n = a_0 \rho_i^n + a_1 \rho_{i-1}^n + a_2 \rho_{i+1}^n + a_3. \tag{2}$$

By a suitable choice of a_0 through a_3, either the local value of ρ_i^n is kept (i.e., for an integrated point $a_0 = 1$; a_1, a_2, $a_3 = 0$) or it is replaced by a boundary condition as given by the appropriate combinations of a_1 through a_3 ($a_0 = 0$). The coefficients a_0 through a_3 are array multiplier masks and for time-invariant boundary conditions can be computed during initialization of the program. Since the array multiplier masks contain the data needed for the selection of the boundary cells, the CSHIFT construct can be used for the NEWS communication to provide fast interprocessor communication between the next and previous values in the spatial arrays. In CM FORTRAN, the equivalent statement of equation (2) using the CSHIFT construct is

$$\rho = a0 * \rho + a1 * cshift(\rho, idim, -1) + a2 * cshift(\rho, idim, +1) + a3 \tag{3}$$

where $idim$ specifies the direction along which the array values are shifted. Note that once the boundary conditions are computed using either of the uncoupled approaches, the value of the integration quantity, ρ^n, is updated using any appropriate CFD algorithm.

2.2. Fully Coupled Methods

The uncoupled UBA eliminates the inefficiencies of local boundary algorithms, due to address calculations, by replacing them with array multiplier masks for selecting the boundary elements. In the fully coupled UBA additional inefficiencies are removed by eliminating superfluous CSHIFT constructs which are used for performing boundary calculations. In essence, the boundary calculations are performed by utilizing the CSHIFT constructs which are required by the CFD algorithm. FCT calculates at the cell interfaces average values and flux differences for the integrated quantities. These expressions, where the boundary conditions must enter into the FCT formulation, are of the form $\rho_i \pm \rho_{i-1}$. In the present work, ρ_i^n and ρ_{i-1}^n are reformulated to include the boundary conditions as follows:

$$\rho_i^n = b_0 \rho_i^n + b_1 \rho_{i-1}^n + b_3 \rho_o \tag{4}$$

and

$$\rho_{i-1}^n = b_2 \rho_i^n + (1 - |b_2|)(1 - b_4)\rho_{i-1}^n + b_4 \rho_o. \tag{5}$$

Thus, the fully coupled form of the UBA applied to the FCT method is given by combining equations (4) and (5) to obtain the expression for $\rho_i \pm \rho_{i-1}$:

$$(\rho_i \pm \rho_{i-1})^n = c_0 \rho_i^n + c_1 \rho_{i-1}^n + c_2 \tag{6}$$

where

$$c_0 = b_0 \pm b_2, \tag{6a}$$

$$c_1 = b_1 \pm (1 - |b_2|)(1 - b_4), \tag{6b}$$

and

$$c_2 = (b_3 \pm b_4)\rho_o. \tag{6c}$$

The array multiplier masks, or coefficients, in equation (6) are precomputed during initialization of the program. The value of these constants depends on the boundary conditions that are applied. Note that the expressions in equations (6a)-(6c) allow the user to define boundary conditions required for supersonic and hypersonic flow calculations such as reflecting walls, specified inflow, and extrapolated outflow. Table 1 shows the values that can be assigned to the constants b_0 through b_4 to construct these boundary conditions.

The fully coupled form of the UBA adds the boundary conditions to the CFD algorithm without additional NEWS communication calls. The boundary conditions do, however, increase the number of floating point operations. For periodic problems, the expression $\rho_i \pm \rho_{i-1}$ requires one floating point operation and one NEWS communication or CSHIFT construct. For nonperiodic problems, the reformulation for the expression $\rho_i \pm \rho_{i-1}$ as given by equation (6) increases the number of floating point operations to four while maintaining the same number of CSHIFT calls. An advantage of the fully coupled UBA is that fewer dead cells are required for the computation of boundary conditions. Dead cells are those cells that are physically allocated for the sole purpose of storing boundary values on the domain. For the uncoupled boundary

Table 1
Array multiplier assignments for supersonic / hypersonic boundary conditions

Boundary condition	b_1	b_2	b_3	b_4
Supersonic inflow				
density ρ	0	0	0 or 1	1 or 0
momentum ρV	0	0	0 or 1	1 or 0
energy ρE	0	0	0 or 1	1 or 0
Supersonic outflow				
density ρ	1 or 0	0 or 1	0	0
momentum ρV	1 or 0	0 or 1	0	0
energy ρE	1 or 0	0 or 1	0	0
Reflecting wall				
density ρ	1 or 0	0 or 1	0	0
momentum ρV_t	1 or 0	0 or 1	0	0
momentum ρV_n	-1 or 0	0 or -1	0	0
energy ρE	1 or 0	0 or 1	0	0
None				
density ρ	0	0	0	0
momentum ρV	0	0	0	0
energy ρE	0	0	0	0

Note: $b_0 = 1$ for integrated cells; $b_0 = 0$ for dead cells

methods used in the present work, these cells are specified along the entire perimeter of the domain. Thus for a two-dimensional grid, where $N_x \times N_y$ cells are integrated, an $(N_x + 2) \times (N_y + 2)$ grid must be allocated. Alternately, for the same two-dimensional problem, the fully coupled UBA requires one extra row and column of dead cells. The dead cells are needed here because the integration of N cells requires calculation and storage of information at N+1 cell interfaces. Thus, the fully coupled approach utilizes half the number of dead cells used by the uncoupled methods.

3. APPLICATIONS

In order to assess the performance of the UBA for compressible, high-speed flows, the method was applied with the FCT algorithm [4] to model the two-dimensional, unsteady Euler equations. The version of FCT implemented in the present work is an explicit, fourth-order, monotone method designed to solve one-dimensional continuity equations with source terms in an accurate, high-resolution, positivity-preserving manner. Two-dimensional problems were solved using both timestep and direction-splitting techniques. The code was developed on the CM2 at the University of Maryland's Institute for Advanced Computer Studies (UMIACS) and the Naval Research Laboratories'

(NRL) CM200. The model problems described below were solved on the NRL CM200, and were chosen to illustrate the ability of the UBA to apply boundary conditions typical of supersonic and hypersonic flow calculations.

The first model problem is a two-dimensional shock tube with streamwise gradients. The diaphragm pressure ratio is $p_4/p_1 = 10$. The diaphragm density ratio is $\rho_4/\rho_1 = 10$ as well, providing an initial uniform temperature in both the driver and driven sections of the tube. A total of 32K grid points was used for the computation, and boundary conditions consisted of reflecting walls at the top and bottom, as well as the right and left ends of the tube. In Figure 1, a time history of the density contours is given as the shock wave moves through the tube. The distinguishing features of the flowfield such as the expansion fan, contact surface, and shock wave are clearly evident. Note that by 1800 timesteps, both the expansion wave and incident shock wave have reflected off the left and right endwalls respectively. This model problem demonstrated the ability of the UBA to successfully handle reflecting boundary conditions.

The next model problem is a supersonic rectangular blast wave in a channel. This problem is very similar to the one described above except that the initial gradients are two-dimensional and the domain in the x-direction is infinite. The initial pressure ratio across the diaphragm is $p_4/p_1 = 30$, and the corresponding density ratio is $\rho_4/\rho_1 = 15$. A channel configuration is chosen to illustrate the outflow and reflecting wall boundary conditions. The axis of the channel is oriented in the horizontal direction with a zeroth-order extrapolation applied to simulate the supersonic outflow condition. When the diaphragm surrounding the high pressure gas is ruptured, a shock wave is formed and propagates forward from the region of the rectangular diaphragm. The contact surface follows behind the shock and with time, the suction created by the low pressure in the center of the blast forces the flow at the open boundaries to subsonic values, for which the supersonic extrapolation is no longer valid. Thus the time history of the blast wave is calculated only for the interval in which the outflow is supersonic. Figure 2 shows the time varying temperature contours for the supersonic blast wave solution on a 32K grid. Note that by 400 timesteps the flow has exited the domain on the left and right boundaries and has reflected off the upper and lower walls of the channel. At the corners of the domain, the UBA correctly incorporated the boundary conditions for the CFD algorithm, allowing the calculation of the complex interaction between the outflowing gas and the reflected shock.

The third model problem is the generation of a Mach 5 regular shock reflection from a plane wall. For this computation 64K grid points were used. At the left boundary a split inflow condition is used to generate the incident oblique shock. Along the lower portion of the inflow plane, a Mach 5 parallel flow at sealevel conditions is specified. The remaining grid points at the inflow plane correspond to the conditions behind a Mach 5, 30° oblique shock. Along the upper boundary the same post shock conditions are given as well. At the exit plane a supersonic outflow condition was specified, and at the lower boundary a solid wall for the shock reflection was imposed. The initial condition for the remaining grid points was the Mach 5 parallel flow. Figure 3 shows density contours detailing the timewise progression of the regular reflection. Because of the discontinuity in velocity along the inflow plane, a sheet of vorticity is generated. This vortex sheet is unstable and begins to roll up forming a starting vortex. Once

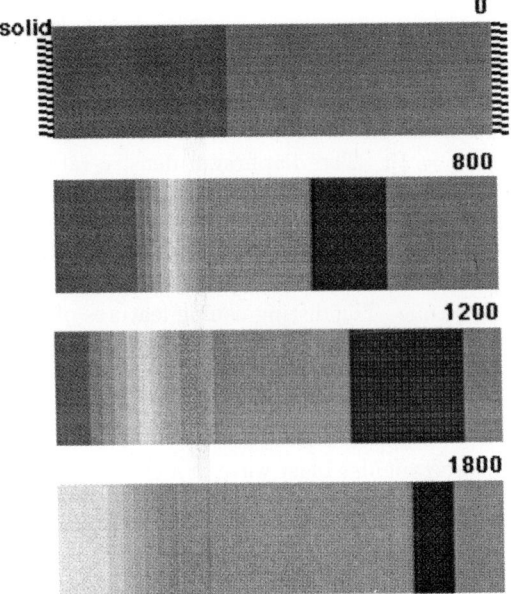

Figure 1. A time history of density contours in a two dimensional shock tube.

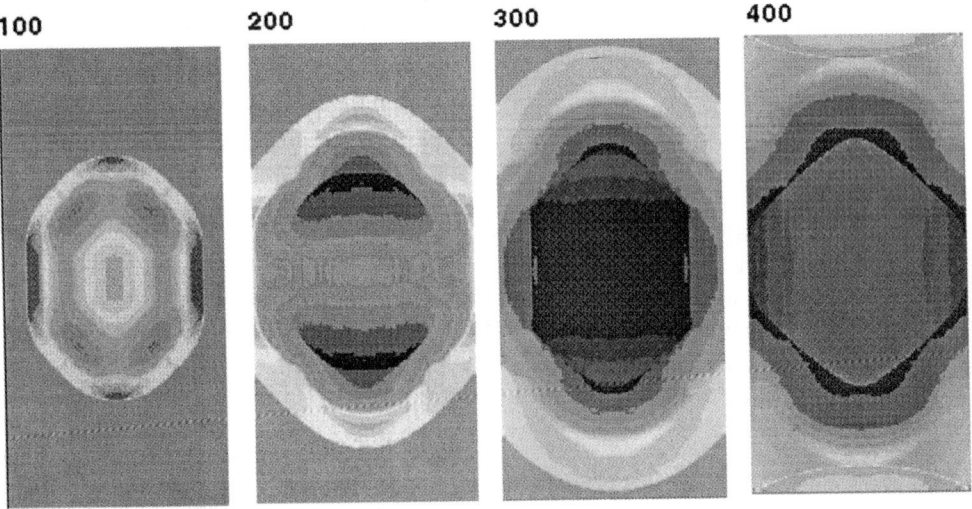

Figure 2. A time history of temperature contours from a supersonic blast wave in a channel.

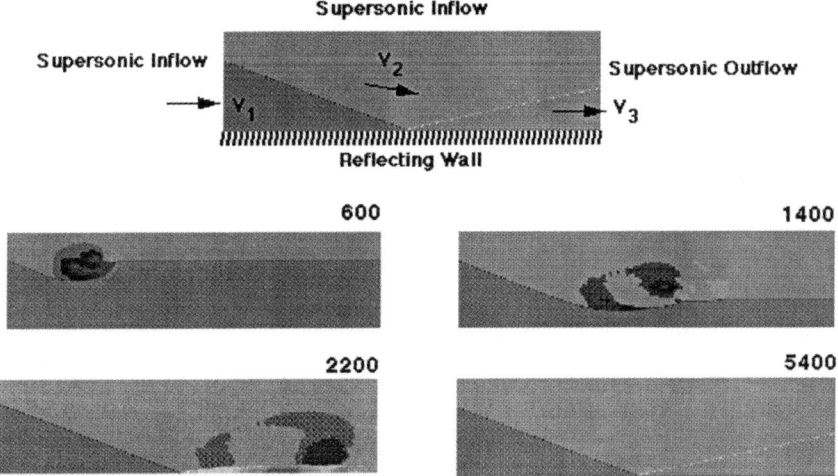

Figure 3. A time history of density contours for a Mach 5 regular shock reflection.

Table 2
Timing comparisons for periodic and nonperiodic boundary conditions (obtained on a CM2 at 7Mhz CM speed)

Grid Points	VP Ratio	PBA (s/timestep)	Coupled UBA (s/timestep)	Uncoupled UBA (s/timestep)
16K	1	.059	.065	.086
32K	2	.098	.107	.141
64K	4	.166	.184	.235
128K	8	.306	.337	.427
256K	16	.562	.623	.774

the incident shock is formed, the vortex follows the formation of the reflected shock. Note that when the flow has reached a steady state at the correct solution the starting vortex has been swept off the domain.

4. PERFORMANCE ANALYSIS

The performance of the coupled and uncoupled versions of the UBA is assessed from the speed, memory requirements, and program flexibility of each method. Performance values for the boundary algorithms were obtained from the UMIACS CM2 using 16K bit serial processors each rated at 7.0 Mhz. The programs were written in CM FORTRAN and executed using PARIS mode since the slicewise compilation mode is not supported for the CM2. Some calculations were performed on the CM200 at NRL using the slicewise mode which is available for CM FORTRAN only. The slicewise mode provided more efficient floating point operations and therefore reduced run times.

4.1. Timing Comparisons

For timing comparisons, a two-dimensional FCT flow solver with periodic boundary conditions was chosen as the baseline algorithm (PBA). As mentioned before, periodic boundary conditions are built into the CM2 hardware by the wrap-around communications network at no added cost to the user. Timings of the baseline code were obtained for problems of different grid sizes. These timings were then compared with equivalent cases where the coupled and uncoupled UBA were implemented to calculate nonperiodic boundary conditions. The results of the timings are given in Table 2 in seconds per timestep.

The timings in Table 2 indicate that for solutions of the nature described in this paper the coupled UBA performs significantly better than the uncoupled UBA. The CPU time for the coupled approach exceeds the run times of the PBA by approximately 10% for all virtual processor (VP) configurations, while the run times of the uncoupled UBA exceed the PBA timings by as much as 45%. The increase in run time by the coupled UBA is due to three added floating point operations required for considering nonperiodic boundary conditions when evaluating the expression given by equation (6). When run on the CM200 to take advantage of the faster clock speed and slicewise execution mode, the PBA speeds are reduced by 55%. Similar improvements can be expected for the coupled UBA approach.

The large cost for computing boundary conditions using the uncoupled UBA was expected due to the added floating point operations and NEWS communication calls. The uncoupled UBA adds two communication calls and six floating point operations each time the boundary conditions are computed as reflected by equation (2). Unlike the coupled UBA where the penalty remains constant for different VP ratios, the penalty incurred by the uncoupled approach decreases to 38% as the VP ratio increases to sixteen. This reduction as the VP ratio is increased stresses the significance of the added communication calls, where physical interprocessor communication for nearest neighbors decreases as the number of virtual processors grows. Finally, the local boundary algorithm described in Section 2.1.1. was used with the FCT code to assess the affects of address calculations on the CPU time of a nonperiodic problem. For a problem size of 32K grid points, the local boundary method was found to be 55% slower than the uncoupled UBA, thus demonstrating the effectiveness of the array multiplier masks implemented by the UBA approaches.

4.2. Memory Requirements and Program Flexibility

Memory requirements are another parameter for assessing the performance of the boundary algorithms. In general, additional memory must be allocated to implement the UBA. The additional memory requirements result from the application of precomputed array multiplier masks and the declaration of dead cells to store boundary values. In the present work, where a direction-split integration technique is employed, eleven array multipliers were added to the PBA when the uncoupled approach was applied, while twenty array multipliers were needed when the fully coupled method was used. Note that while there may be a savings in the number of array multipliers, the uncoupled UBA requires twice as many dead cells as the coupled boundary algorithm. The number of arrays presented here is by no means optimum and is subject to the ingenuity of the programmer and the complexity of the program. Li, for example, has suggested that as few as four array multipliers can be applied with the UBA. Storage requirements can also be reduced by using integer or bit arrays, however the bit arrays are possibilities only for PARIS mode execution or programs written in C*.

In both the coupled and uncoupled versions of the UBA, the coefficients used to compute the boundary conditions are arrays themselves and their values can be varied from one grid point to the next. Thus, the possibility of simulating complex internal boundaries is available at no extra cost to the programmer. This is in direct contrast to local boundary algorithms that select boundary cells using array indexing. For these methods, internal boundaries add directly to the computational cost and the complexity of the program. Thus, in addition to the faster run times, the array multiplier masks are an attractive feature of the UBAs since complex, internal geometries can be handled with the same ease as the problems investigated in the present work.

5. CONCLUSIONS

The Uniform Boundary Algorithm is a method designed to efficiently implement nonperiodic boundary conditions for fluid dynamics problems on massively parallel SIMD computers. In the present work, two versions of the Uniform Boundary Algorithm are incorporated with the Flux-Corrected Transport method in CM FORTRAN to simulate high speed, compressible flows. In the first approach, the Uniform Boundary Algorithm is independent of the routine used to integrate the governing equations. This uncoupled method reduces inefficiencies that are present when traditional serial approaches to calculating boundary conditions are applied on parallel machines. However, the approach introduces two additional NEWS communications and six additional floating point operations each time the boundary conditions are employed. In the second approach, the form of the Uniform Boundary Algorithm is directly linked to the integration routine. For the FCT algorithm, the Uniform Boundary Algorithm adds the boundary conditions directly to the expressions calculating flux differences and cell-interface averages. This has the advantage of introducing no new communications to the scheme and, for the approach discussed in this work, adds only four floating point operations to the FCT method each time the boundary conditions are considered.

The Uniform Boundary Algorithm was successfully applied to compute the boundary effects on a two-dimensional shock-tube, a supersonic blast wave, and a Mach 5 regular reflection. The performance of the Uniform Boundary Algorithms was then assessed from the speed, memory requirements and flexibility of each approach. It was found that the coupled Uniform Boundary Algorithm increased the CPU time of existing periodic codes by 10%, while the uncoupled approach incurred penalties ranging from 38-45%, depending on the virtual processor configuration. The superior performance of the coupled approach was anticipated due to the reduced number of floating point operations and elimination of superfluous NEWS communications.

Increased memory requirements for the Uniform Boundary Algorithm result from the application of precomputed array multiplier masks and the declaration of dead cells to store boundary values. The number of array multipliers can be reduced by additional efforts to combine the coefficients for the stencils used in the direction-split integration. It has been suggested that as few as four array multipliers are necessary to implement the Uniform Boundary Algorithm with the FCT method. Storage requirements can be further reduced by using integer or bit arrays wherever possible. An attractive feature of the Uniform Boundary Algorithm is that array multiplier masks serve as coefficients in the finite-difference stencils such that complex internal geometries can be modeled with the same ease as external boundaries. In conclusion, it has been found for the FCT integration method that the coupled Uniform Boundary Algorithm provides an efficient and flexible approach for calculating nonperiodic boundary conditions on massively parallel computers.

6. ACKNOWLEDGEMENTS

The authors would like to thank the University of Maryland Institute for Advanced Computer Studies (UMIACS) and the parallel computing staff for providing access and support to the CM2. We are also grateful to Robert Whaley of Thinking Machines Corporation, and Mike Young and the CMF staff at NRL for their guidance and expertise with the CM200. We would like to acknowledge Lyle Long at Penn State University for the conversations we shared on boundary condition algorithms. Funding for this work was sponsored in part by the U.S. Army Research Office through the NDSEG Fellowship Program, the Minta Martin Research Fund, and the Naval Research Laboratory through ONR and through the DARPA Applied and Computational Mathematics Program.

7. REFERENCES

1. E.S. Oran, J.P. Boris, R.O. Whaley, and E.F. Brown, Supercomput. Rev., May, (1990) 52.
2. J. Myczkowski, M. Bromley, D. McCowan, Paper No. AIAA-91-0436, AIAA, Washington, DC (1991).
3. C. Li, E.S. Oran, and J.P. Boris, to appear, Parallel CFD, H.D. Simon, (ed.) MIT Press, Cambridge (1992).
4. J.P. Boris and D.L. Book, Meth. Comput. Phys., 16 (1976) 85.

Parallelising explicit and fully implicit Navier-Stokes solutions for compressible flows

X.Xu, N.Qin and B. E. Richards.

Department of Aerospace Engineering, University of Glasgow, G12 8QQ, UK.

Abstract

The paper describes two studies involved with the parallelisation of algorithms for the numerical calculation of hypersonic viscous flows over generic vehicle configurations by solving the Navier-Stokes equations. One involved a scalable explicit formulation that achieved high parallel efficiency on 32 processors of an Intel iPSC860 Hypercube when calculating the 3-dimensional flow over a blunt delta wing at high incidence, the other involved a fully implicit formulation using a Newton-like procedure with a GMRES solver with preconditioning when high efficiency was achieved when run on 8 transputers of a Meiko Computing Surface computer in order to calculate the flow over a cone at high incidence. This latter approach, although more complex, has potential in providing more rapid convergence characteristics, hence efficiency, than the explicit scheme. High order upwind discretisation was used in each case in order to achieve high resolution of important shock and viscous phenomenon within the flow field. The work reported contributes to the aim of a wider programme of work, a summary of which is included, to provide the computational tools to calculate accurately and efficiently steady and unsteady viscous compressible flows over complex aerospace configurations.

1. INTRODUCTION

Following many studies by the Computational Fluid Dynamics (CFD) Team at Glasgow University on the adaption of current state of the art CFD techniques towards predicting hypersonic viscous flows, it is apparent that with increasing complexity of application, there are difficulties in convergence to sufficient accuracy using explicit and approximate implicit schemes without using massive amounts of time on even the most powerful of national supercomputing facilities. A brief commentary on these studies is given in the next section. There appears to be a contemporary train of thought that with the advent of more powerful computers these difficulties will be easily overcome and that the present algorithms will be sufficient. This view however ignores several general points. The past history of rapid developments in CFD was achieved not only by the availability of powerful hardware but also by the development of clever new algorithms. Secondly, the predictions of continuing rapid advances in the future have assumed that the opportunities offered by new architectures will be fully grasped. It is with this in mind that the team has explored, for hypersonic and transonic applications, the characteristics of CFD discretisation techniques and algebra solvers and from this experience either chosen those that give the best accuracy or developed new techniques, such as acceleration procedures or algebra solvers, where these are deficient. Both vector processors and parallel architectures have been used in these developments. The developments to date in parallel computing will be the main focus of this paper.

2 BACKGROUND

In this work a cell-centred finite volume formulation is used to reduce sensitivity to the highly stretched/skewed meshes generated in the types of problem to be tackled. Structured grids are used for spatial discretisation, but unstructured grids are not precluded. An upwind approach, namely Osher's flux difference splitting (FDS) method, has been chosen to be used following demonstrations [1,2] that it provides low numerical dissipation in dealing with viscous/ turbulent flows. By using a third order MUSCL scheme, good resolution of shocks, boundary layers and wakes is achieved so that the number of grid meshes can be reduced for a particular problem. These demonstations have been done on the corner flow problem of two intersecting 8 degree wedges at right angles and $30°$ sweep and at $M = 12.75$, $Re= 5 \times 10^6$/m, $T = 38.73$ K, wall $T = 300$ K [1] and the Tracy [3] flow of a $7°$ half-angle cone at $0°$ angle of attack, $M = 7.95$, $Re = 4.2 \times 10^6$m, $T = 55.4$ K, wall $T = 309.8$ K [2].

The high order explicit finite volume Osher FDS approach has been deployed to numerically solve the fully 3D-NS equations. An explicit multi-stage Runge-Kutta method was employed in the iterative time integration in the 3D-NS part using local time stepping. The full NS model needs to be used for forebody calculations and for separated flows due to upstream influence, but many afterbody flows including viscous effects can be tackled adequately by using 3D-P(arabolised)NS modelling. In a 3D-PNS code developed, the unsteady term and the viscous derivatives in the marching direction were omitted and the streamwise pressure gradient terms treated using Vigneron's approach. In the resulting marching method an implicit treatment in the marching plane was employed which required the iterative solution of a non-linear system at each station. A combined programme has been used successfully [4] to calculate the flow field around a generic vehicle configuration representing the ESA Hermes which was in the shape of a blunt delta wing as illustrated in Figures 1 (the grid used) and 2 (results of the iso-Mach no. lines). Despite the success in calculation, however, overall convergence was slow. This is attributed to solution dependence on cell Reynolds number, necessitating the use of the very stretched grids to resolve the high gradients in thin shear layers near walls and the use of the sophisticated high order upwind scheme.

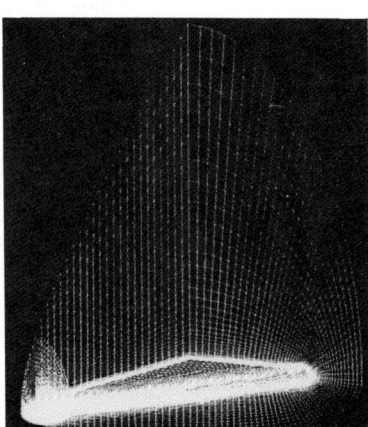
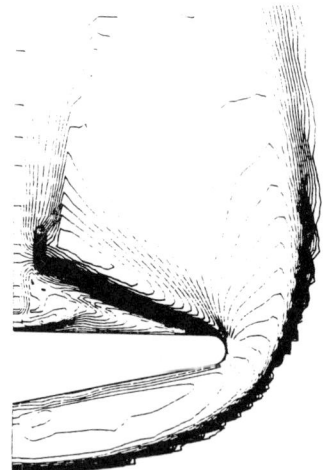

Fig 1. Grid around blunt leading-edged delta-wing. Symmetry plane and rear surface.

Fig 2. Mach number contours at cross sectional plane at 50% chord of delta wing.

Current means of overcoming this convergence difficulty have been through developments in multigrid methods aimed mainly (but not exclusively) at explicit methods and through implicit time marching approaches which have been shown to achieve a degree of success in Euler solvers for inviscid compressible flows. However these developments have been disappointing for complex problems in that the efficiency of multigrid techniqes have been found to be significanly reduced on these highly stretched grids. Furthermore implicit procedures involve linearization which, if exact, implies analytical calculation of the Jacobian of the non-linear system. This feature then creates difficulties in dealing with the inherent non-linearity of the Navier-Stokes equations using high resolution schemes. The result is that the majority of implicit methods use an approximate linearization, using for example only first order upwind inviscid terms in the implicit operator, which generally results again in poor convergence for Navier-Stokes solvers in this case because of the unbalanced left and right hand sides.

A fully implicit approach enables virtual elimination of the stability problems that cause these comparatively slow convergences in explicit and approximate implicit approaches which bear also upon their ability to tackle efficiently complex steady and time-accurate unsteady flows. The application of the sparse finite difference Newton (SFDN) and sparse quasi-Newton (SQN) methods, developed for CFD applications at Glasgow [5], to the evaluation of the Jacobian, provides the ability to tackle highly non-linear problems generated for example by transonic behaviour, turbulence and real gas effects and provides fast convergence (quadratic or superlinear, respectively).

Adopting these approaches, however, does pose the dual problem of dealing efficiently with the solution of the resulting large 13-point (for 2-d, 21 for 3-d cases) ill-conditioned sparse non-symmetric linear system and providing for the large memory required in this approach. The efficient linear solver recently devised in this continuing research [6] to tackle the first problem is a multi-level iterative method using the generalised minimum residual (GMRES) approach with a diagonalised preconditioner and a damping factor (α, giving rise to the name α-GMRES solver).

The overall scheme has been tested, with excellent comparison, on the case due to Tracy [3] of the laminar flow over a $10°$ cone at $24°$ incidence in a Mach 7.95 flow with Reynolds number of 4.2×10^6, wall temperature of 309.8 K and freestream temperature of 55.4 K. The flow is a demanding case including massive flow separation, bow and flow embedded shocks and very high temperature gradient in the windward boundary layer as illustrated in Figure 3.

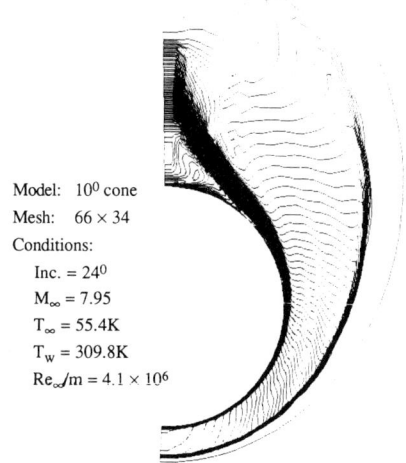

Fig 3. Flowfield temperature contours at a cross section of the cone.

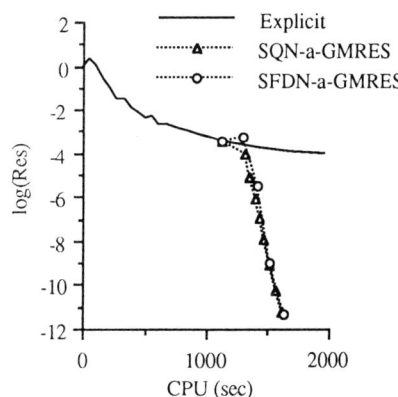

Fig 4. Convergence of SFDN and SQN methods using α-GMRES solver for NS solution.

However since the flow is nearly conical then it can be well modelled by the locally conical Navier-Stokes equations thus reducing the dimensions of the problem by one. Mesh sizes of 34 x 34 and 66 x 34 have been used. Fig. 4 demonstrates using an IBM RS6000, the fast convergence at low processor cost of the SFDM-α-GMRES and SQN-α-GMRES methods, once the calculation had been started by an explicit method. This starting solution shown used the time dependent approach with Runge-Kutta integration and local time stepping. This approach is robust for starting the solution from freestream conditions, but slow in convergence. An approximate eigenvalue analysis [7] using Arnoldi's method revealed the favourable effects of preconditioning and damping on the eigenvalue spectra which explains the successful convergence characteristics of the GMRES method.

In the overall research programme numerical solutions of the Reynolds' averaged 3-d N-S equations have been made using the Baldwin-Lomax turbulence model [8]. The non-equilibrium models of Johnson-King and K~ε models which have been deployed before in earlier 2-d studies are presently being incorporated. Some fast subroutines for including equilibrium air [9] and eventually non-equilibrium and frozen air for very high gas energies have been developed to tackle high enthalpy cases. Additions of this nature to a Navier-Stokes model make the systems to be solved more non-linear which thus make solution convergence even more difficult to achieve.

It is to explore this magnitude, in terms of processing and also particularly memory requirements, of problem associated with viscous aerodynamic design support, that algorithms are now being developed by the Research Team to make use of the new generation of scalable distributed memory parallel computers. The paper then describes two pieces of work involved with the parallelization of algorithms for the numerical calculation of hypersonic viscous flows over generic vehicle configurations. Efficient message passing and balanced use of the multi-processors has been a theme of the research. Chapter 3 describes a scalable explicit formulation that achieved high parallel efficiency on 32 processors of an Intel iPSC860 Hypercube for solving the full 3-dimensional Navier-Stokes equations. Chapter 4 describes a fully implicit formulation using a Newton-like procedure and a GMRES solver with pre-conditioner in which high parallel efficiency was achieved on 8 transputers of a Meiko Computing Surface in solutions of the locally conical Navier-Stokes equations.

3. PARALLELISING THE EXPLICIT FINITE VOLUME SOLVER

In this section, we present some aspects of parallelisation of the explicit three dimensional finite volume Navier-Stokes solver on a hypercube parallel computer. Aspects of the parallelisation of the implicit operator will be dealt with in the next section.

Parallel processing introduces several concepts into the design of the software, that are unique to parallel programming. These are the concepts of "load balancing", "communication" and "scaling". After a brief description of the hypercube parallel computer in Sec.3.1, we discuss the parallelisation of the three dimensional finite volume Navier-Stokes solver in the following sections according to these basic principles.

3.1 The Intel iPSC/860 Hypercube Computer

The INTEL iPSC/860 hypercube is a parallel computer that makes concurrent computation available at a relatively low cost. It consists of a set of independent i860 processors, each with its own memory and capable of operating on its own data. Each processor has its own program to execute and processors are linked together by communication channels. Due to the "hypercube" connectivity, the number of the processors is always of the form 2^d, where d is an integer and known as the dimension of the hypercube.

The hypercube architecture has a number of desirable features. The network provides a good balance between the requirement for a high degree of connectivity between the processing nodes and the engineering requirements of ease of construction and minimal cost. In addition, the hypercube provides a "fixed" architecture (in the conceptual sense), which simplifies the design of software and allows the rescaling of an algorithm to larger numbers of nodes to occur in a simple way.

Communication between nodes is handled by generic routines supplied by the manufacturer. These routines permit the user to send and receive messages of specified length between specified nodes.

3.2 Domain Decomposition and Load Balance

Domain decomposition is a technique well suited for CFD problems, where the fundamental data structures can be decomposed by splitting up the physical domain of the computation. For the present three dimensional Navier-Stokes calculation, the domain decomposition has been achieved through subdividing the three dimensional block into slices in the streamwise direction.

The discretisation used in the Navier-Stokes solver requires a 21-point stencil. At the interfaces of each subdomain, the information at two grid surfaces in the neighbouring subdomains is required to complete the update of the solution at all the cell centres in the current subdomain. Therefore communication between nodes has to be arranged.

It is important that the total workload is equally distributed among the nodes to avoid wasting the computing resources of the system. Thus the above subdivision is carried out in such a fashion that each subdomain will have the same amount of grid points and therefore the same amount of workload apart from a small difference due to the two boundary slices.

3.3 Scalability

Another of the programming principles is to make the code scalable; that is, to program it so it can be executed independent of the number of nodes currently available to be used. A special effort has been made to make the parallel code scalable. The code adjusts the array size according to the size of the subcube allocated to the node and is made independent of the size of the cube.

3.4 Accuracy and Efficiency

It is obvious that the accuracy and the iterative convergence of the sequential code should be maintained with the current parallel approach. To study the efficiency of the parallel approach, computation has been carried out to solve the hypersonic viscous flow around a blunt delta wing related to the European Hermes space programme using a C-O grid of the size 65x33x33 similar to that in Fig 1 (earlier). The results are identical to that illusttrated in Fig 2 (earlier). The computing time, speedup and efficiency are shown in the following table using 1 to 32 processors.

d	No. of Processors (p)	Computing Time (T_p, Second/Iteration)	Speedup ($S_p=T_1/T_p$)	Efficiency ($E_p=S_p/p$)
0	1	129.5	1.00	1.00
1	2	65.3	1.98	0.99
2	4	33.0	3.92	0.98
3	8	17.6	7.35	0.91
4	16	9.0	14.39	0.90
5	32	5.4	23.98	0.75

In the present computation, the surface communication time is comparatively shorter than the computation time in the subdomain, which yields in a relatively high performance on the medium grain parallel computer. It is clear from the table that the efficiency reduces as the number of the processor increases. This is because in the present computation the total computation per iteration is constant and the communication time for each subdomain (or processor) is also constant since the same amount of surface data has to be sent or received no matter how large or small is the subdomain. However the computation time in each subdomain is shorter if the subdomain is smaller when using more processors. Thus the ratio of the computation time to communication time becomes smaller and smaller if more and more processors are used. The performance given in the table above is compatible with a communication time, T_c, of around 1.1 seconds. It can be seen that T_c is small when compared with a computing time of 65.3 seconds for the 2 processor case, but significant when compared with 5.4 seconds for the 32 processor case.

The number of iterations required for reasonable convergence in this implicit approach is high (about 8000 iterations). Multigrid acceleration will reduce processor time, however the authors feel that a more efficient approach is through the parallelised fully implicit scheme outlined in the following section.

4. PARALLELISING THE FULLY IMPLICIT SOLVER

In the Background Section in which the fast convergence characteristics of the fully implicit approach was illustrated we also referred to the large memory requirements of the method. This memory problem that becomes critical for very large grids is being tackled through the use of distributed memory multi-processors. These provide the features of fast processing, high memory and low cost enabling these large problems to be tackled. Parallelisation is then used both to evaluate the Jacobian and solve the linear system

First we discuss the linear solver. We denote the linear system by

$$A x = b \qquad (4.1)$$

where the structure of A depends on the spatial discretisation scheme used. Typically we consider the following system resulting from a second or third order high resolution scheme using a structured grid for a two-dimensional Navier-Stokes solution. The linear system will be a block 13-point diagonal matrix which can be denoted as

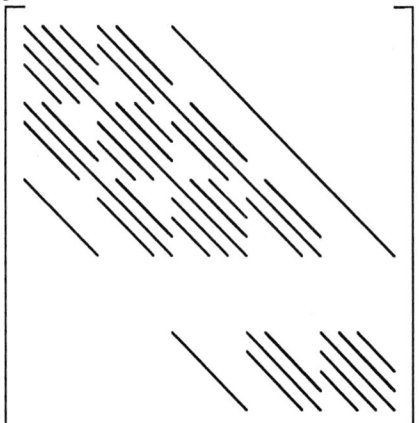

4.1. The α-GMRES Method

For the large sparse non-symmetric linear system $A x = b$, the α-GMRES method [6] is written as

$$(\alpha I + D^{-1} A) x^{n+1} = D^{-1} b + \alpha x^n. \qquad (4.2)$$

Given x^n, the above equation is solved for x^{n+1} using the GMRES method [10]. This procedure is continued until the sequence x^n is converged, and the convergent vector is the solution of the original linear system. Here D is a block diagonal matrix of A, I is a unit matrix, and $\alpha > 0$.

4.2. The Parallel α-GMRES Method

It is assumed there are P processors available.

4.2.1. Data Distribution

The matrix A can be written in columns as $A = [A^1, A^2, ..., A^P]$, where A^p are N×L submatrics, p=1,2,...,P. The transposition of vector v can be written as $v^T = [(v^1)^T, (v^2)^T, ..., (v^P)^T]$ where v^p is vector of order L corresponding to A^p, p=1,2,...,P. A^p and v^p are stored in each processor p. The distribution of the matrix data in columns does not increase the data storage compared to the sequential case.

Remark: The L in the order of N×L may be not the same in different processors.

4.2.2. Parallel Algorithm

Let ε_1 be the convergence criterion of the GMRES algorithm and ε_2 be the convergence criterion of the α-GMRES algorithm. In processor p, we perform the following calculations and communications.

Step 1: Initialisation

Set an initial guess x^p_0, we have

$$A^p x^p_0 = \bar{r}^p_0 ,$$
$$r^p = b^p - \bar{r}^p , \qquad (*)$$

and

$$\|r_0\| = \sqrt{\sum_{p=1}^{P} (r^p_0, r^p_0)} . \qquad (**)$$

Let $\delta = \|r_0\|$ and $w^p_0 = x^p_0$.

Remarks:

(*): Matrix-vector multiplication can be generated as follows:

$$A v_i = (A^1, A^2, ..., A^P) \left(\begin{pmatrix} v^1_i \\ 0 \\ \vdots \\ 0 \end{pmatrix} + \begin{pmatrix} 0 \\ v^2_i \\ \vdots \\ 0 \end{pmatrix} + \cdots + \begin{pmatrix} 0 \\ 0 \\ \vdots \\ v^P_i \end{pmatrix} \right) = A^1 v^1_i + A^2 v^2_i + \cdots + A^P v^P_i$$

$$= \begin{pmatrix} * \\ * \\ \vdots \\ * \end{pmatrix} + \begin{pmatrix} * \\ * \\ \vdots \\ * \end{pmatrix} + \cdots + \begin{pmatrix} * \\ * \\ \vdots \\ * \end{pmatrix} \Rightarrow \begin{pmatrix} \bar{v}^1_i \\ 0 \\ \vdots \\ 0 \end{pmatrix} + \begin{pmatrix} 0 \\ \bar{v}^2_i \\ \vdots \\ 0 \end{pmatrix} + \cdots + \begin{pmatrix} 0 \\ 0 \\ \vdots \\ \bar{v}^P_i \end{pmatrix}$$

where "⇒" indicates the communication of data among different processors to form $\bar{v}_i$. In this way, the task of calculating $A v_i$ for P processors is divided by calculating $A^p v^p_i$ on processor p. This is the main calculation in the GMRES method. The resulting vector $\bar{v}_i$ is again distributed to the P processors. The only communication required in the calculation is in the formation of $\bar{v}_i$. Due to the sparsity of the matrix A, this communication is only of a limited nature.

(**): Here requires the collection of the partial inner products carried out on each processor.

Step 2: Calculate $B = (\alpha I + D^{-1}A)$ and $c = D^{-1}b$

We can write B in columns as $B = [B^1, B^2, ..., B^P]$, which has the same stencil as matrix A. Parts of D^{-1} are calculated in each processor separately. Thus $D^{-1}A$ can be performed in each processor provided that appropriate communications are arranged. Then the α is added in diagonal elements in each processor so that we have B^P in each processor. $c = D^{-1}b$ can be performed in each processor without any communication.

Step 3: Calculation

Let $e^p_0 = c^p + \alpha w^p_0$ and we have

$$B^P w^P_0 = \bar{f}^P_0,$$
$$f^P_0 = e^P_0 - \bar{f}^P_0,$$

and then set

$$\hat{v}^P_1 = f^P_0,$$

so we have

$$\|f_0\| = \sqrt{\sum_{p=1}^{P} (f^P_0, f^P_0)}.$$

$$v^P_1 = \frac{f^P_0}{\|f_0\|}.$$

Let $\delta_1 = \|f_0\|$ in the first iteration.

Step 4: For i=1 to k

$$B^P v^P_i = \bar{v}^P_i,$$

the elements of the Hessenberg matrix are calculated using

$$\beta_{i+1,j} = \sum_{p=1}^{P} (\bar{v}^P_i, v^P_j).$$

We then calculate

$$\hat{v}^P_{i+1} = \bar{v}^P_i - \sum_{j=1}^{i} \beta_{i+1,j} v^P_j,$$

and

$$\|\hat{v}_{i+1}\| = \sqrt{\sum_{p=1}^{P} (\hat{v}^P_{i+1}, \hat{v}^P_{i+1})},$$

and normalise the base vector as follows

$$v^P_{i+1} = \frac{\hat{v}^P_{i+1}}{\|\hat{v}_{i+1}\|}.$$

After k steps, the Hessenberg matrix is

$$H_k = \begin{pmatrix} \beta_{2,1} & \beta_{3,1} & \cdots & \beta_{k+1,1} \\ \|\hat{v}_2\| & \beta_{3,2} & \cdots & \beta_{k+1,2} \\ 0 & \|\hat{v}_3\| & \ddots & \vdots \\ \vdots & \vdots & \ddots & \beta_{k+1,k} \\ 0 & 0 & \cdots & \|\hat{v}_{k+1}\| \end{pmatrix}_{(k+1)\times k}.$$

<u>Step 5</u>: Uses a Q-R algorithm to find y such that

$$\| \delta_2 e_1 - H_k y \| = \min_{y_0 \in R^k} \| \delta_2 e_1 - H_k y_0 \| \quad ,$$

where $y = (y_1, y_2, ..., y_k)^T$, $e_1 = (1, 0_1, ..., 0_k)^T$ and $\delta_2 = \|f_0\|$, so we have $w^p = w^p_0 + \sum y_k v^p_k$.

<u>Step 6</u>: Calculation

$$B^p w^p = \bar{f}^p ,$$
$$f^p = e^p_0 - \bar{f}^p ,$$

and

$$\|f\| = \sqrt{\sum_{p=1}^{P} (f^p, f^p)} \quad .$$

If $\|f\| < \delta_1 \times \varepsilon_1$ then we go to next step, or else let $w^p_0 = w^p$ and go to step 3.

<u>Step 7</u>: Calculation

$$A^p w^p = \bar{r}^p ,$$
$$r^p = b^p - \bar{r}^p ,$$

and

$$\|r\| = \sqrt{\sum_{p=1}^{P} (r^p, r^p)} \quad .$$

If $\|r\| < \delta \times \varepsilon_2$ then the algorithm is stopped, otherwise we let $w^p_0 = w^p$ and go to step 3.

4.3. The Parallel N-S Solver

The method of distributed storage of the matrix data results in the corresponding geometric domain decomposition in solving N-S equations which is illustrated in Fig.5.

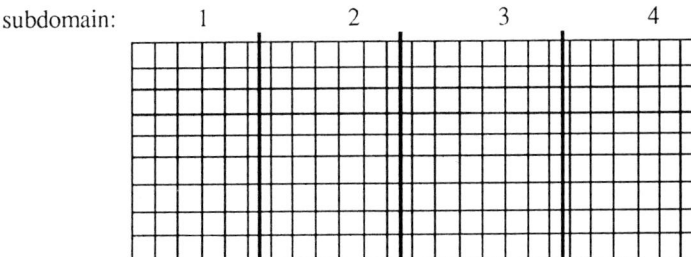

Fig 5. The domain decomposition.

The discretised N-S equations can be written as follows
$$F(Q) = 0 \tag{4.3}$$
The general Newton's method is
$$\left(\frac{\partial F}{\partial Q}\right)^n \Delta^n Q = -F^n(Q) \tag{4.4}$$
By forming $F(Q)$ and the Jacobian $J = \partial F / \partial Q$ at the known nth iterate, the increment $\Delta^n Q$ is then found by solving the linear system. The value of Q at the new iterate is given by
$$Q^{n+1} = Q^n + \Delta^n Q \tag{4.5}$$
In the parallel case the $F(Q)$ can be formed in each required subdomain.

The approach is to replace the analytic Jacobian with a numerically approximated Jacobian [5]. In the sequential computation case: Let h be square root of machine epsilon, if J is a band matrix of band-width $m = 2\rho - 1$ then the difference $F_i(Q + h \, e_j) - F_i(Q)$ is zero if $|i - j| \geq \rho$; it follows that we may find simultaneously approximations to columns $j + km$, $k = 0, 1, 2, ...$, of J from the difference $F_i(Q + \Sigma \, h \, e_{j+km}) - F_i(Q)$. Here the sum Σ is for the subscript k. In this way the total number of subroutine calls needed can be reduced from n+1 to m+1. This strategy positively minimizes the total number of function evaluations in view of the number of unknown coefficients in each row of J. In the parallel case: Because the property of forming the Jacobian is according to columns the Jacobian J can be generated in each subdomain. This procedure should deal with a relative large subdomain compared with the original subdomain, which is divided according to the matrix storage.

After the Jacobian was generated in a parallel form we can use the parallel α-GMRES algorithm to form the completely parallel algorithm. The implicit iterative method begins from a relative 'good guess' and include three iterative loops: the inner GMRES iterative loop, the middle α-GMRES iterative loop, and the outer Newton iterative loop.

In the resulting developed parallel algorithm (described fully in [11]), communication is kept to a minimum and it leads to an efficient geometric domain decomposition type solver to complete this fully implicit overall approach. Fig 6 illustrates the convergence history using the Meiko M40 Computing Surface, and Fig 7 the speed-up achieved using 8 T800 transputer nodes providing a parallel efficiency of 83%. Fig. 8 demonstrates the reduction in memory size per node achieved.

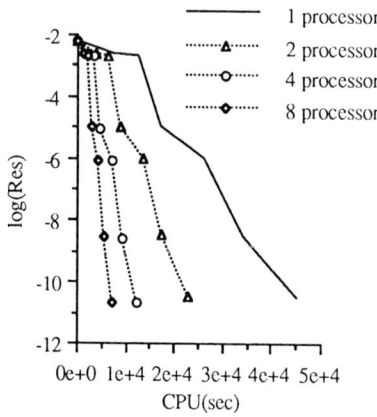

Fig 6. Convergence of N-S solution with different number of processors.

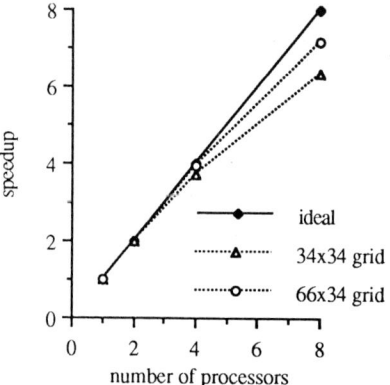

Fig 7. Speed-up acheived using 8 T800 transputer nodes.

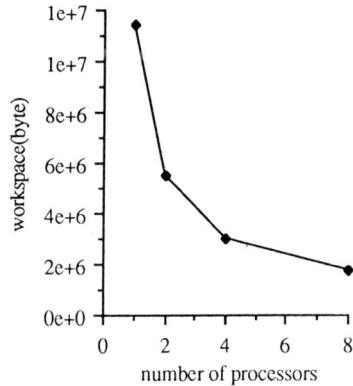

Fig 8. Memory needed for N-S solution with different number of processors

It is planned to extend the work to research the ability of the scheme to tackle more complex multi-dimensional problems using DL's Intel iPSC/860 Hypercube with faster processors and larger memory. The very preliminary work so far done has not brought the efficiency up to the level of that on the Meiko, because the processing on one i860 node is much faster than that on a T800 transputer. Steady state hypersonic and time-accurate unsteady transonic flow cases are planned in the future. The solver could have application in other areas.

5 CONCLUSIONS

Two studies of the parallelisation of algorithms to support the numerical calculation of hypersonic viscous flows over generic vehicle configurations by solving the Navier-Stokes equations have been made. High order Osher flux difference splitting upwind discretisation was used in each case to achieve high resolution of important shock and viscous phenomenon within the flow field. A scalable parallel explicit formulation on an Intel iPSC860 Hypercube computer was developed for calculating the 3-dimensional flow over a blunt delta wing at high incidence. This was demonstrated to achieve high parallel efficiency using up to 32 processors. A fully implicit formulation using a Newton-like procedure with a GMRES solver with pre-conditioning was applied to the calculation of the flow over a cone at high incidence. Again high efficiency was achieved using 8 transputers in a Meiko Computing Surface multi-processor. This implicit approach, although more complex, has potential in providing more rapid convergence characteristics, hence efficiency, than the explicit scheme and it is shown to be parallelisable. The work reported contributes to a programme of work to assist the development of computational tools to calculate accurately and efficiently steady and unsteady viscous compressible flows over complex aerospace configurations.

The first author was supported by University of Glasgow and ORS scholarship awards.

6. REFERENCES

1. N. Qin and B.E. Richards, Aeronautical Journal, 1991 pp 152-160.
2. N. Qin, Presentation to SERC CC-CFD Workshop, November, 1990. University of Glasgow Aero Report 9120.
3. R. Tracy, California Institute of Technology Report, GALCIT Memorandum 62, 1962.
4. N. Qin and B.E. Richards, Proceedings of the Workshop in Hypersonic Flows for Reentry Problems, Part I, 1991, Springer-Verlag 1992.
5. N. Qin and B.E. Richards, Notes in Num. Fluid Mechanics, Vol. 29, Vieweg, 1990.
6. X. Xu, N. Qin and B. E. Richards, University of Glasgow Aero Report 9110. Int. J. of Numerical Methods for Fluid Dynamics (in press).
7. N. Qin, X. Xu and B. E. Richards, University of Glasgow Aero Report 9228. J. of Computing Systems in Engineering (in press).
8. N. Qin and B. E. Richards, Proceedings of the Workshop in Hypersonic Flows for Reentry Problems, Part II, 1991. (in press, Vieweg 1992).
9. J. Anderson, University of Glasgow Aero Report 9206, 1992.
10. Saad, Y and Schultz, M.H. *SIAM J. Stat. Comp.*, Vol. 7, No. 3, 1986, pp. 856-869.
11. X.Xu, N. Qin and B. E. Richards, University of Glasgow Aero Report 9210, April 1992.   Journal of Parallel and Distributed Computing (under review).

Parallelization of KIVA-II on The iPSC/860 Supercomputer

Osman Yasar and Christopher J. Rutland

College of Engineering, University of Wisconsin-Madison, 1500 Johnson Drive, Madison, WI 53706

Abstract

KIVA-II was originally developed at Los Alamos National Laboratory for the numerical simulation of transient, chemically reactive fluid flows with sprays in two- and three-dimension. It was developed for applications of internal combustion engines, but its modularity and generality make it applicable to a wide variety of multidimensional problems in fluid dynamics. Its wide-use in the auto-industry and academia and the impact of massively parallel machines in the computational sciences have drawn our attention to the parallelization of such a computer code. When analyzed with FORGE, an optimization tool for the iPSC/860, the unparallelized version of the code is found to be spending 70 % of its time on several major DO-loops over the spatial mesh. Decomposition of these loops in the z-direction seems favorable to the other directions in terms of the ease of modification and reduction of communication overhead. Hence, a slave-slave control structure was used with a decomposition of the computational work in the z-direction. The parallel version of KIVA-II achieves a speedup of 2.7 on four processors and a speedup of 5 on sixteen processors for a test problem with no piston movement. Yet, only moderate speedups (2.6 on 16 nodes) are obtainable in engine applications with piston movement. This is due to the decreasing grid size in the z-direction during compression. The number of z-layers and thus the computational load per processor decreases during compression, causing the communication/computation ratio to go higher.

1 Introduction

KIVA-II [1] is a time-dependent 3-D CFD code that solves equations of motion of turbulent and chemically reactive mixture of ideal gases and sprays. Its is a finite-difference code and uses either first order (partial donor cell) or quasi-second-order upwind schemes in a control volume approach which largely preserves the local conservation properties of the differential equations.

The spatial domain is divided in cells, the corners of which are the vertices. A typical finite-difference cell and the labels for its vertices are shown in Fig. 1. One can identify vertices with their (i,j,k) coordinates but an alternative notation has been added to KIVA-II to number everything in sequence and with a single variable. Hence, as shown

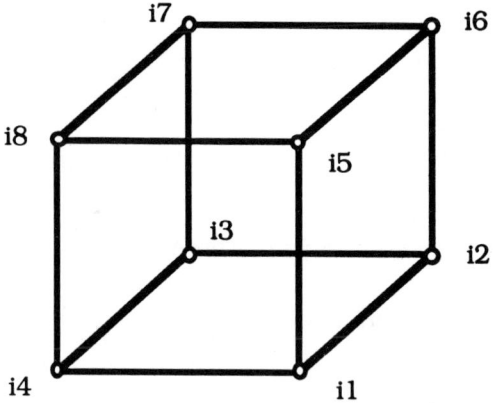

Figure 1. A typical finite-difference cell and its indexing notation.

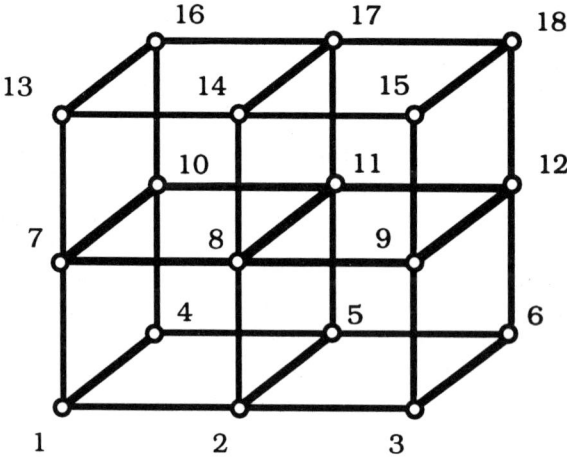

Figure 2. A 3-D KIVA mesh structure and its indexing notation for 1-D array storage.

in Fig. 2, the vertices are numbered as one moves first in x-direction, then y-direction, and z-direction. By doing so, one rather sweeps the whole domain with one DO-loop rather than three nested DO-loops that degrade vectorization on CRAY computers. The same numbering sequence is also used for the cells with the understanding that the sequence of vertex $i4$ is the same as cell (i, j, k). This sequence, as in Fig. 2, is used to store the three-dimensional flow quantities in one-dimensional arrays.

The loop over vertices goes from one to the total number of vertices, called $ijkall$ in the code. When one iterates over the cells, the loop count goes beyond the total number of real cells which may seem odd to the user. When each cell number is matched with its $i4$ vertex, the cells that correspond to the boundary vertices become fake and should also be looped over to reach the last cell without breaking the loop. For a mesh structure in Fig. 2, the loop count for vertices ($ijkall$) is 18, whereas the loop count for cells ($ijkvec$) is 8 including the fake cells 3, 4, 5, 6 and excluding those after the last cell 8. Since we included four fake cells, the real number of cells is 4 as one can tell from the picture.

The temporal differencing is the first-order approximation to the time derivative. Each time cycle has two phases: a lagrangian phase where computational cells follow the fluid (no convection), and a rezoning phase where the flow field is frozen and rezoned. The first phase involves implicit solutions of flow fields whereas the second one (convection) uses explicit solutions. The convection time step is a multiple of the main timestep and does satisfy the Courant condition.

2 Timing and Dependency Analysis

KIVA-II was analyzed using FORGE [2] for both timing and dependency analysis. FORGE is an interactive program with a graphics interface on Sun workstations. It aids the user in understanding the global data and control flow of an existing sequential Fortran program. By providing a powerful and concise display of information, FORGE eliminates the need to dig through stacks of paper listings trying to understand the program.

The most time consuming parts of the KIVA-II program are those that use implicit methods to solve pressure (psolve), velocity (vsolve), temperature (tsolve), turbulence kinetic energy (kesolv), and mass diffusion (ysolve). The implicit method is a two-step iterative procedure (predicted-corrected). In fact, one starts with a predicted value of pressure (linear extrapolation using previous two cycles) and calculates velocity and temperature fields respectively. Based on these, the solution for corrected pressure field is found, which involves simultaneous solutions of cell face velocities, cell volumes, and isentropic equation of state that relates the corrected pressures and volumes. The predicted and corrected pressure fields are then compared to see if convergence has been attained. If so, then most recently predicted velocities and temperatures are carried over to the next step. As one may expect from this solution recipe, among these, the pressure solution should be the most time consuming one. Our timing study indicates that 44.5 % of the execution time is spent in *psolve* whereas other implicit routines altogether account for only 30 %. Another outcome of the timing study is that this 70 % execution time is spent only in seven major DO-loops that iterate over the spatial domain.

These time consuming implicit solvers are concentrated together in one part of the main program, indicating that a large granularity can be obtained for parallel processing. Although obtaining a coarse-grained parallelism is an advantage of distributed-memory parallel machines, one has to deal with determining the data dependencies between the processors which makes programming lengthy and difficult compared to compiler-aided shared-memory parallel environments. It may become combersome to port large-size existing codes if the porter is not very familiar with the code. Nevertheless, the availability of tools like FORGE is increasing [3] and distributed-memory parallel architectures are now being supplemented with hard-working operating systems that offer the easeness of shared-memory parallel programming.

3 Parallel Structure of KIVA

One obvious decomposition strategy for KIVA, as we have been stressing so far, is the domain decomposition where each processor is responsible for some parts of the spatial domain. One can perhaps try functional (control) decomposition where each processor simulates different functions of the parts of the system. With domain decomposition, one can follow two different ways: master-slave and slave-slave cases. In the master-slave case, a master node can run the sequential part of the code and then distribute the tasks when things can be done in parallel. One has to realize that the slave nodes stay idle for some time when the sequential part is running. Also the need for data communication has to be handled explicitly with direct message passing. The second approach is that each node deals only with some given domain during the whole simulation and thus needs to communicate with others as data is needed. To ease the burden of message passing, one can have every processor run exactly the same program including the whole physical domain except where parallelism is an alternative. Although there is some degree of redundancy here as processors do the same calculations for the sequential part, the need for message passing is decreased. In any of these cases, one still has to determine how to decompose the mesh between the processors when calculations are done in parallel.

A typical 3-D structure of finite-difference cells for KIVA is shown in Fig. 2. The vertices are numbered as one moves first in x-direction, then y-direction, and z-direction. As far as the communication with neighbor points is concerned, the number of references in the loop bodies for all directions is about the same. However, the fact that vertices are numbered in ascending order in the z direction makes it eaiser to modify the loops for decomposition. The layers in the z direction are divided between processors and in case of data dependency between these layers one has to have each processor include a layer up or down or both in calculations to avoid message passing. There may appear to be a trade-off between calculation overhead and communication overhead, and thus one needs to check the performance for both cases. Also the number of layers each processor gets should be far more than the number of overlapped layers (maximum of 2) to come up with a low ratio of communication/computation.

4 Parallel Performance

The parallelization was done in two different ways as suggested before. The first choice was to have a master-slave paradigm with master node doing sequential and slave nodes doing parallel computations when reqiured. The second choice was to run the same program on all nodes including the sequential part as well, so that much of the communication is eliminated. Although there is some degree of redundancy here, the pay-off is much better in terms of speedup.

In the master-slave case, we started the parallelization only on the pressure iteration (*psolve* routine) to get a sense about the speedup. This routine takes 44.5 % of scalar run time. Initial results showed a significant performance degradation compared to a single-node run. The parallel code on four nodes becomes 1.52 times slower (one master and three slaves) than the one node version. For every time-cycle master node executes psolve 1.47 times on the average and psolve awakens slave nodes with a 200 KB data. Slave nodes receive the data, assumes float point operations and sends back 30 KB data to the master node. If convergency test is failed, then master node calls back the slave nodes again and this process is repeated 24 times on the average until convergency is achieved. Given the fact that the master node has to send data back and forth to all three nodes so often per convergency, a communication bottleneck may have been encountered and the communication/computation ratio must have been larger than 1 for the computation body in subroutine *resp*. In fact, this ratio is already larger than one per floating point operation ($2.8 \mu sec/25 nsec$) for iPSC/860 system with 2.8 MB/sec bandwidth and 40 MFLOP node performance. So, ideally one would like to have a large granularity to reduce this ratio for the parallel portion of computations.

Considering the first choice non-promising, we have switched to the second one where we run the same code on all processors. The communication bottleneck mentioned above is absent here because every node has the required data in its own storage. The only communication is to update some vector quantitites with global concatenation calls (gcolx). Also, the inter-computation communication is eliminated by having the nodes include in their domains a layer up and/or down. By parallelizing only the psolve branch here, we achieve a 20 % faster run on two nodes. With the inclusion of ysolve, tsolve, vsolve, kesolv running in parallel, the two-node parallel version becomes 1.36 times faster (26 % faster).

Considering the second choice as a promising way, we have added more loops to the parallel mode. The new speedup on two nodes becomes 1.49 with 74 % processor utilization. Figure 3 shows how the the speedup changes with the number of parallel processors. Results are obtained for two different cases with different physics and grid size. The less realistic case (grid: 20x1x79), which we call the *test* problem, has a larger grid in the z-direction (80 mesh points) to explore the scalability of the parallel algorithm. Also it assumes a fixed grid size for the whole duration of the simulation. The second case (20x1x39), which we call the *Los Alamos engine* problem, has a smaller grid size (40 mesh points) and it involves piston movement. The piston movement towards the fixed cylinder head reduces the grid size in the axial direction (z-direction) and thus it results in a less computational load for each processor.

The communication overhead comes from both the *gcolx* global communication and

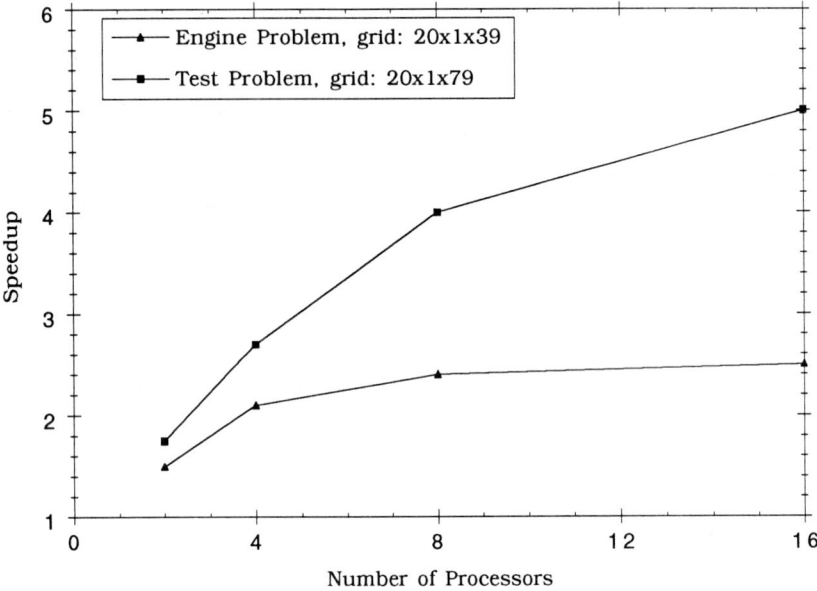

Figure 3: Performance of initial port of KIVA onto Intel iPSC/860 Hypercube.

the overlapping computation. We know that the number of overlapped layers is at least 2 per processor and the total number of layers per processor has to be significantly larger than this for us to be able to gain any benefit from parallel processing. In the test case, the number of layers per processor changes from 40 to 5 when the number of processors go from 2 to 16. Notice then that the processor utilization goes from 75 % down to 30 %. In the engine problem case, though, the number of layers per processor is in the range of 2-20. In fact there might be some processors running idle during part of the calculations when grid size is less than 40. Appearently, the speedup in this case is far below the *test* case, and it also holds no great promises for problems with this grid size on large number of processors. If the patform of choice is a small size parallel machine then perhaps parallelism is a definite plus. However, the gain on larger machines will depend on the size of the problem. In one sense, we can say the algorithm is scalable so long as there is enough computation per processor to compensate for the communication and overlapped-computation overhead.

5 Conclusion

Porting KIVA fluid dynamics code to iPSC/860 has been an important task for us as we indicated before. The initial results are encouraging. KIVA-II now runs in parallel on a massively parallel MIMD machine and the fact that speedup gets better with the problem size is a good indication of its scalability and potential for better improvement. Although decomposition in the z-direction is not suitable for engine applications, it may find usage in areas such as numerical propulsion systems. One can argue however how much more improvement we can get. One can seek decomposition in other directions for engine applications, however, it may introduce more communication overhead since noncontiguous sets of data (i.e., x-z or y-z planes) would have to be passed between processors.

The current parallelization of KIVA-II is only partial. Approximately 50 % of the code is still not parallelized and the data was not split up among the processors. The largest task to completing the parallelization is the data decomposition. This would permit much larger problems to run, but may require a significant increase in communication overhead. One could rework some of the algorithms to increase parallelization and reduce communications overheads. Some parts of the code will not parallelize without major changes to the algorithms (for example the Lagrangian spray calculations). Also, for example, an explicit time differencing could be used instead since it is easier to parallelize due to its comunication requirement only at the end of each time step.

References

[1] A. A. Amsden, P. J. O'Rourke, T. D. Butler, "KIVA-II: A Computer Program for Chemically Reactive Flows with Sprays," Los Alamos National Laboratory Report LA-11560-MS, May 1989.

[2] Forge User's Guide, Intel Corporation, Part # 311866-001, May 1990.

[3] Technology Focus Conference, Intel Supercomputer Systems Division University Partnership Program, April 1991.

[4] Erol Gelenbe, *Multiprocessor Performance*, John Wiley & Sons, New York 1989.

On the Implementation Issues of Domain Decomposition Algorithms for Parallel Computers

Jianping Zhu

NSF Engineering Research Center, P. O. Box 6176, Mississippi State University, Mississippi State, MS 39762, USA

Abstract

It is known that interprocessor communications can have significant impact on algorithms implemented on distributed memory parallel computers. This paper addresses three aspects of implementation issues of domain decomposition algorithms on Intel iPSC/860 parallel computers: Different domain decompositions, nearest neighbor communications and bandwidth improvement by using forced message type. It is found that the domains involved in the computation should be decomposed in such a way as to have sizes in all dimensions as close as possible, and that the interprocessor communication rate can be improved significantly by using forced message type.

1 Introduction

Domain decomposition algorithms [1,2,3,4] are very popular in CFD computations because it introduces natural parallelism on parallel computers. The basic idea is to decompose the original domain where the problem is defined into subdomains and assign different subdomains to different processors for parallel processing. The correct physical relations between subdomains are retained during the solution process by inter-processor communications. The efficiency of an algorithm on a parallel computer depends very much on the implementation details of the algorithm[1], like how the domain is decomposed and how the inter-processor communication is scheduled.

For example, a 2-D rectangular domain can be decomposed into strips as shown in figure 1(a), or into patches as shown in figure 1(b). The scheme in figure 1(a) corresponds to decomposing the domain in the horizontal dimension only, while the scheme in figure 1(b) corresponds to decomposing the domain in two dimensions, both the vertical and the horizontal dimensions. There are even more possibilities for 3-D domains.

2 Different Domain Decomposition Schemes

A spatial domain can be decomposed into subdomains in many different ways. But most of the time, the domains are decomposed into subdomains with simple and regular geometries for simplicity of algorithm implementations. The domains discussed here

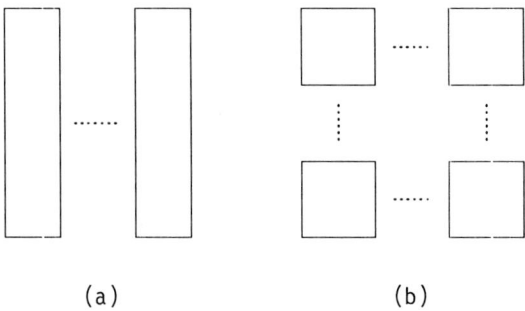

(a)   (b)

Figure 1: two domain decomposition schemes for a 2-D domain

are all rectangular or cubic domains. For more complicated domains, numerical grid generation schemes can be used to map the complex physical domains to the regular computational domains.

We first discuss two ways to decompose a rectangular 2-D domain, i.e. the strip decomposition (D_{21} as shown in figure 1(a)) and the patch decomposition (D_{22} as shown in figure 1(b)).

A subdomain needs to exchange a message with all adjacent subdomains. It is assumed that the number of single precision data in a message is equal to the number of grid points on the boundary.

For the 2-D $m \times m$ domain in figure 1, with strip decomposition algorithm each subdomain needs to exchange with two subdomains a message of length m. The communication time will be

$$T_{D_{21}} = 2(\sigma + 4\beta m) \qquad (1)$$

where σ is the communication start-up time and β is the time required to send one byte of data. The communication (send and receive) subroutines on iPSC/860 require that the data to be exchanged locate in contiguous memory units. If the grid data is stored in an array $A(m, l)$ where $l = m/P$ and P is the number of processors, the data item corresponding to the two boundaries of the subdomain are stored in $A(1:m, 1)$ and $A(1:m, l)$, which are the first and last columns of the array respectively. With FORTRAN, data items in an array is stored by columns, therefore the two data sets to be exchanged are located in the contiguous memory units and can be exchanged directly. If other programming languages are used which map array elements to memory units by rows, the grid data of the subdomain should be transposed and stored in $A(l, m)$ to secure that the data to be exchanged are in the contiguous memory locations.

The disadvantage of this decomposition scheme is that the length of message to be exchanged does not decrease as the number of processors increases. The message length is always m regardless how many processors are used in the computation. On the other hand, the amount of numerical computation is proportional to the number of grid points m^2/P in the subdomain which is decreasing as the number of processors P increases.

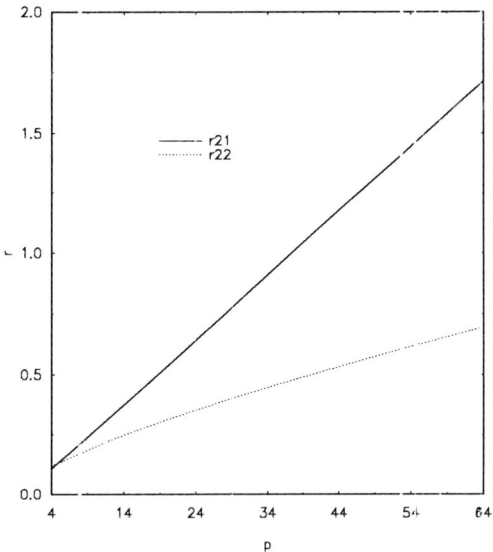

Figure 2: communication/computation ratios vs. the number of processors P for decomposition schemes D_{21} and D_{22}

If the Jacobi iteration is used, the amount of numerical computation will be $4m^2/P$. Therefore the communication/computation ratio is

$$r_{21} = \frac{2P(\sigma + 4\beta m)}{4m^2 \tau} =: \frac{P(\sigma + 4\beta m)}{2m^2 \tau}$$

which increases proportionally to the number of processors P used for the computation. τ is the time required for a floating point operation.

For decomposition algorithm D_{22}, assume $\sqrt{P}$ is an integer and m is divisible by $\sqrt{m}$. The processors can be arranged as a $\sqrt{P} \times \sqrt{P}$ grid and a subdomain will have dimensions $\frac{m}{\sqrt{P}} \times \frac{m}{\sqrt{P}}$. A subdomain needs to communicate with four adjacent subdomains, except the subdomains at the boundary of the original domain which communicate with two or three subdomains. The message length in each message exchange is $\frac{m}{\sqrt{P}}$. The communication time is

$$T_{D_{22}} = 4(\sigma + 4\beta \cdot \frac{m}{\sqrt{P}})$$

In contrast to $T_{D_{21}}$, we can see that $T_{D_{22}}$ decreases as the number of processors increases. The amount of numerical computation is the same as in algorithm D_{21} since a strip and a patch have the same number of grid points. Therefore the communication/computation ratio of algorithm D_{22} is

$$r_{22} = \frac{P(\sigma + 4\beta \frac{m}{\sqrt{P}})}{m^2 \tau} =: \frac{\sigma P + 4\beta m \sqrt{P}}{m^2 \tau}.$$

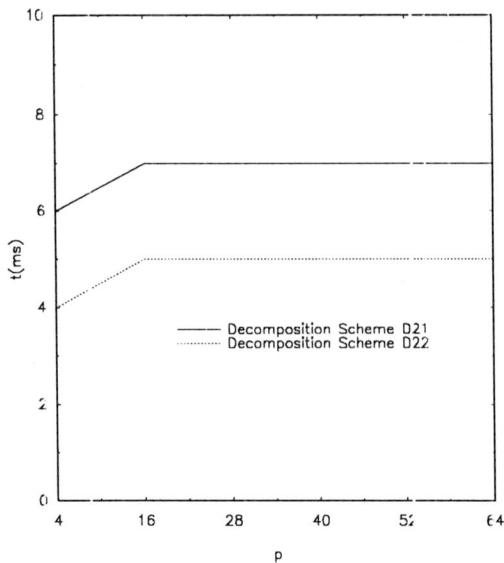

Figure 3: timing curves of decomposition algorithms D_{21} and D_{22} for exchanging messages of a 512×512 domain

It can be seen that one term which corresponds to the start-up time in the communication/computation ratio r_{22} increases proportionally to P. For long messages, the communication time is dominated by the second term $4\beta m\sqrt{P}$ which is a very slowly increasing function.

Figure 2 shows the curves representing r_{21} and r_{22} vs. the number of processors used in the computation. The parameter values are $m = 512$, $c = 100$, $\beta = 0.5$ and $\tau = 0.08$. The machine parameters are based on numerical experiments like those done in [5], not on peak performance values. From the figure it is clear that algorithm D_{22} has a lower communication/computation ratio which increases much more slowly than does r_{21}.

The disadvantage is that an interior subdomain needs to communicate with four adjacent subdomains, as opposed to two subdomains with algorithm D_{21}. If the grid data of a subdomain is stored in an array $A(l,l)$ where $l = \frac{m}{\sqrt{P}}$, then the first and last columns and the first and last rows of A need to be exchanged with the four adjacent subdomains respectively. As discussed before, the first and last columns of A can be exchanged directly by calling communication subroutines on the iPSC/860 since they are stored in the contiguous memory locations. The data items of the first and last rows of A, however, are not stored in the contiguous memory units with FORTRAN. Therefore, they have to be packed into one dimensional arrays before communication, typically using the statements like

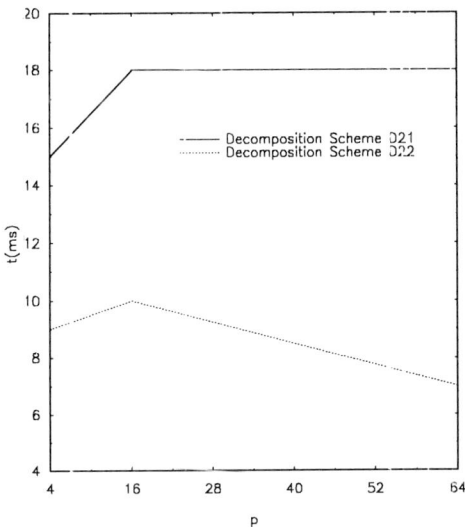

Figure 4: timing curves of decomposition algorithms D_{21} and D_{22} for exchanging messages of a 2048 × 2048 domain

$$\begin{aligned}&do\quad i = 1,\, l\\&\quad buff1(i) := A(1,i)\\&\quad buff2(i) := A(l,i)\\&end\quad do.\end{aligned} \quad (2)$$

To avoid doing this, the programmer must store the boundary data separately from the interior data of the subdomain.

The curves in figure 3 show the timing results of message exchange for a 512 × 512 2-D domain with algorithms D_{21} and D_{22} respectively. With algorithm D_{21}, processors are configured as a chain and each processor needs to communicate with two processors a message of 512 words, except the first and last processors in the chain which communicate with only one processor. When algorithm D_{22} is used, the P processors are arranged as a $\sqrt{P} \times \sqrt{P}$ grid, i.e. 2 × 2 (4 processors), 4 × 4 (16 processors) and 8 × 8 (64 processors) grids. In the more general case when $\sqrt{P}$ is not an integer, the P processors can be configured as a $n_1 \times n_2$ grid with $P = n_1 \times n_2$. The message length is 256, 128 and 64 words for 4, 16 and 64 processors, respectively.

Figure 3 shows that algorithm D_{22} takes less time to exchange messages between subdomains. However, the time used by algorithm D_{22} does not decrease as the number of processors increases. In the case of 4 processors arranged as a 2 × 2 grid, each processor needs to communicate with actually only two processors with a message length of 256 words. When 16 processors are used (arranged as a 4×4 grid), the communication time is dominated by the processors that have to exchange message with 4 adjacent processors. Although the message length is only 128, the reduction of the communication time from the shorter message length is not enough to offset the increase of the start-up times of 4 communications, therefore the total communication time actually went up in the case of 16 processors.

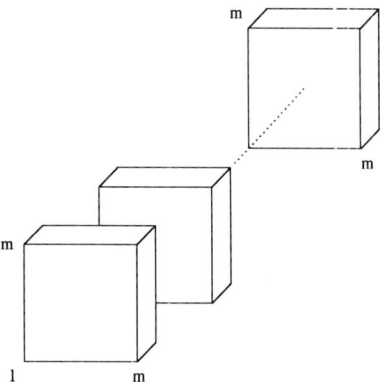

Figure 5: slice decomposition of a 3-D $m \times m \times m$ domain

The curves in figure 4 represent the same thing as those in figure 3, except a much larger domain is used (2048 × 2048). The decomposition scheme D_{22} used much less communication time than did scheme D_{21}. For the same reason discussed for figure 3, it takes more time to exchange messages for 16 processors than for 4 processors. However, since a much larger domain is involved here the reduction of the communication time from shorter message length by using more processors eventually offsets the additional communication overhead incurred as the number of processors increases. Therefore, the total communication time decreases as the number of processors increases.

For three dimensional domains, three decomposition schemes are discussed here, i.e. the slice decomposition (D_{31} shown in figure 5), the column decomposition (D_{32} shown in figure 6) and the cubic decomposition (D_{33} shown in figure 7).

With algorithm D_{31}, the situation is pretty much the same as with algorithm D_{21} in the case of a 2-D domain, except the message length is now m^2 and the amount of computation is proportional to $6\frac{m^3}{P}$. The total communication time is

$$T_{D_{31}} = 2(\sigma + 4\beta m^2)$$

and the communication/computation ratio is

$$r_{31} = \frac{2(\sigma + 4\beta m^2)}{6\frac{m^3}{P}\tau} = \frac{P(\sigma + 4\beta m^2)}{3m^3\tau}.$$

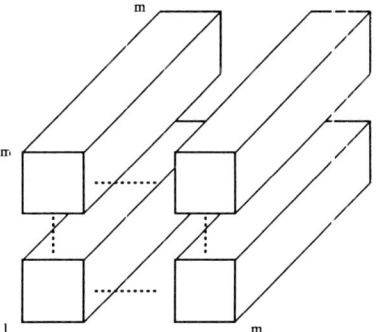

Figure 6: column decomposition of a 3-D $m \times m \times m$ domain

If the grid data of the subdomain are stored in an array $A(m,m,l)$ where $l = m/P$, the boundary data to be exchanged are all in contiguous memory locations and can be transferred directly. The ratio r_{31} increases proportionally to the number of processors as in the case of r_{21}.

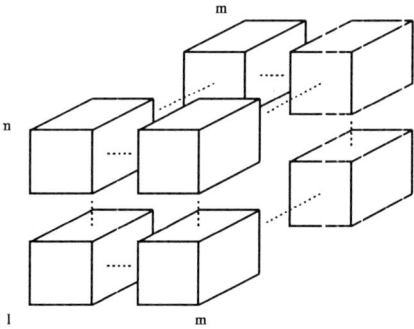

Figure 7: block decomposition of a 3-D $m \times m \times m$ domain

If the domain is decomposed in 2 dimensions and the third dimension remains intact, as shown in figure 6, each processor needs to exchange with four adjacent columns in general, except the processors on the boundary. If the P processors are arranged as a $\sqrt{P} \times \sqrt{P}$ grid, then the message length of each communication is $\frac{m^2}{\sqrt{P}}$. The total communication time is then

$$T_{D_{32}} = 4(\sigma + 4\beta \frac{m^2}{\sqrt{P}})$$

and the communication/computation ratio

$$r_{32} = \frac{4(\sigma + 4\beta \frac{m^2}{\sqrt{P}})}{6\frac{m^3}{P}\tau} = \frac{2\sigma P + 8\beta m^2 \sqrt{P}}{3m^3 \tau}.$$

This is similar to the case when algorithm D_{22} is applied to the 2-D domain. The term in r_{32} corresponding to the start-up time grows proportionally to P, while the other term grows as a function of $\sqrt{P}$.

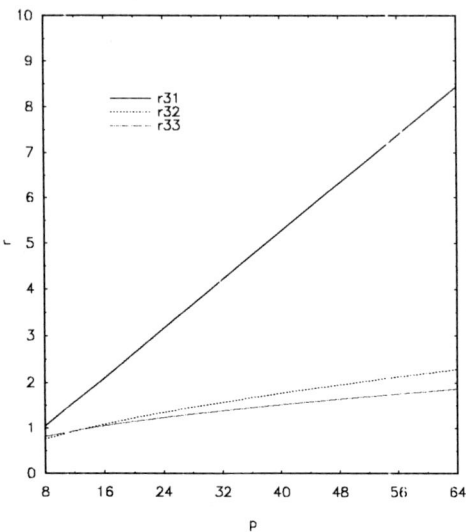

Figure 8: communication/computation ratios vs. the number of processors P for decomposition schemes D_{31}, D_{32} and D_{33}.

Figure 7 shows the block decomposition algorithm D_{33} in which all three dimensions of the original domain is decomposed. Assume that the processors are arranged as a $P^{\frac{1}{3}} \times P^{\frac{1}{3}} \times P^{\frac{1}{3}}$ grid, the subdomains then have dimensions $\frac{m}{\sqrt[3]{P}} \times \frac{m}{\sqrt[3]{P}} \times \frac{m}{\sqrt[3]{P}}$. A processor holding an interior subdomain needs to communicate with 6 processors holding adjacent blocks a message of length $\frac{m^2}{\sqrt[3]{P^2}}$. The total communication time is then

$$T_{D_{33}} = 6(\sigma + 4\beta \frac{m^2}{\sqrt[3]{P^2}})$$

and the communication/computation ratio is

$$r_{33} = \frac{6(\sigma + 4\beta \frac{m^2}{\sqrt[3]{P^2}})}{6\frac{m^3}{P}\tau} = \frac{\sigma P + 4\beta m^2 \sqrt[3]{P}}{m^3 \tau}.$$

We can see that the second term in r_{33} grows as a cubic root function of P.

Figure 8 shows the curves of r_{31}, r_{32} and r_{33} vs. the number of processors P. It is clear that algorithm D_{33} has the lowest communication/computation ratio as P increases.

Similar to the situation when algorithm D_{22} is applied to a 2-D domain, with algorithms D_{32} and D_{33}, some boundary data to be exchanged must be packed into a one dimensional array before communication, as expressed by (2). The influence of this data packing is hard to quantify analytically. Figure 9 and 10 show the timing curves of the three decomposition algorithms for a $64 \times 64 \times 64$ and a $128 \times 128 \times 128$ grids. For algorithm D_{32}, processors are arranged as 4×4 and 8×8 grids. For algorithm D_{33}, processors are arranged as $2 \times 2 \times 2$ and $4 \times 4 \times 4$ grids respectively. It can be seen from figure 9 and figure 10 that algorithm D_{33} takes the least amount of time to exchange boundary data of the subdomains, even though a processor has to communicate with 6 processors and pack some boundary data.

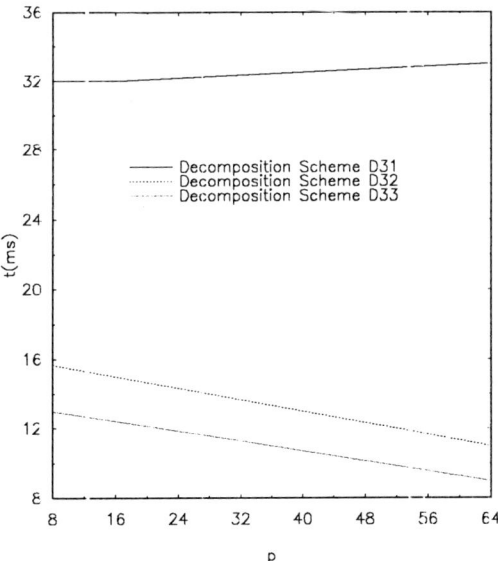

Figure 9: timing curves of decomposition algorithms D_{31}, D_{32} and D_{33} for exchanging messages of a $64 \times 64 \times 64$ domain.

3 Bandwidth Improvement by Forced Message Type

On an iPSC/860 hypercube, a processor can send a message to another processor at the rate of 2.8Mbytes/second. If a message arrives at a processor and there is no memory allocated for it, the message will go to the system buffer. To avoid this, the receiving processor can post a non-blocking *irecv* to allocate the storage for the incoming message.

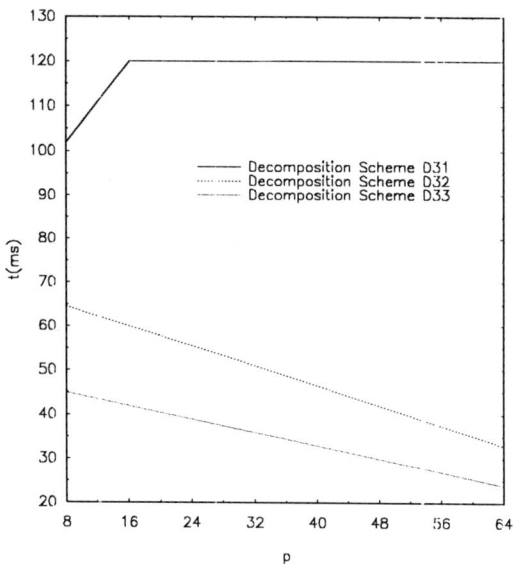

Figure 10: timing curves of decomposition algorithms D_{31}, D_{32} and D_{33} for exchanging messages of a $128 \times 128 \times 128$ domain.

To make sure both processors involved in the communication have allocated storages, a zero byte message is exchanged. After a processor received the signal message, it can then send the real data to the destination processor by using the forced message type which makes the message go directly to the storage allocated on the destination processor, rather than following the usual 4-step message passing process. The algorithm on one processor is:

{ 1. *post non-blocking irecv to allocate memory for incoming message*}

{ 2. *send a zero-byte message to signal the allocation of memory*}

{ 3. *wait for the zero-byte message from the other processor*}

{ 4. *send the real message using forced message type*}

For long messages, the above algorithm can take full advantage of the bi-directional communication channels between two processors. If there are more than two processors involved in the communications, the processors should be grouped into pairs to take advantage of bi-direction channels.

We take algorithm D_{21} as an example. A 2-D domain is decomposed into strips and assigned to 8 processors which are configured as a chain. A processor needs to exchange a message with adjacent processors. To improve the bandwidth by using forced message type, the communication process is divided into 2 stages as shown in figure 11. In each stage, forced message type is used for pairwise communications.

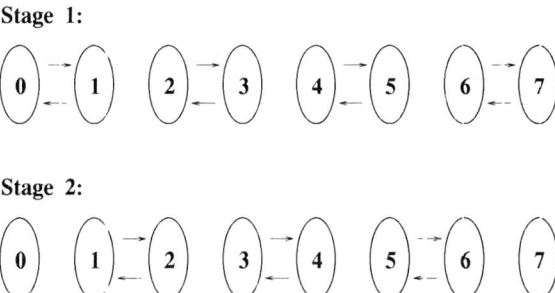

Figure 11: two stages of the communication process of a chain in which each processor exchanges a message with neighboring processors.

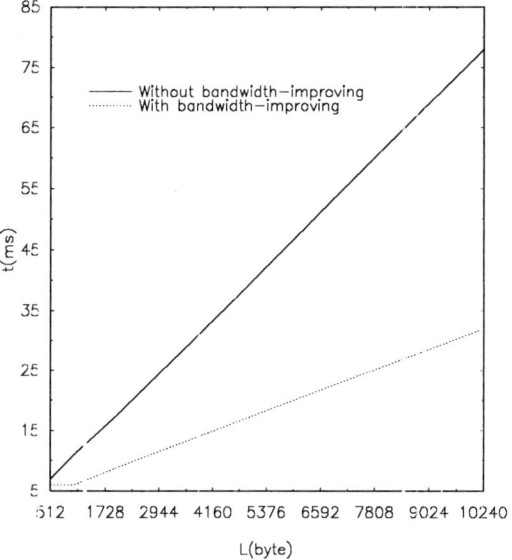

Figure 12: timing curves for a 8-processor chain to exchange message with and without bandwidth improvement by forced message type

Figure 12 shows the time used by the two communication methods for all processors in the chain to exchange messages with neighboring processors. Each processor exchanges with the neighboring processors a message of length L bytes. We can see that the difference between the two methods increases significantly as the message length increases. For long messages, the bandwidth-improving technique reduces the communication time to less than half of the original value. Therefore, it is well worth the extra effort to use the bandwidth-improving technique in scheduling inter-processor communications.

4 Effect of the Nearest Neighbor Communications

All experiments reported in figure 3, 4, 9, 10 and 12 are performed with and without the nearest neighbor communications (using gray code), the timing results are almost identical, which shows that the effect of non-nearest neighbor communication is not significant in one, two and three dimensional problems. This will simplify the implementation of algorithms and the mapping of subdomains to different processors.

5 Conclusions

In summary, for large scale CFD computations, the original domain should be decomposed into subdomains with sizes in all dimensions as close to each other as possible. This often requires decomposing the domain in all dimensions as the number of processors increases. As the routing technology improves, the nearest neighbor communication is not as important as it used to be on iPSC/1 hypercubes, which simplies implementations and mappings of parallel algorithms. Also, the effective bandwidth of communication channels for long messages (> 512 bytes) can be increased significantly by using forced message type.

6 References

1 W. Gropp and E. Smith, Computational Fluid Dynamics on Parallel Processors, Computer & Fluids, 18(1990), 289.
2 W. Gropp and D. Keyes, Domain Decomposition on Parallel Computers, In: Domain Decomposition Methods (Edited by T. Chan et al.), SIAM, Philadelphia (1989).
3 T. Chan, Domain Decomposition Algorithms and Computational Fluid Dynamics, The International Journal of Supercomputer Applications, 2(1988), 72-83.
4 L. Kang, Parallel Algorithms and Domain Decomposition. Wuhan, China: Wuhan University Press, 1987.
5 M. Heath, G. Geist and J. Drake, Early Experience with the Intel iPSC/860 at Oak Ridge National Laboratory, J. Supercomputer Applications, 5(1991), 10-26.